Intellectual Property and Biotechnology

Biological Inventions

Matthew Rimmer

Australian Research Council Future Fellow
Associate Professor, The Australian National University
College of Law
Associate Director, ACIPA

Edward Elgar
Cheltenham, UK • Northampton, MA, USA

© Matthew Rimmer 2008
© Michael Kirby, Foreword, 2008

All rights reserved. No part of this publication may be reproduced, stored in a retrieval system or transmitted in any form or by any means, electronic, mechanical or photocopying, recording, or otherwise without the prior permission of the publisher.

Published by
Edward Elgar Publishing Limited
The Lypiatts
15 Lansdown Road
Cheltenham
Glos GL50 2JA
UK

Edward Elgar Publishing, Inc.
William Pratt House
9 Dewey Court
Northampton
Massachusetts 01060
USA

Paperback edition 2011

A catalogue record for this book
is available from the British Library

Library of Congress Cataloguing in Publication Data

Rimmer, Matthew.
　Intellectual property and biotechnology : biological inventions / by Matthew Rimmer.
　　p. cm.
　Includes bibliographical references and index.
　1. Intellectual property.　2. Biotechnology—Law and legislation.　I. Title.
　K1519.B54R56　2007
　346.04'8—dc22
　　　　　　　　　　　　　　　　　　　　　　　　　　　2007030342

ISBN 978 1 84542 947 8 (cased)
ISBN 978 0 85793 370 6 (paperback)

Printed and bound by MPG Books Group, UK

Contents

Foreword The Hon Justice Michael Kirby AC CMG	vi
Preface	xi
Preface to this edition	xv

Introduction		1
1	Anything under the sun: patent law and micro-organisms	24
2	Franklin barley: patent law and plant breeders' rights	50
3	The human chimera patent initiative: patent law and animals	82
4	The storehouse of knowledge: patent law, scientific discoveries and products of nature	110
5	The book of life: patent law and the human genome project	138
6	The dilettante's defence: patent law, research tools and experimental use	164
7	The Utah saints: patent law and genetic testing	187
8	The alchemy of junk: patent law and non-coding DNA	216
9	Still life with stem cells: patent law and human embryos	248
Conclusion. Blue sky research: patent law and frontier technologies		280
Bibliography		308
Index		363

Foreword

The Hon Justice Michael Kirby AC CMG

I became aware of the subjects of this book almost by accident. In the early 1980s, when HIV/AIDS so unexpectedly came upon the world, I was invited by that fine epidemiologist turned international civil servant, Dr Jonathan Mann, to join the World Health Organisation inaugural Global Commission on AIDS.

This experience threw me into close contact with some of the leaders of medical science at the time, including Robert Gallo and Luc Montagnier, the two scientists who first isolated the virus that causes AIDS. I was soon attending meetings with leading biomedical experts and hearing them describe their experiments, their dreams and hopes.

How clearly I remember the predictions of those days that we would have a vaccine against HIV transmission within a decade or so and a cure within twenty years. Despite all the talent and the investment of great resources, the world still has no safe vaccine. There is no cure, although remarkable advances have occurred in the development of antiretroviral drugs, some of them actually produced earlier and for other purposes but put to work in the battle against AIDS, often with remarkable efficacy.

Looking at those conferences from the outside, as a non-scientist, I could not help but contrast the two moods that were often present in the debates. I do not refer to the moods of optimism and pessimism, although we alternated between hope and despair as one product after another looked promising but then dashed our expectations. The contrast in moods to which I refer was between those scientists of the old school who preached that the pandemic was a great moral challenge for our species and that advances would best be secured by endeavours of pure science, working by serendipity with free sharing of knowledge and research. And those of the new school who saw the hope of progress as lying in huge investments in scientific experimentation which, they assured us, would ultimately produce the vaccine and cure and deliver a couple of Nobel prizes into the bargain.

The foremost proponent of the pure science theory was a young American biochemist, David Baltimore. A decade and more before HIV burst upon the world, he had begun investigating a rare simian retrovirus that existed in

African chimpanzees. When the human retrovirus we now know as HIV appeared, it was David Baltimore's research that cut a decade off the time of the ensuing investigations. He had not conducted his research for the glittering prizes of financial gain and investment profits. I do not believe that he was even motivated by the hope of a Nobel Prize, although that was duly awarded to him. His basic motivation was human curiosity. He was intrigued by the peculiarities and cleverness of the virus that he studied.

Baltimore's story provides an important antidote to those who think that the greatest leaps of science are always made in committees like that of the Manhattan Project and as a result of huge capital investments. On the contrary, sometimes the biggest leaps in scientific knowledge, essential to the most important technological breakthroughs, come about just because human beings are puzzled and want to get to the bottom of an intriguing problem.

At the HIV meetings, scientists began to speak of the biotechnology revolution that was underway in the United States following the closely divided decision of the Supreme Court of that country in *Diamond v Chakrabarty*, with which Dr Rimmer begins this book. That decision was announced by the Supreme Court in 1980. By five Justices to four, the Court found that Ananda Chakrabarty's patent application in respect of an oil-eating bacteria, constituted either a manner of manufacture or a composition of matter and was therefore patentable under United States law.

That decision was one of those turning points in legal history, like *Donoghue v Stevenson* (1932) (on the law of negligence), *Brown v Board of Education* (1954) (on equal rights for racial minorities), or the *Engineers Case* (1920) (on the literalist interpretation of the *Australian Constitution* 1901).

It is interesting, but futile, to speculate on what might have happened for the subjects of this book if Chief Justice Burger, who wrote the majority opinion of the Court, or one of those Justices who concurred with him, had slipped on an oily substance whilst climbing the beautiful marble stairs to his chambers in the Supreme Court building, momentarily distracted by the aspirational legend: 'Equal Justice Under Law'. If the Court had been evenly decided or if the vote had affirmatively gone the other way, the momentum of which the scientists spoke in those early AIDS colloquia might have turned out quite differently.

In the curious manner of these things, my encounter with the international scientific, legal and public health experts working on HIV/AIDS led to subsequent appointments that kept me in close touch with these fascinating experimental scientists. In quick succession, I was added to the Ethics Committee of HUGO (the Human Genome Organisation) and to the International Bioethics Committee of UNESCO (IBC).

This was an exciting time to be working with HUGO. It stood on the brink of the completion of the map of the entire human genome. That was an achievement that came to pass in 2001, suitably enough, just in time for a new millennium. In the meetings of the HUGO Ethics Committee, and of the UNESCO IBC, the participants were challenged by new developments that had arisen in the United States, possibly stimulated by the outcome in Dr Chakrabarty's case.

One of these developments was the enactment of new federal laws, proposed by the Reagan administration, obliging American institutions, funded by federal subventions, to secure intellectual property protection for their original work as the price for the support of American public money. How many times I heard leading scientists lament the demise of the previous culture of unrestricted scientific exchange in the fields of biomedicine. Instead, now, they and their institutions were required by law to install intellectual property protection. With federal gold came obligations to defend what was increasingly seen as a crucial source of America's national income. Coinciding with the developments in the United States, the moves in the World Trade Organisation, the negotiation of the *TRIPS Agreement* (1994) and the *Doha Declaration on Public Health and the TRIPS Agreement* 2001 sought new ways to regularise and internationalise the technological and legal culture that flowed in the wake of *Diamond v Chakrabarty*.

At meetings with participants from developing countries, both in the context of international responses to the AIDS pandemic (by now the responsibility of UNAIDS) and in the context of HUGO and the IBC, developments of intellectual property law in Western countries were vehemently denounced. For the civil society organisations representing the poor, the infected and the sick, the new developments of intellectual property protection of biological inventions were not exciting means to promote scientific investment and experimentation that would help cure the world's ills. Instead, they were condemned as a new form of Western hegemony.

The old Empires might have faded away. But at conference after conference I heard delegates from poorer countries proclaim that intellectual property law, as it was advancing in the world, would strangle the poorer nations. It would put them in perpetual thrall to the pharmaceutical corporations of the wealthy states. Moreover, those states would invest their capital not in the diseases that afflicted most of humanity but in the products that would quickly recoup the largest financial returns. As it was often put: 'Face creams before malaria'. For the critics, intellectual property law had become the medium to divert the erstwhile noble dream of medical inquiry into a debased handservant of global capital movements, many of them flowing in the direction of the United States under free trade agreements which were insistent in this respect.

In 2001, just before the preliminary draft of the sequence of the human genome was published, UNESCO convened an international symposium in Paris on the topic of Ethics, Intellectual Property and Genomics. I chaired the concluding session. Many of the debates, outlined above, came to a head. The differences seemed irreconcilable. In the outcome, the Director-General of UNESCO invited the IBC to draft a new *Universal Declaration on Bioethics and Human Rights*. I chaired the drafting committee. The object of the project was to attempt a reconciliation of the ancient discipline of medical bioethics (initiated by Hippocrates and his equivalents in ancient times and by the medical and scientific professions since) and universal human rights (largely developed by lawyers in the wake of the devastating events of the Second World War and its aftermath).

Eventually this *Declaration* was adopted by the IBC. It was modified by governmental committees to reflect political and economic concerns. As so modified, it was adopted unanimously by the General Conference of UNESCO in October 2005. Some of the provisions of the *Declaration* reflect biological debates that emerged in the early days of HIV/AIDS and later as the Human Genome Project moved its conclusion.

This is not the place to explain the principles that were endorsed in the *Declaration*. However, the headings will indicate the guiding rules which the international community accepted in principle. Thus, Article 3 insists on respect for human dignity and human rights. Article 4 demands a balance between benefits and risks of harm. Article 8 insists on respect for human vulnerability and personal integrity. Article 10 asserts the fundamental equality of all human beings and the demand that they be treated justly and equitably. Article 11 expresses the principle of non-discrimination and non-stigmatisation. Article 12 reflects the need for respect for cultural diversity and pluralism. Several articles (13, 15 and 16) are concerned with human solidarity and cooperation across borders; the obligation to share benefits of science and technology; and the need to protect future generations. Article 14 insists on the obligation of science to respect social responsibility and to advance human health. Article 17 demands protection of the environment, the biosphere and biodiversity.

There are many other provisions in the *Declaration* that are worthy of attention. They grow out of the recognition, reflected in Dr Rimmer's book, that we stand on the brink of amazing and exciting developments of science and technology that, overwhelmingly, will be for the benefit of humanity. We must ensure that these developments occur and go forward in a world that understands and cherishes the essential unity of the human species and its interdependence with other living things in a biosphere, itself a living phenomenon.

In a sense, human beings are trustees for all living things. Law is ultimately a servant of our species. At the present moment in human history, it is unfortunate that we have not had the time, the will or the imagination to think freshly about the intellectual property regimes that would be suitable for the astonishing advances that are occurring about us. Instead, beginning with *Diamond v Chakrabarty*, we have built on the old legal regimes that were originally created for the age of sailing ships, wheels and cogs and machinery. Some developments in the applicable law have occurred. They are described in these pages. However, the fundamental ethical questions remain those debated in *Diamond v Chakrabarty* and reflected in the UNESCO *Universal Declaration on Bioethics and Human Rights*.

Dr Rimmer's book is a marvellous introduction to a crucial topic of our time. He writes engagingly, provocatively and always with good humour. A highly technical and complex area of law has been reduced to clear descriptions and searching analysis. Truly, this is an important book on an essential topic that will help define the ethics of a future that includes nothing less than the future of our species.

Michael Kirby
Canberra, 1 October 2007

Preface

furphy *n.* (*pl.*furphies). 1 a false report or rumour. 2 an absurd story.
adj. (furphier, furphiest) absurdly false, unbelievable:
that's the furphiest bit of news I ever heard. (*Australian Oxford Paperback Dictionary*, 1996)[1]

In the nineteenth century, patent law provided exclusive rights to inventors in respect of mechanical inventions, but it did not extend such protection to biological inventions. My mother's family hail from Shepparton in the Goulburn Valley in Victoria, Australia. In 1873, the blacksmith, John Furphy, set up a forge in the town, and produced a range of farm machinery. He was awarded a Victorian colonial patent in respect of a 'grain stripping machine' in 1882.[2] The invention won first prize at the Grand National Show in 1884, and enjoyed great popularity at agricultural fairs. The Furphy Foundry became most famous for the Furphy Water Cart, with its catchy advertising slogan, 'Good, better, best/ Never let it rest/ Until your good is better/ And your better best.' After the Water Cart was used by the Australian army in World War I, the word 'furphy' became a byword for gossip, idle rumour and tall stories. John's brother, Joseph Furphy, wrote the classic work of Australian literature, *Such is Life*, while working at the foundry.[3] The Furphy family were inventive in both the arts and the sciences.

Since the time of John and Joseph Furphy, patent law has become unrecognizable. With federation, the Australian Federal Government gained the exclusive power to make laws with respect to intellectual property, including patents of invention. Moreover, the Australian *Patents Act* 1990 (Cth) has been heavily influenced by international treaties, such as the *TRIPS Agreement* 1994, and the *Australia–United States Free Trade Agreement* 2004. Once the province of mechanical inventions and chemicals, patent law has expanded in its scope to cover all sorts of biological inventions, including micro-organisms, plants and animals; methods of human treatment, pharmaceutical drugs and research tools; and human genes, stem cells and tissues. No doubt, some of these inventions would seem to be farfetched and incredible 'furphies'. The mechanical engineers of the ilk of John and Joseph Furphy have been joined by new species of inventors: micro-biologists, plant and animal breeders, genetic engineers, stem cell scientists and nanotechnology developers. This book considers how the patent system, a product of the industrial revolution, has accommodated and adapted to the recent developments in the life sciences.

I have been fortunate to have received such support and mentoring from a number of teachers of intellectual property. Professor Peter Drahos at Regnet, at the Australian National University, first sparked my interest in intellectual property and biotechnology. My doctoral supervisor, Professor Kathy Bowrey, of the University of New South Wales, provided the sage advice that 'patents could be fun' and taught the art of writing about intellectual property in an accessible way. Professor Jill McKeough of the University of Technology Sydney has always been a stalwart supporter. Professor Brad Sherman from the University of Queensland has enlarged my vision of patent law, with his historical vision of biological property.

This book was written at the Australian National University College of Law. I am grateful for the academic freedom that I have been given by the leadership of this institution, including the Dean, Professor Michael Coper and the Head of School, Professor Stephen Bottomley, and his predecessor, the late great, Professor Phillipa Weeks. I have also appreciated the insights of Dr Don Anton, Dr Thomas Faunce and Matthew Zagor whose work intersects with my own. A small army of research assistants and Summer Research Scholars have worked with me over the years, including Katrina Gunn, Ishtiaque Omar, Elsa Gilchrist, Jessica Graham, Christine Henry and Paul Clarke.

This book has been written in the research centre, the Australian Centre for Intellectual Property in Agriculture (ACIPA), which is based at the Australian National University, the Griffith University and the University of Queensland. I am grateful for the help and support of all the researchers and administrators who have worked under its banner. In particular, I am indebted to Antony Taubman for providing such a good introduction to issues associated with gene patents, access to genetic resources and traditional knowledge. My knowledge of intellectual property and biotechnology has been augmented by friends of the centre, such as Geoff Budd and John Lovett of the Grains Research and Development Corporation. I have also learnt much from visiting keynote speakers to the ACIPA conferences, especially Dr Mildred Cho of Stanford University, Professor Mark D. Janis of the University of Iowa College of Law, Dr Margaret Llewelyn of the University of Sheffield and Dr Kate Murashige of Morrison & Foerster. I have also been grateful for the intellectual insights of fellow travellers, Dr Dianne Nicol of the University of Tasmania, and Dr Janet Hope of Regnet at the Australian National University.

My understanding of intellectual property and biotechnology has been enriched by conversations and dialogues with a number of scientists, researchers, geneticists and technology transfer managers, including

Dr Vijoleta Braach-Maksvytis of the Commonwealth Scientific and Industrial Research Organisation (CSIRO); Dr Hugh Dawkins of the Genomics Directorate of the Department of Health in Western Australia; Professor Simon Easteal of the John Curtin Medical Research School at the Australian National University; Professor Wayne Hall of the Institute for Molecular Biosciences at the University of Queensland; Professor John Mattick, co-director of the Institute for Molecular Biosciences at the University of Queensland; Professor Nicos Nicola of the Walter and Eliza Hall at the University of Melbourne; Dr Peter O'Leary of the Genomics Directorate of the Department of Health in Western Australia; Professor Ron Quinn, director of Astra Zeneca R&D; Professor Rodney Scott of the John Hunter Hospital; Professor John Shine of the Garvan Institute of Medical Research; Professor Grant Sutherland, the director of the Cytogenetics department of the Women's and Children's Hospital and the former chairman of the Human Genome Organisation (HUGO); Dr Kathy Tucker of the Prince of Wales Hospital; and Associate Professor Paul Waring of the Peter MacCallum Cancer Institute. I have also been kept up-to-date with the latest developments in intellectual property and biotechnology by a number of Australian journalists, including Jonathon Holmes, Danny Kingsley, Leigh Dayton, Judy Skatssoon and Deborah Smith.

The construction of this book has also been aided by conversations and dialogues with members of government agencies and institutions. I have been much assisted by Dr Doug Waterhouse of the Plant Breeders' Rights Office; Geoff Burton of the Department of the Environment and Heritage; and Dr Ian Heath of IP Australia. I have consulted members of the law reform bodies, Australian Law Reform Commission and the Canadian Biotechnology Advisory Committee. This research has been supported by an Australian Research Council Discovery Project, 'Gene patents in Australia: options for reform' (2003–05) and an Australian Research Council Linkage Project, 'The protection of botanical inventions' (2003).

I am grateful for the productive dialogues that I had in Canada with Professor Michael Geist, Jeremy de Beer and Marcus Bornfreund of the University of Ottawa Law and Technology Programme, Yann Joly of the University of Montreal, Professor Margaret Ann Wilkinson of the University of Western Ontario and Professor Myra Tawfik of the University of Windsor. I have also learnt much from conversations in Scandinavia with Eva Hemmungs Wirten of Uppsala University, Mathias Klang of Göteborg University, and Lee Davis of the Biotech Business School of Copenhagen University.

I am also obliged for the support of a number of academics from other institutions, including Dr Kirsten Anker of McGill University, Dr Livio

Dobrez of the Australian National University, Associate Professor Andrew Kenyon of the University of Melbourne and Professor Carolyn Sappideen of the University of Western Sydney. I am also indebted to my friends for their sage advice, who include Dr Rachel Bacon, Dr Alastair Blanshard, Kevin Boreham and Edwin Ho, Helen and James Chisholm, Janine Lapworth, Dr Simone Murray, Dr Kristin Natalier and Al King, Dan Neidle, Dr Mark Nolan, Tanya Richards-Pugh and Ivan Sun.

I am most grateful for the support and help of the publisher, Edward Elgar, and his team, including Luke Adams, Nep Elverd and Kate Pearce, from the eureka moment of inspiration through to the long, hard process of publication.

I am grateful to my parents, Professor Peter Rimmer and Dr Susan Rimmer, for providing me with such good genetic stock, and nurturing my scholarship. My grandmother, Joane Ford, has been an inspiring correspondent. My siblings, Joe Rimmer and Rachel Rimmer, have offered great support over the years. My children, Marina Rimmer and Joshua Rimmer, have provided me with much joy and distraction. I am also most grateful to all the child-care workers at the University Pre School and Child Care Centre for looking after them so well, while I have been writing this book. As always, my wife, Susan Harris Rimmer, has provided great love, forbearance and inspiration. Her suspicions about biotechnology have been a perfect foil to my own enthusiasm for the miracles of modern science.

NOTES

1. Ludowyk, F. (ed.) (1997), 'Ozwords: Furphy', http://www.anu.edu.au/andc/ozwords/November_97/, November.
2. Furphy, J. (1882), 'Grain stripping machine', Victorian Patent No: 3297.
3. Furphy, Joseph (1903), *Such is Life: Being the Diary of Certain Extracts from the Diary of Tom Collins*, Melbourne: Oxford University Press.

Preface to this edition

Since the publication of *Intellectual Property and Biotechnology: Biological Inventions* in 2008, the debate over gene patents has intensified further.

In the United States, the American Civil Liberties Union and the Public Patent Foundation filed a lawsuit in 2009 against the validity of patents related to BRCA1 and BRCA2 held by Myriad Genetics Inc. and the University of Utah Research Foundation. A district court judge, Sweet J, was sympathetic to the challenge, expressing reservations about taking a broad approach to patentable subject matter.[1] The judge was particularly concerned about the impact of gene patents upon patient care, medical research, and the administration of health-care: 'The resolution of the issues presented to this Court deeply concerns breast cancer patients, medical professionals, researchers, caregivers, advocacy groups, existing gene patent holders and their investors, and those seeking to advance public health.'[2] The judge held that 'the patents at issue directed to "isolated DNA" containing sequences found in nature are unsustainable as a matter of law and are deemed unpatentable subject matter under 35 U.S.C. § 101.'[3] The matter is currently under appeal in the United States Court of Appeals for the Federal Circuit.

There are grave doubts as to whether the decision is consistent with the broad approach taken to patentable subject matter in previous precedents of superior courts in the United States. In the case of *Bilski* v. *Kappos*, the Supreme Court of the United States held that a patent application for method of hedging risk in field of commodities trading in the energy market was an unpatentable abstract idea.[4] Nonetheless, in the lead judgment, Kennedy J declined to restrict the breadth of patentable subject matter, reflecting that 'the patent law faces a great challenge in striking the balance between protecting inventors and not granting monopolies over procedures that others would discover by independent, creative application of general principles'.[5] In the remanded case of *Prometheus Laboratories Inc.* v. *Mayo Collaborative Services*, the United States Court of Appeals for the Federal Circuit applied the ruling in *Bilski* v. *Kappos* and held that Prometheus's asserted medical treatment claims were patentable subject matter.[6]

In the European Union, there has also been a similar reappraisal of gene patents, with ongoing conflicts in the European Patent Office.[7]

In Australia, there has been ongoing controversy over gene patents. In June 2010, Cancer Voices Australia and Yvonne D'Arcy brought an action in the Federal Court of Australia arguing that claims 1–3 of Australian patent 686,004 relating to the BRCA1 was invalid, on the grounds that 'genes and the information represented by human gene sequences are products of nature universally present in each individual'.[8] The respondents included Myriad Genetics Inc., the Centre de Recherche du Chul, the Cancer Institute of Japan and Genetic Technologies Limited.

In November 2010, the Senate Community Affairs Reference Committee released its long awaited report on gene patents.[9] The majority of the committee recommended a number of procedural and substantive reforms to improve the quality of patents granted – particularly in the area of biotechnology. However, a minority of the committee were of the view that there should be a prohibition on gene patents altogether. A private members' bill has been introduced into the Australian Parliament, entitled the *Patent Amendment (Human Genes and Biological Materials) Bill* 2010 (Cth), which proposes a broad prohibition: 'The following are not patentable inventions: (a) human beings, and the biological processes for their generation; and (b) biological materials including their components and derivatives, whether isolated or purified or not and however made, which are identical or substantially identical to such materials as they exist in nature.' It is doubtful that such a bill will win the support of the major parties in the Australian Parliament.

2011

NOTES

1. *Association for Molecular Pathology v. United States Patent and Trademark Office, Myriad Genetics Inc., and the directors of the University of Utah Research Foundation* 702 F. Supp. 2d 181 at 185 (2010).
2. Ibid.
3. Ibid.
4. *Bilski* v. *Kappos* 130 S. Ct. 3218 (2010).
5. Ibid., at 3228.
6. *Prometheus Laboratories, Inc.* v. *Mayo Collaborative Services*, No. 2008-1403, slip op. (Fed. Cir. 2010).
7. Hawkins, N. (2010), 'Human gene patents and genetic testing in Europe: a reappraisal', *Script-ed*, 7(3), 453, http://www.law.ed.ac.uk/ahrc/script-ed/vol7-3/hawkins.asp.
8. *Cancer Voices Australia* et al. v. *Myriad Genetics Inc.* et al. (2010), Federal Court of Australia, Statement of Claim, 8 June.
9. Senate Community Affairs Reference Committee (2010), *Gene Patents*, Canberra: Australian Parliament, http://www.aph.gov.au/senate/committee/clac_ctte/gene_patents_43/report/index.htm.

Introduction

In a witty satire of prevailing patenting practices, the English poet and part-time casino waitress, Donna MacLean, sought a patent application – GB0000180.0 – in respect of herself.[1] She explained that she had satisfied the usual patent criteria in that she was 'novel', displayed an 'inventive step', and was eminently 'useful':

> It has taken 30 years of hard labor for me to discover and invent myself, and now I wish to protect my invention from unauthorized exploitation, genetic or otherwise. I am new: I have led a private existence and I have not made the invention of myself public. I am not obvious.[2]

MacLean quipped that she had many industrial applications: 'For example, my genes can be used in medical research to extremely profitable ends – I therefore wish to have sole control of my own genetic material.'[3] She explained the serious motives that lay behind her stunt: 'There's a kind of unpleasant, grasping, greedy atmosphere at the moment around the mapping of the human genome... I wanted to see if a human being could protect their own genes in law.'[4] The episode raises larger questions about the philosophy, ethics and politics of 'patenting lives'.[5]

The contemporary debate over patent law and biological inventions is not new. There has been a long-standing controversy over the grant of monopolies in respect of scientific inventions and technologies. In the sixteenth century, English monarchs granted monopoly privileges to inventors and imports of new technology in return for the payment of royalties to the Crown.[6] The courts objected to the Crown rewarding political patronage with trading monopolies.[7] The English Parliament sought to constrain the exercise of such royal prerogatives. The first modern patent legislation, the *Statute of Monopolies* 1623 (UK), limited the grant of monopolies to the 'first and true inventors' of 'any manner of new manufactures of the realm', so long as they were 'not contrary to the law, nor mischievous to the state, by raising prices of commodities at home, or hurt of trade, or generally inconvenient'. As it first developed, there was no clear procedure for the grant of patents. The process of obtaining patent protection was slow, expensive and cumbersome. In the midst of the industrial revolution, the English Parliament sought to reform the administration of patents.[8] In particular, patent applicants were required to define their claims to an invention in written documents known as specifications. The *Paris Convention for the Protection of Industrial Property* 1883 established an international union for the protection of industrial property – including protection for patents, trademarks and designs. Since that time, there have been a number of national, regional and international legal developments, which have created the modern network of patent offices.

Patent law grants exclusive economic rights in respect of the use and exploitation of inventions, in order to benefit society through encouraging innovation, and promoting the disclosure of scientific knowledge.[9] Binnie J of the Supreme Court of Canada has described the 'patent bargain' in these terms:

> A patent, as has been said many times, is not intended as an accolade or civic award for ingenuity. It is a method by which inventive solutions to practical problems are coaxed into the public domain by the promise of a limited monopoly for a limited time. Disclosure is the *quid pro quo* for valuable proprietary rights to exclusivity which are entirely the statutory creature of the *Patent Act*. Monopolies are associated in the public mind with higher prices. The public should not be expected to pay an elevated price in exchange for speculation, or for the statement of 'any mere scientific principle or abstract theorem', or for the 'discovery' of things that already exist, or are obvious. The patent monopoly should be purchased with the hard coinage of new, ingenious, useful and unobvious disclosures.[10]

Members of the World Trade Organization (WTO) are required to provide patent protection for 'any inventions, whether products or processes, in all fields of technology, provided that they are new, involve an inventive step and are capable of industrial application'.[11] The extent of patent protection is further limited in terms of territory and temporality. Patent protection is limited to the jurisdiction within which the grant was made. Nation states must provide protection of patents for 20 years from the filing date.[12] In certain exceptional circumstances, pharmaceutical drug patents may obtain an additional extension of the patent term for up to five extra years.

Furthermore, there are a number of legal doctrines which facilitate access to patented inventions. Members of the WTO can provide 'limited exceptions to the exclusive rights conferred by a patent', such as a defence of experimental use, and a safe harbour for research in respect of pharmaceutical drugs.[13] Moreover, nation states can allow for use of the subject matter of a patent without the authorization of the right holder, including use by the government or third parties authorized by the government. There is also scope for competition measures 'to prevent the abuse of intellectual property rights by right holders'.[14] Countries also have the capacity to exclude from patentability inventions on the grounds of public order and morality.[15]

Patent law has become a sprawling empire, exercising dominion over a wide range of scientific fields and technologies, with few limits or boundaries. Over the last century, Parliaments, Courts and the Patent Offices round the world have progressively and incrementally expanded the limits of patentable subject matter, until 'anything under the sun that is made by man' has been considered to be patentable. Initially, patent offices granted patents in respect of micro-organisms, such as yeasts, moulds, fungi, bacteria, algae, cell lines, viruses and protozoa. Then, intellectual property rights were incrementally extended to plants: the *Plant Patent Act* 1930 (US) provided protection in respect of asexually reproduced varieties of plants; plant breeders' rights offered exclusive rights in respect of sexually reproducing plants; and finally patent protection was granted in respect of traditionally bred plants, hybrid plants and genetically modified crops. Patent law also enveloped the animal kingdom: after it was recognized that polyploid oysters could constitute patentable subject matter, patents were sought in respect of the Harvard oncomouse, model organisms, such as drosophila, mice and zebra fish, and even methods to clone animals, such as Dolly the Sheep.

The prohibition against patenting methods of human treatment has been lifted in a number of Western jurisdictions. Patents have thus been sought in respect of medical devices, surgical techniques and diagnostic tests, as well as research tools, pharmaceutical drugs and personalized medicine. More recently, of course, patents have been granted in respect of human tissues, genes, stem cells and somatic nuclear cell transfer, so-called 'therapeutic cloning'. There remain few taboo inventions under patent law: perhaps only human cloning and animal–human hybrids remain clearly outside the scope of patentable subject matter. The limits of patentable subject matter have even been stretched to accommodate frontier technologies, such as bioinformatics, proteomics, pharmacogenomics and nanotechnology.

Patent loyalists – lawyers, patent attorneys and policy makers, as well as members of the pharmaceutical and biotechnology industries – have defended the expansion of the patent system to include biological inventions.[16] They have maintained that the patent system has achieved its objectives of encouraging innovation, boosting investment in research and development, and facilitating access to scientific information. The peak body, the Biotechnology Industry Organization (BIO), is exemplary in its defence of the extension of patent protection in respect of biotechnological inventions:

> For over 200 years the carefully crafted intellectual property laws have been the driving force for innovation and progress in the United States. The U.S. patent system fosters the development of new products and discoveries, new uses for

old products and employment opportunities for millions of Americans. Nowhere is this more apparent than in the biotechnology arena. The biotechnology industry as we know it did not exist prior to the landmark Supreme Court decision of *Diamond v. Chakrabarty* in 1980, where the court held that anything made by the hand of man was eligible for patenting. Since this decision, the biotechnology industry has flourished and continues to grow. Strong intellectual property protection is essential to the success, and in some instances to the survival, of the over 1,200 biotechnology companies in this country. For these companies, the patent system serves to encourage development of new medicines and diagnostics for treatment and monitoring of intractable diseases, and agricultural and environmental products to meet global needs.[17]

BIO emphasizes that patent protection is an invaluable incentive for both capitalists and scientists alike: 'Enticed by the prospect of the market exclusivity afforded by U.S. patent protection, U.S. entrepreneurs and scientists expend great resources to develop and produce cutting edge biotechnology products.'[18] The peak body emphasizes: 'Patents provide the needed assurance for investors to risk the capital necessary in the long development process; e.g. that his/her investment cannot only be recouped but also generate a profit.'[19] The group believes that the model of the United States patent system should be emulated by other countries. 'BIO takes an active role in educating policymakers, opinion leaders and the public at large, both in the U.S. and abroad, about the value of the biotechnology sector.'[20]

By contrast, a number of commentators argue that the patent system can accommodate new technologies within its framework, but only through the flexible use of patent doctrines and administrative guidelines. Dan Burk and Mark Lemley maintain that, under a façade of technology neutrality, the patent system is technologically specific in the way that it deals with new technologies:

> This seeming paradox – a monolithic legal incentive for wildly disparate industries – is resolved by the realization that, despite the appearance of uniformity, patent law is actually as varied as the industries it seeks to foster. Closer examination of patent law demonstrates that it is unified only in concept. In practice the rules actually applied to different industries have shown increasing divergence. The best examples of such divergence are found in biotechnology and computer software cases, where the courts have applied the common legal standards of obviousness, enablement, and written description in ways that differ radically in result. As a practical matter, it appears that although patent law is technology-neutral in theory, it is technology-specific in application.[21]

These authors question whether patent law should explicitly attempt to tailor protection to the needs of specific industries, as many have suggested. Instead, they suggest that Patent Offices and courts should make use of existing policy levers within patent law to address and respond to new technologies: 'The great flexibility in the patent statute presents an opportunity for courts to take account of the needs and characteristics of different industries.'[22] Burk and Lemley, in particular, mention a number of existing doctrines, such as prohibition against the patenting of abstract ideas; the level of skill of a person skilled in the art; secondary considerations of inventiveness; the criteria of utility; written description requirements; various indicia of patent infringement; and the defence of experimental use. They also identify a number of other potential policy levers, such as the presumption of validity; anti-trust considerations; and the use of remedies, such as injunctions.

Law reformers have recommended that the patent system could be reformed and improved, so that it is better adapted to the unique problems presented by gene patents. They have made both recommendations for procedural reform, in terms of patent administration and examination standards, as well as substantive reform, such as raising the threshold of patent criteria in respect of novelty, inventive step and utility, and expanding the exceptions to patent infringement. In its inquiry into gene patenting and human health, the Australian Law Reform Commission (ALRC) commented:

> The ALRC has adopted a nuanced approach to reform, which recognises both the generality and longevity of the patents system, on the one hand, and the new challenges generated by human genetic science and technology, on the other. There are many different points at which the patent system might be reform to address the actual and anticipated problems posed by the patenting of genetic materials and technologies. This does not mean that reform must be sought at every point, but rather that intervention – where needed – should be directed to those areas in which it will be most effective . . . The Report makes important recommendations for reform but it does not suggest any radical overhaul of the patent system.[23]

The Commission was exemplary of a model of a minimalist, liberal and rational law reform, with its ideals of technology neutrality, regulatory flexibility and legislative compromise. Similar approaches were taken in other jurisdictions, by sister reform bodies such as the Canadian Biotechnology Advisory Committee,[24] the Nuffield Council on Bioethics,[25] The New Zealand Royal Commission on Genetic Modification,[26] the National Academy of Sciences,[27] and the National Research Council.[28]

In a classic paper that captured the zeitgeist, Michael Heller and Rebecca Eisenberg speculated that biomedical research suffered from the 'tragedy of the anticommons'.[29] The authors contended: 'A proliferation of intellectual property rights upstream may be stifling life-saving innovations further downstream in the course of research and product development'.[30] Heller and Eisenberg elaborated:

> Thirty years ago in *Science*, Garrett Hardin introduced the metaphor 'tragedy of the commons' to help explain overpopulation, air pollution, and species extinction. People often overuse resources they own in common because they have no incentive to conserve. Today, Hardin's metaphor is central to debates in economics, law, and science and is a powerful justification for privatizing commons property. Although the metaphor highlights the cost of overuse when governments allow too many people to use a scarce resource, it overlooks the possibility of underuse when governments give too many people rights to exclude others. Privatization can solve one tragedy but cause another. Since Hardin's article appeared, biomedical research has been moving from a commons model toward a privatization model.[31]

Heller and Eisenberg concluded: 'An anticommons in biomedical research may be more likely to endure than in other areas of intellectual property because of the high transaction costs of bargaining, heterogenous interests among owners, and cognitive biases of researchers.'[32]

There have been a number of empirical studies, which have investigated the impact of gene patents upon scientific research, communication and innovation.[33] The evidence has been inconclusive. Reviewing the available empirical evidence, the eminent panel of Tim Caulfield, Robert Cook-Deegan, F. Scott Kieff and John Walsh questioned whether the phenomenon of the anti-commons had materialized:

> The evidence regarding the anticommons and restricted access concerns is clearer. The empirical research suggests that the fears of widespread anticommons effects that block the use of upstream discoveries have largely not materialized. The reasons for this are numerous and are often straightforward matters of basic economics. In addition to licensing being widely available, researchers make use of a variety of strategies to develop working solutions to the problem of access, including inventing around, going offshore, challenging questionable patents and using technology without a license.[34]

There is an important gap, though, between the opinions of the interpretative community of lawyers, patent attorneys, business managers and policy makers, and wider public opinion about patenting life forms. R. Stephen Crespi noted in his report for the Organisation for Economic Co-operation and Development (OECD) that there is 'a large gap between the views of experts and public opinion about problems engendered by the patenting of genetic inventions'.[35] Empirical evidence suggests that there is widespread community concern about genetic patents.[36] Technocrats may wishfully like to think that the debate about the patentability of genes has been conclusively resolved; however, the question is still very much an open-ended subject of passionate debate in the wider community.

Bioethicists have maintained that ethical considerations are and should be relevant in assessing applications for gene patents.[37] The current manner of manufacture test is not sufficient to accommodate such considerations. An independent body should become relevant in assessing ethical considerations related to assessing applications for gene patents. In an article in the *Lancet*, Richard Gold and Timothy Caulfield argue that the patent system can address ethical concerns in biotechnology: 'The patent system provides a useful mechanism by which to address ethical and social concerns in biotechnology, not because patents are necessarily the cause of concern, but because the system for granting them provides a practical way to regulate compliance with ethical and social values.'[38] The Canadian academics propose that patents for inventions that present social and ethical questions should be subject to suspension by an independent, transparent and responsible tribunal made up of specialists in ethics, research and economics. This suspension should be reversible so that, when the social or ethical concerns have been addressed in appropriate manner, the suspension can be lifted. Although controversial, such a flexible mechanism would assist governments and industry in enhancing public support for patents in the biotechnology area. The political philosopher, Francis Fukuyama, has called for greater regulation of genetic engineering.[39]

A number of commentators believe that sui generis regimes of intellectual property should be minted to accommodate new technologies and scientific developments.[40] Special legislative schemes have been developed to deal with plant breeders' rights, and access to genetic resources. Sui generis regimes have been mooted for all manner of other subject matter, including scientific discoveries, animal breeders' rights, genetic databases and the protection of traditional knowledge. However, such an approach seems increasingly unrealistic, given the broad expansion of patentable subject matter in national jurisdictions, and the ratcheting up of minimum obligations under international treaties.

There are also a number of patent abolitionists who contend that biological inventions should not be eligible for protection as patentable subject matter. Jeremy Rifkin has been a long-time opponent of biotechnology, generally, and gene patents, more particularly. He was involved in the *Diamond v Chakrabarty* case as a friend of the court and has supported the Human Chimera Patent Initiative as a means of critiquing the administration of the United States Patent and Trademark Office (USPTO). In *The Biotech Century*, Jeremy Rifkin summarized his concerns about the commercialization of life forms:

> A handful of global corporations, research institutions, and governments could hold patents on virtually all 100,000 genes that make up the blueprints of the human race, as well as the cells, organs, and tissues that comprise the human

body. They may also own similar patents on tens of thousands of microorganisms, plants, and animals, allowing them unprecedented power to dictate the terms by which we and future generations will live our lives.[41]

Rifkin warns that 'multinational corporations and governments are already scouting the continents in search of the new "green gold", hoping to locate microbes, plants, animals, and humans with rare genetic traits that might have future market value'.[42]

Following the lead of Jeremy Rifkin and the Peoples Business Commission, a number of non-government organizations have expressed a range of ethical and moral objections to the patenting of genes.[43] Peter Drahos and John Braithwaite have noted the increasing participation of civil society groups in policy debates over the intellectual property rights: 'The decline of moral respectability of intellectual property rights has been accompanied by increasing levels of transnational activism against the use and extension of intellectual property regimes.'[44] Scientists and researchers, such as John Sulston, have contended that genes should not be patented because they are scientific discoveries, products of nature and the common heritage of human kind. Folk heroes such as Percy Schmeiser and farmers' collectives, such as the Network of Concerned Farmers, have expressed concerns that plant patents could undermine farmers' rights to save seed, and engage in traditional agricultural activities.[45] Animal rights' activists, such as the American Anti-Vivisection Society, have protested that it is unethical and immoral to patent animals, because they are sentient beings.[46] Environmental groups have objected to the patenting of plants, animals and human genes, complaining about the commodification of life forms. Greenpeace, for instance, has declared: 'Greenpeace opposes all patents on genes, plants, humans and parts of the human body and regards the biodiversity of this planet as the common heritage of humankind.'[47]

Consumer organizations, such as Ralph Nader's Consumer Project on Technology, have campaigned for access to knowledge and access to essential medicines.[48] Anti-biotechnology activists, such as the ETC Group, have protested against the creation of monopolies in respect of new biological technologies.[49] Health activists, such as the Institut Curie, Médecins Sans Frontières and the Treatment Action Campaign, have contended that patents have undermined access to essential medicines.[50] Religious denominations have objected to the patenting of genes and stem cells on the basis that life is sacred. Similarly, Indigenous communities and peak bodies like the Indigenous Peoples Council on Biocolonialism, have complained that they have been the victims of biopiracy through the assertion of patent rights and other related forms of intellectual property.[51] Anti-globalization groups have objected to the impact of gene patents on developing countries, noting that research dollars and the beneficial effects of patented products are concentrated in developed countries.[52]

This book contends that there is a need to reform intellectual property and biotechnology in order to better accommodate scientific and technological developments. Sagely, Lester Thurow observed that the patent system has become rigid and inflexible in its 'technology-neutral' approach:

> Fundamental shifts in technology and in the economic landscape are rapidly making the current system of intellectual property rights unworkable and ineffective. Designed more than 100 years ago to meet the simpler needs of an industrial era, it is an undifferentiated, one-size-fits all system. Although treating all advances in knowledge in the same way may have worked when most patents were granted for new mechanical devices, today's brainpower industries pose challenges that are far more complex.[53]

It is submitted that the boundaries of patentable subject matter need to be better demarcated and delimited, so as to preserve the public domain and the scientific commons. The thresholds for the patent criteria of novelty, inventive step and utility should be raised, so as to require more than merely follow-on innovation. There should be an expansion of defences and exceptions to patent infringement, especially in respect of experimental use, farm-saved seed and medical treatment. Innocent bystanders should not be the subject of patent infringement actions. Bioethical concepts of informed consent and benefit sharing should inform the operation of the patent system. Moreover, there should be greater scope for the flexible use of compulsory licensing, Crown use and competition law. Furthermore, patent law needs to recognize the global nature of scientific inquiry, commonly featuring 'Big Science' projects, which involve collaborations between the public and private sectors.

In analysing intellectual property and biotechnology, this book draws upon a mixture of methodologies, including the history of science,[54] the sociology of science[55] and a comparative analysis of patent law, policy and practice.[56]

First, this book is part of a larger project of seeking to document the historical origins of the biotechnology industry. The Oral History Office of Bancroft Library at the University of California has been conducting interviews with scientists, entrepreneurs and university administrators who were involved in the development and commercialization of the life sciences.[57] Drawing upon this work, Sally Smith Hughes has written a dazzling case study of the Cohen-Boyer patent in respect of recombinant DNA.[58] She argues that the patent was a turning point in the commercialization of molecular biology and a harbinger of the social and ethical issues associated with biotechnology today. Stephen Hall made an early attempt to document the race to synthesize a human gene, focusing upon Genentech, Biogen, Eli Lilly and the University of California.[59] Daniel Kevles has written about key moments in the history of intellectual property and biotechnology.[60] The anthropologist Paul Rabinow has also told a number of stories about the history of biotechnology. He has written accounts of the polymerase chain reaction (PCR),[61] the French genomics project[62] and the Icelandic genomics project by DeCODE Genomics.[63] Similarly, the sociologist Alberto Cambrosio and his collaborators have written a history of scientific research into monoclonal antibodies.[64] There is much to be learned from such historical case studies.

Second, this text explores whether patent law, and allied rights, have an impact on the social norms of scientific communities: in particular, Robert Merton's key values of universalism, communism, disinterestedness and organized scepticism.[65] Rebecca Eisenberg explores the potential negative impact of patent rights on scientific norms in the field of biotechnological research: 'By providing such broad exclusive rights, patent law may aggravate pre-existing conflict between scientific norms and the reward structure of science.'[66] Her collaborator, Arti Rai, supports this claim:

> Legal rules and social norms are powerful and interdependent institutions for shaping behaviour. Law-and-norms analysis represents a valuable tool for determining how these institutions should be deployed. Applying an efficiency-focused variant of law-and-norms analysis to basic research in molecular biology reveals that the federal government's past efforts to displace information-sharing norms with intellectual property rights have failed to recognize those contexts in which invention and development goals are promoted more effectively through the public domain than through privatization.[67]

By contrast, F. Scott Kieff is a naysayer who argues that intellectual property rights are consistent with the norms of science: 'It is not even clear that the pre-1980 basic biological research community had a prescriptive norm that specifically rejected patents, as distinct from other forms of intellectual property.'[68] Such arguments need to be grounded in historical and sociological work about the understanding of intellectual property by scientists at that time.

Third, this book considers how the legal problems in respect of biological inventions have been addressed in a number of key jurisdictions, including the United States, the European Union, Canada, Australia and New Zealand. There have been noticeable tensions and rivalries between the USPTO, the Court of Appeals for the Federal Circuit and the Supreme Court of the United States. The United States Congress has debated a number of legislative proposals in respect of biological inventions, such as *The Genomic Research and Diagnostic Accessibility Act* 2002 (US), *The*

Genomic Science and Technology Innovation Act 2002 (US) and the *Genomic Research and Accessibility Act* 2007 (US). However, the United States Government has been somewhat reluctant to implement such measures. The Supreme Court of Canada has been divided between supporters of gene patents and a cohort of naysayers who have ethical qualms about biological inventions. The Canadian Parliament has been noticeably slow to adopt the recommendations of the Canadian Biotechnology Advisory Committee.[69]

By contrast, the European Parliament passed the comprehensive *European Union Directive on the Legal Protection of Biotechnological Inventions* 1998 (EU). Nonetheless, there has been much debate amongst member states over the implementation of this Directive. There has been discord on the issues of gene patents and stem cell patents between the European Parliament, the European Patent Office and specialist law reform advisory bodies. Canada presents a striking hybrid of British, European and North American influences on patent law.

In Australia, IP Australia, the Federal Court of Australia, and the High Court of Australia have had to grapple with a number of frontier technologies. The Australian Law Reform Commission conducted an extensive inquiry into gene patenting and human health; however, the Australian Government has shown little inclination to implement its minimalist recommendations.[70] The New Zealand Government has commissioned policy papers on genetic engineering, and more particularly on the impact of gene patents on human health.[71]

There has also been much debate about biological inventions in a number of international forums. The *Paris Convention for the Protection of Industrial Property* 1883 established a multilateral regime for the protection of various forms of industrial property – including patents, trademarks and designs. The *UPOV Convention* 1961, and its successors, the *UPOV Convention* 1978 and the *UPOV Convention* 1991, provided a blueprint for the development of a sui generis regime protection for plant breeders' rights. The *Patent Cooperation Treaty* 1970 was designed to enable the filing of an international patent application, which can be assessed for novelty by a search of the prior art. The *Budapest Treaty* 1977 provided international recognition of the deposit of micro-organisms for the purposes of patent disclosure. The *Rio Convention on Biological Diversity* 1992 established a framework for access to genetic resources of sovereign nation states on the basis of prior informed consent and benefit sharing.[72]

The *TRIPS Agreement* 1994 clarified the existing criteria for granting a patent, and also confined the nature of the exclusions to patentable subject matter that can be applied in national patent laws.[73] There has been much debate about access to essential medicines under the *TRIPS Agreement*

1994. The *Doha Declaration on the TRIPS Agreement and Public Health* 2001 recognized that nation states could take measures under patent law to protect public health. The *WTO General Council Decision* 2003 acknowledged that member states could export pharmaceutical drugs to developing countries.[74] The United States has sought to increase the level of patent protection through the means of bilateral agreements such as the *Australia–United States Free Trade Agreement* 2004, and regional agreements like the proposed *Free Trade Area of the Americas*. The World Intellectual Property Organization (WIPO) has hosted policy debates over intellectual property and development. The *UNESCO Universal Declaration on Bioethics and Human Rights* 2005 has promoted the principles of informed consent and benefit sharing in respect of biomedical research.[75] The United Nations Permanent Forum On Indigenous Issues has supported a rights-based approach to the protection of Indigenous cultural heritage in order to provide better protection of traditional knowledge.[76]

Chapter 1 investigates the progressive extension of patent protection to micro-organisms. In *Diamond v Chakrabarty*, the majority of the Supreme Court of the United States held by a majority of five to four that a new strain of bacteria produced artificially by bacterial recombination was a useful patentable invention.[77] The decision was of wider significance. The Supreme Court stated that 'anything under the sun made by man' was patentable subject matter. It opened the way for the USPTO to take a broad approach to statutory subject, and grant patents in respect of micro-organisms and other biotechnological inventions. Without the sound and fury of the Supreme Court of the United States decision, other jurisdictions, such as Australia, the United Kingdom and Canada, came to similar conclusions that micro-organisms could indeed be patentable subject matter.[78]

Chapter 2 considers the relationship between patent law and plant breeders' rights in light of modern developments in biotechnology. It examines how a number of superior courts have sought to manage the tensions and conflicts between these competing schemes of intellectual property protection. The chapter considers the High Court of Australia case of *Grain Pool of Western Australia v the Commonwealth*, dealing with Franklin barley.[79] It also examines the significance of the Supreme Court of the United States decision in *JEM Ag Supply Inc v Pioneer Hi-Bred International Inc* with respect to utility patents and hybrid seed.[80] The chapter considers the Supreme Court of Canada case of *Monsanto Canada Inc. v Schmeiser*, in which a Saskatchewan canola farmer was sued for infringing a patent on glyphosate-resistant canola.[81] It considers the implications of the decision for patent protection of agricultural products, farmers' rights and the position of innocent bystanders.

Chapter 3 explores the legal, commercial and ethical debate over the patenting of animals. There has been great litigation over the Harvard oncomouse, a transgenic animal designed to be genetically predisposed to develop cancerous tumours. The USPTO granted a patent for 'Transgenic Non-Human Mammals' on the 12 April 1988.[82] The European Patent Office granted a similar patent on the Harvard oncomouse on 13 May 1992.[83] However, there has been continuing litigation over the Harvard oncomouse in the European Patent Office. The European Patent Office modified claim number 1 to include only 'transgenic rodents' rather than 'transgenic non-human mammals'.[84] By contrast, in *Harvard College v the Commissioner for Patents*, the Supreme Court of Canada ruled by a five to four majority that the Harvard oncomouse was not patentable subject matter.[85] In the leading judgment for the majority, Bastarache J emphasizes that Parliament must give an express legislative direction to authorize the patenting of higher life forms: 'I believe that the best reading of the words of the Act supports the opposite conclusion – that higher life forms such as the oncomouse are not currently patentable in Canada.'[86] There has also been much controversy over the patenting of polyploid oysters, model animals and cloned animals. Of particular note has been Jeremy Rifkin and Stuart Newman's patent application in respect of a human–animal chimera.[87] This application challenged the USPTO to consider the morality of certain biological inventions.[88]

Chapter 4 considers the ramifications of the ruling of the Supreme Court of the United States in *Laboratory Corp. of America Holdings v Metabolite Laboratories Inc* for scientific discoveries, natural principles, abstract ideas and methods of human treatment.[89] The case involved a patent application, which claimed a process for helping to diagnose deficiencies of two vitamins, folate and cobalamin. A majority of five judges of the Supreme Court of the United States ruled that the writ of certiorari had been improvidently granted, and dismissed the action. This decision reflected the view of the judges that the written record had been insufficiently developed to consider the question of patentable subject matter. Nonetheless, Breyer J wrote a dissenting judgment, with the support of Stevens J and Souter J. His Honour emphasized: 'Patent law seeks to avoid the dangers of overprotection just as surely as it seeks to avoid the diminished incentive to invent that underprotection can threaten.'[90] The dissenting judges ruled that the patent application should have been ruled invalid because it sought to claim natural principles and scientific discoveries. Breyer J emphasized that the position of the majority 'threatens to leave the medical profession subject to the restrictions imposed by this individual patent and others of its kind'.[91] Indeed, he observed: 'Those restrictions may inhibit doctors from using their best medical judgment; they may force doctors to spend unnecessary time and energy to enter into license agreements; they may divert resources from the medical task of health care to the legal task of searching patent files for similar simple correlations; they may raise the cost of healthcare while inhibiting its effective delivery.'[92]

Chapter 5 analyses recent litigation over patent law and expressed sequence tags. In the matter of *In re Fisher*, the agricultural biotechnology company Monsanto sought to patent express sequence tags in maize plants.[93] The USPTO examiner and Board of Appeals rejected such claims on the grounds of lack of utility and enablement. Monsanto appealed to the United States Court of Appeals for the Federal Circuit, arguing that the Board applied a heightened standard for utility in the case of express sequence tags. For the majority, Michel CJ held that the claimed invention lacked a specific and substantial utility, and the application did not enable a person skilled in the art to use the invention. His Honour rejected the argument that express sequence tags were analogous to research tools, such as a microscope. Rader J dissented, saying that the claimed ESTs have such a utility, at least as research tools in isolating and studying other molecules. His Honour responded: 'These research tools are similar to a microscope; both take a researcher one step closer to identifying and understanding a previously unknown and invisible structure.'[94] There is a discussion of various attempts by representatives of the United States Congress to reform patent law to provide greater access to genetic inventions for scientists and patients alike.

Chapter 6 considers whether patent law should have a defence for research use and, if so, what its scope should be. It will explore the impact of such an exemption upon a number of important industries, such as agriculture, biotechnology and health care. It will also examine the repercussions of such a defence for universities, research organizations and educational institutions. In the United States, there has been much controversy over the decision of the Federal Circuit in *Madey v Duke University* over patent law and experimental use.[95] Gajarsa J held that the common law defence of experimental use was circumscribed: 'Regardless of whether a particular institution or entity is engaged in an endeavor for commercial gain, so long as the act is in furtherance of the alleged infringer's legitimate business and is not solely for amusement, to satisfy idle curiosity, or for strictly philosophical inquiry, the act does not qualify for the very narrow and strictly limited experimental use defense.'[96] The judge concluded that the district court attached too great a weight to the non-profit, educational status of Duke, 'effectively suppressing the fact that Duke's acts appeared to be in accordance with any reasonable interpretation of Duke's legitimate business objectives'.[97] In *Merck KGaA v Integra Lifesciences I, Ltd.*, the Supreme Court of the United States considered the safe harbour for pharmaceutical drugs, the so-called 'Bolar' exception.[98] The case concerned whether uses of patented inventions in preclinical research, the results of which are not ultimately included in a submission to the Food and Drug Administration, are exempted from infringement by 35 U.S.C. §271(*e*)(1). In a pithy, leading decision, Scalia J observed that the safe harbour is to be read broadly: 'Properly construed, §271(*e*)(1) leaves adequate space for experimentation and failure on the road to regulatory approval.'[99]

Chapter 7 considers the litigation and controversy over the patents held by the Utah biotechnology firm, Myriad Genetics, in respect of genetic diagnostic testing for BRCA1 and BRCA2, which are related to breast cancer and ovarian cancer. In France, the Institut Curie initiated a number of opposition procedures against the patents lodged by Myriad Genetics in respect of genetic tests for breast cancer and ovarian cancer. The Institut Curie and its supporters have challenged Myriad Genetics' patents – EP 699 754, EP 705 902 and EP 705903 (patents relating to BRCA1) and EP 785 216 (the patent relating to BRCA2).[100] The European Patent Office has revoked one of its patents dealing with BRCA1, narrowed the scope of a couple of its patents dealing with BRCA1 and awarded Michael Stratton and Cancer Research UK a patent dealing with BRCA2.[101] Myriad Genetics has transferred some of its rights to the University of Utah Research Foundation. Myriad may well appeal against such decisions and may also rely upon its licence from GTG to commercialize the patents with respect to non-coding DNA. This chapter considers the ramifications of this dispute for the *European Union Directive on the Legal Protection of Biotechnological Inventions* 1998 (EU) and its implementation by member states.

Chapter 8 examines the related debate over patents in respect of non-coding DNA and genomic mapping. The firm Genetic Technologies Limited (GTG) was able to obtain broad patents on a range of scientific inventions arising out of the work of Malcolm Simons. Most significantly, the USPTO awarded U.S. Patent No. 5,612,179 to GTG for an invention entitled 'Intron sequence analysis method for detection of adjacent and remote locus alleles as haplotypes'.[102] Furthermore, the USPTO also issued U.S. Patent No. 5,851,762 to GTG for an invention entitled 'Genomic mapping method by direct haplotyping using intron sequence analysis'.[103] GTG has embarked upon an ambitious licensing programme. Most significantly, GTG has obtained an exclusive licence from Myriad Genetics to use and exploit its medical diagnostics in Australia, New Zealand, and the Asia-Pacific region. In the United States, GTG brought a legal action for patent infringement against the Applera Corporation and its subsidiaries.[104] In response, Applera has counter-claimed that the patents of GTG were invalid because they fail to comply with the requirements of US patent law, such as novelty, inventive step and written specifications. In New Zealand, the Auckland District Health Board brought legal action in the High Court, seeking a declaration that the patents of GTG were invalid, and that the Board has not in any case infringed them.[105] This matter was settled. The New Zealand Ministry of Health and the Ministry of Economic Development have reported to Cabinet on the issues relating to the patenting of genetic material.[106] Similarly, the Australian Law Reform Commission has also engaged in an inquiry into gene patents and human health; and the Advisory Council on Intellectual Property has considered whether there should be a new defence in respect of experimental use and research.[107]

Chapter 9 explores the ethical and political controversy over patents relating to stem cell research, so-called 'therapeutic cloning' (nuclear transfer) and human cloning. It highlights concerns about commercialization, access to essential medicines and bioethics. The chapter questions the meaning of section 18(2) of the *Patents Act* 1990 (Cth), which provides that 'Human beings, and the biological processes for their generation are not patentable inventions.' It considers the interpretation of section 18(2) of the *Patents Act* 1990 (Cth) in two key decisions by the Deputy Commissioner of Patents: *Fertilitescentrum AB and Luminis Pty Ltd* and *Woo-Suk Hwang*.[108] This chapter examines the strong patent protection secured by the Wisconsin Alumni Research Foundation and Geron Corporation in respect of stem cell research in the United States. It considers the challenge to the validity of such patents in the USPTO by the California-based Foundation for Taxpayer and Consumer Rights, and the New York-based Public Patent Foundation.[109] This chapter investigates the marginal position of stem cell research under the *European Union Directive on the Legal Protection of Biotechnological Inventions* 1998 (EU). It examines a number of decisions of the European patent office in respect of the 'Edinburgh patent', a Wisconsin Alumni Research Foundation patent application, and a California Institute of Technology Patent Application.[110] It also considers the inquiry of the European Group on Ethics in Science and New Technologies into 'The Ethical Aspects of Patenting Inventions Involving Human Stem Cells',[111] as well as the practice of the United Kingdom Patent Office.[112]

The Conclusion considers how the patent regime will accommodate frontier technologies – in light of substantial investment in the areas of genomics, bioinformatics, proteomics, pharmacogenomics and nanotechnology. Such new scientific advancements will no doubt test the flexibility of patent law and practice. There has been much debate as to whether such new technologies can be accommodated within the framework of current law, or if they require new examination guidelines and legislative reforms.

NOTES

1. MacLean, D. (2000), 'Myself', United Kingdom Patent Application No: GB0000180.0.
2. MacLean, D. (2000), 'I am that I am', *Harper's Magazine*, **301** (1803)18.
3. Ibid.
4. Meek, J (2000), 'Poet attempts the ultimate in self-invention – patenting her own genes', *The Guardian*, 29 February.
5. Gibson, Johanna (2005–2007), 'Patenting lives', http://www.patentinglives.org/ and http://patentinglives.blogspot.com/.
6. Boehm, Klaus with Aubury Silberston (1967), *The British Patent System*, Cambridge: Cambridge University Press.
7. *Darcy v Allen* (1602) 11 Co Rep 84 (the Case of Monopolies).
8. Sherman, Brad and Lionel Bently (1999), *The Making Of Modern Intellectual Property Law: The British Experience 1760–1911*, Cambridge: Cambridge University Press.
9. For a survey of patent law and doctrine, see Philip Grubb (2004), *Patents for Chemicals, Pharmaceuticals and Biotechnology*, 4th edn, Oxford: Oxford University Press. For an introduction to United States patent law, see Janice Mueller (2006), *Introduction to Patent Law*, 2nd edn, Frederick (MD): Aspen Publishers; and Robert Merges, Peter Mennell and Mark Lemley (2006), *Intellectual Property in the New Technological Age*, Frederick (MD): Aspen Publishers. For summaries of Australian patent law, see Jill McKeough, Kathy Bowrey and Philip Griffith (2007), *Intellectual Property: Commentary and Materials* (4th edn), Sydney: Law Book Co; Jill McKeough, Andrew Stewart and Philip Griffith (2004), *Intellectual Property in Australia*, 3rd edn, Sydney: LexisNexis Butterworths; and Sam Ricketson and Megan Richardson (2005), *Intellectual Property: Cases and Materials*, 3rd edn, Sydney: LexisNexis Butterworths. For a survey of Canadian patent law, see Daniel Gervais and Elizabeth Judge (2005), *Intellectual Property: The Law in Canada*, Toronto: Thomson Carswell; and for surveys of United Kingdom patent law, see William Cornish and David Llewelyn (2003), *Intellectual Property: Patents, Copyright, Trade Marks and Allied Rights*, 5th edn, London: Sweet & Maxwell; and Lionel Bently and Brad Sherman (2004), *Intellectual Property*, 2nd edn, Oxford: Oxford University Press.
10. Binnie J in *Apotex Inc. v Wellcome Foundation Ltd.*, [2002] 4 S.C.R. 153, 2002 SCC 77.
11. Article 27.1 of the *TRIPS Agreement* 1994.
12. Article 33 of the *TRIPS Agreement* 1994.
13. Article 30 of the *TRIPS Agreement* 1994.
14. Article 8(2) of the *TRIPS Agreement* 1994.
15. Article 27(2) of the *TRIPS Agreement* 1994.
16. Crespi, S. (2001/2002), 'Patenting and ethics – a dubious connection', *Bio-Science Law Review*, **5**(3), 71–8.
17. Biotechnology Industry Organization, 'The importance of intellectual property', http://www.bio.org/ip/.
18. Ibid.
19. Ibid.
20. Ibid.
21. Burk, D. and M. Lemley (2003), 'Policy levers in patent law', *Virginia Law Review*, **89**, 1575–1696 at 1577.
22. Ibid., 1641.
23. Australian Law Reform Commission (2004), *Genes and Ingenuity: Gene Patenting and Human Health, Report 99*, Sydney: Australian Commonwealth, http://www.austlii.edu.au/au/other/alrc/publications/reports/99/, June, 14.
24. Canadian Biotechnology Advisory Committee (2002), *Patenting Of Higher Life Forms*, Ottawa: Canadian Biotechnology Advisory Committee, June; and Canadian Biotechnology Advisory Committee (2003), *Advisory Memorandum: Higher Life Forms And The Patent Act*, Ottawa: Canadian Biotechnology Advisory Committee, 24 February; and Canadian Biotechnology Advisory Committee (2006), *Human Genetic Materials,*

Intellectual Property, and the Health Sector, Ottawa: Canadian Biotechnology Advisory Committee, http://cbac-cccb.ca/epic/internet/incbac-cccb.nsf/en/ah00578e.html

25. Nuffield Council on Bioethics (2002), *The Ethics of Patenting DNA, A Discussion Paper*, London: Nuffield Council on Bioethics, http://www.nuffieldbioethics.org/go/ourwork/patentingdna/publication_310.html.

26. New Zealand (2000), *Royal Commission on Genetic Modification*, http://www.mfe.govt.nz/issues/organisms/law-changes/commission/; New Zealand (2002), *Report of the Royal Commission on Genetic Modification*, http://www.mfe.govt.nz/publications/organisms/royal-commission-gm/index.html, Chapter 10, http://www.mfe.govt.nz/publications/organisms/royal-commission-gm/chapter-10.pdf; New Zealand, Minister of Health and Associate Minister of Commerce (2003), *Implications of the Granting of Patents Over Genetic Material*, Cabinet Policy Committee, available at *Internet Archive*, <http://web.archive.org/web/*/http://www.med.govt.nz/buslt/int_prop/genetic-material/cabinet/implications/implications.pdf, November/; and New Zealand, Ministry of Health and Ministry of Commerce (2004), *Memorandum to Cabinet Policy Committee: Report Back with Recommendations and Options for Addressing Genetic Material Patents*, http://www.med.govt.nz/templates/Multipage-DocumentTOC___1148.aspx, May, 1.

27. National Academy of Sciences (2004), *A Patent System for the 21st Century*, Washington, DC: National Academy of Sciences, http://books.nap.edu/catalog.php?record_id=10976.

28. National Research Council Committee on Intellectual Property Rights in Genomic and Protein Research and Innovation (2005), *Reaping the Benefits of Genomic and Proteomic Research: Intellectual Property Rights, Innovation and Public Health*, *Washington DC:* National Academies Press, Washington, DC.

29. Heller, M. and R. Eisenberg (1998), 'Can patents deter innovation? The anticommons in biomedical research', *Science*, **280**, 698–701.

30. Ibid.

31. Ibid.

32. Ibid., at 701.

33. Walsh, J.P., A. Arora and W.M. Cohen (2003), 'Working through the patent problem', *Science*, **299**, 1021; Merz, J., A. Kriss, D. Leonard and M. Cho (2002), 'Diagnostic testing fails the test: the pitfalls of patenting are illustrated by the case of haemochromatosis', *Nature*, **415**, 577–9; M. Cho et al. (2003), 'Effects of patents and licenses on the provision of clinical genetic testing services', *Journal of Molecular Diagnostics*, 5(1), 3–8; D. Nicol and J. Nielsen (2003), 'Patents and medical biotechnology: an empirical analysis of issues facing the Australian industry', Occasional Paper No. 6, Centre for Law and Genetics, University of Tasmania, http://www.ipria.org/publications/workingpapers/BiotechReportFinal.pdf; J. Strauss (2002), 'Genetic inventions and patents: a German empirical study', Organisation for Economic Co-operation and Development Workshop, http://www.oecd.org/dataoecd/36/22/1817995.pdf, 24–5 January; B. Verbeure, G. Matthijs and G. van Overwalle (2005), 'Analysing DNA patents in relation to diagnostic genetic testing', *European Journal of Human Genetics* 14(1), 26–33; E. Campbell et al. (2002), 'Data withholding in academic genetics: evidence from a national survey', *Journal of the American Medical Association*, 287(4), 473–80; K. Jensen and F. Murray (2005), 'Intellectual property landscape of the human genome', *Science*, **310**, 239–40; Organisation for Economic Co-operation and Development (2002), *Genetic Inventions, Intellectual Property Rights and Licensing Practices: Evidence and Policies*, Paris: Organisation for Economic Co-operation and Development, http://www.oecd.org/dataoecd/42/21/2491084.pdf.

34. Caulfield, T., R.C. Cook-Deegan, F.C. Kieff and J. Walsh (2006), 'Evidence and anecdotes: an analysis of human gene patenting controversies', *Nature Biotechnology*, 24(9), 1091–4, at 1093.

35. Organisation for Economic Co-operation and Development (2002), *Genetic Inventions, Intellectual Property Rights and Licensing Practices: Evidence and Policies*, Paris: Organisation for Economic Co-operation and Development, http://www.oecd.org/dataoecd/42/21/2491084.pdf, p. 79.

36. As part of a quantitative empirical study conducted in 2004 and 2005, I asked 199 undergraduate science students to rate on a seven-point scale whether patents should be granted upon a range of inventions. Responses ranged widely: strongly disagree (1), disagree (2), moderately disagree (3), no opinion (4), moderately agree (5), agree (6) and strongly agree (7). The results showed a spectrum of attitudes to patents across a range of technologies.

There was strong opposition to patents being granted in respect of human genes (2.28), stem cell lines (2.80) and human cloning (2.65). There was a notable dislike of patents being granted in respect of animals (2.64). There was significant resistance to patents being granted in respect of biomedical research, such as methods of human treatment (2.84), genetic diagnostic tests (2.99) and express sequence tags (3.32). There was moderate opposition to patents being granted in respect of micro-organisms (3.46), plants (3.17) and bioprospecting (3.85). There was strong support for patents being granted in respect of mechanical inventions (6.28), chemicals (5.49), research tools (4.44) and pharmaceutical drugs (4.44). Interestingly, the subjects were relatively unconcerned about patents being granted in the emerging field of nanotechnology (4.93). The findings demonstrate that community attitudes are not uniform across all biological inventions. The results demonstrate a preference towards patents being granted in respect of traditional subject matter, such as mechanical inventions, chemical and pharmaceutical drugs. The study shows a clear bias against the patenting of higher life forms, with a lesser concern about the patenting of lower life forms.
37. See, for instance, David Magnus, Arthur Caplan and Glenn McGee (eds) (2002), *Who Owns Life?*, Amherst (NY): Prometheus Books; and J. Merz and M. Cho (2005), 'What are gene patents and why are people worried about them?', *Community Genetics*, **8**, 203–8.
38. Gold, E.R. and T. Caulfield (2002), 'The moral tollbooth: a method that makes use of the patent system to address ethical concerns in Biotechnology', *The Lancet*, **359**, p. 2268.
39. Fukuyama, Francis (2002), *Our Posthuman Future: Consequences of the Biotechnology Revolution*, London: Profile Books.
40. Ellinson, D. (1988), 'The patent system – time to reflect', *Law Institute Journal*, **292–3**; and Luigi Palombi (2004), 'The patenting of biological materials in the context of the agreement on trade related aspects of intellectual property', PhD thesis, University of New South Wales, Sydney http://cgkd.anu.edu.au/menus/PDFs/PhDThesisFinal.pdf.
41. Rifkin, Jeremy (1998), *The Biotech Century: Harnessing the Gene and Remaking the World*, New York: Penguin Putnam Inc, p. 2.
42. Ibid., 37.
43. Hindmarsh, Richard and Geoffrey Lawrence (eds) (2001), *Altered Genes II: The Future?*, Melbourne: Scribe Publications; Brian Tokar (ed.) (2001), *Redesigning Life? The Worldwide Challenge to Genetic Engineering*, London: Zed Books; and Rachel Schurman, Dennis Doyle and Takahashi Kelso (2003), *Engineering Trouble: Biotechnology and its Discontents*, Berkeley and Los Angeles: the University of California Press.
44. Drahos, Peter and John Braithwaite (2002), *Information Feudalism*, London: Earthscan Publications, p. 16.
45. Network of Concerned Farmers, http://www.non-gm-farmers.com/.
46. The American Anti-Vivisection Society, 'Stop Animal Patents!', http://www.stopanimalpatents.org/.
47. Greenpeace, 'Patents on Life', http://www.greenpeace.org/international/campaigns/genetic-engineering/ge-agriculture-and-genetic-pol/patents-on-life.
48. Consumer Project on Technology, http://www.cptech.org/.
49. ETC Group, http://www.etcgroup.org/en/.
50. Institut Curie, http://www.curie.fr/index.cfm/lang/_gb.htm, Médecins Sans Frontières http://www.accessmed-msf.org/; and Treatment Action Campaign, http://www.tac.org.za/.
51. Indigenous Peoples Council on Biocolonialism, http://www.ipcb.org.

52. Shiva, Vandana (2000), *Tomorrow's Biodiversity*, London: Thames and Hudson; and Ikechi Mgbeoji (2006), *Global Biopiracy: Patents, Plants, and Indigenous Knowledge*, Vancouver: University of British Columbia Press.
53. Thurow, L. (1995), 'Needed: a new system of intellectual property rights', *Harvard Business Review*, September–October 95.
54. Smith Hughes, S. (2001), 'Making dollars out of DNA: the first major patent in biotechnology and the commercialization of molecular biology, 1974–1980', *Isis*, **92**, 541–75: Robert Cook-Deegan (1994), *The Gene Wars: Science, Politics, and the Human Genome*, WW Norton, New York and London; Paul Rabinow (1996), *Making PCR: A Story of Biotechnology*, Chicago: The University of Chicago Press; Paul Rabinow (1998), *French DNA: Trouble In Purgatory*, Chicago: The University of Chicago Press; and Donna Haraway (1997), *Modest Witness: Feminism And Technoscience*, Routledge: New York.
55. Biagioli, Mario (ed.) (1999), *The Science Studies Reader*, New York: Routledge; and Mario Biagioli and Peter Galison (eds) (2003), *Scientific Authorship: Credit and Intellectual Property in Science*, Routledge: New York.
56. Heller, M. and R. Eisenberg (1998), 'Can patents deter innovation? The anticommons in biomedical research', *Science*, **280**, 698–701; and David Magnus, Arthur Caplan and Glenn McGee (eds) (2002), *Who Owns Life?*, Amherst (NY): Prometheus Books.
57. The Bancroft Library, Bioscience and Biotechnology: Resources for Historical Research, 1999–2002, http://sunsite.berkeley.edu:2020/dynaweb/teiproj/oh/science/.
58. Smith Hughes, S. (2001), 'Making dollars out of DNA: the first major project in biotechnology and the commercialization of molecular biology, 1974–1980', *Isis*, **92**, 541–78.
59. Hall, Stephen (1987), *Invisible Frontiers: The Race To Synthesize A Human Gene*, London: Sidgwick and Jackson.
60. Kevles, D. (1994), 'Ananda Chakrabarty wins a patent: biotechnology, law and society, 1972–1980', *Historical Studies in the Physical and Biological Sciences*, **25**, 111–35; Daniel Kevles (1998), '*Diamond v Chakrabarty* and beyond', in Arnold Thackray (ed.), *Private Science: Biotechnology and the Rise of the Molecular Sciences*, Philadelphia: University of Pennsylvania Press, pp. 65–79; D. Kevles (2002), 'Of mice and money: the story of the world's first animal patent', *Daedalus*, **131**(2), 78–88; and D. Kevles (2002), 'A history of patenting life in the United States with comparative attention to Canada and Europe', European Group on Ethics in Science And New Technologies To The European Commission, 12 January.
61. Rabinow, Paul (1996), *Making PCR: A Story of Biotechnology*, Chicago: The University of Chicago Press.
62. Rabinow, Paul (1998), *French DNA: Trouble In Purgatory*, Chicago: The University of Chicago Press.
63. Rabinow, P. (2000), 'Learning to fly: a new page in history', *GeneLetter*.
64. Cambrosio, Alberto and Peter Keating (1998), 'Monoclonal antibodies: from local to extended networks', in Arnold Thackray (ed.), *Private Science: Biotechnology and the Rise of the Molecular Sciences*, Philadelphia: University of Pennsylvania Press, pp. 165–81.
65. Merton, Robert (1973), *The Sociology of Science*, Chicago: University of Chicago Press; see also Peter Drahos (1994), 'Decentring communication: the dark side of intellectual property', in Tom Campbell and Wojciech Sadurski (eds), *Freedom Of Communication*, Aldershot: Dartmouth, pp. 249–79.
66. Eisenberg, R. (1987), 'Proprietary rights and the norms of science in biotechnology research', *Yale Law Review*, **97**, 177–223.
67. Rai, A. (1999), 'Regulating scientific research: intellectual property rights and the norms of science', *Northwestern University Law Review*, **94**, 77–152, 151–2.
68. Kieff, F.S. (2001), 'Facilitating scientific research: intellectual property rights and the norms of science – a response to Rai and Eisenberg', *Northwestern University Law Review*, **95**, 691–706.
69. Nuffield Council on Bioethics (2002), *The Ethics of Patenting DNA, A Discussion Paper*, London: Nuffield Council on Bioethics, http://www.nuffieldbioethics.org/go/ourwork/patentingdna/publication_310.html.

70. Australian Law Reform Commission (2003), *Gene Patenting and Human Health, Issue Paper 27*, Sydney: Australian Commonwealth, http://www.austlii.edu.au/au/other/alrc/publications/issues/27/, July; Australian Law Reform Commission (2004), *Gene Patenting and Human Health, Discussion Paper 68*, Sydney: Australian Commonwealth, http://www.austlii.edu.au/au/other/alrc/publications/dp/68/, February; and Australian Law Reform Commission (2004), *Genes and Ingenuity: Gene Patenting and Human Health, Report 99*. Sydney: Australian Commonwealth, http://www.austlii.edu.au/au/other/alrc/publications/reports/99/, June.
71. New Zealand (2000), *Royal Commission on Genetic Modification*, http://www.mfe.govt.nz/issues/organisms/law-changes/commission/; New Zealand (2002), *Report of the Royal Commission on Genetic Modification*, http://www.mfe.govt.nz/publications/organisms/royal-commission-gm/index.html, Chapter 10, http://www.mfe.govt.nz/publications/organisms/royal-commission-gm/chapter-10.pdf; New Zealand, Minister of Health and Associate Minister of Commerce (2003), *Implications of the Granting of Patents Over Genetic Material*, Cabinet Policy Committee, available at *Internet Archive*, <http://web.archive.org/web/*/http://www.med.govt.nz/buslt/int_prop/genetic-material/cabinet/implications/implications.pdf, November/; and New Zealand, Ministry of Health and Ministry of Commerce (2004), *Memorandum to Cabinet Policy Committee: Report Back with Recommendations and Options for Addressing Genetic Material Patents*. http://www.med.govt.nz/templates/MultipageDocumentTOC____1148.aspx, May, 1.
72. The intersection of intellectual property and access to genetic resources is beyond the scope of this book. For further analysis, see Kerry Ten Kate and Sarah Laird (1999), *The Commercial Use Of Biodiversity: Access to Genetic Resources and Benefit-Sharing*, London: Earthscan; and M. Rimmer (2003), 'Blame it on Rio: biodiscovery, native title and traditional knowledge', *The Southern Cross University Law Review*, 7, 1–49.
73. Gervais, Daniel (2003), *The TRIPS Agreement: Drafting History and Analysis*, 2nd edn. London: Sweet and Maxwell; and Nuno Pires de Carvalho (2003), *The TRIPS Regime of Patent Rights*, 2nd edn, The Hague: Kluwer Law International.
74. A discussion of patent law and access to essential medicines is beyond the limits of this book. For further analysis, see M. Rimmer (2004), 'The race to patent the SARS virus: the TRIPS agreement and access to essential medicines', *Melbourne Journal of International Law*, **5**(2), 335–74; and M. Rimmer (2005), 'The Jean Chrétien pledge to Africa Act: patent law and humanitarian aid', *Expert Opinion on Therapeutic Patents*. **15**(7), 889–909.
75. The inter-relationship between patent law, informed consent and benefit sharing is further explored in D. Gitter (2004), 'Ownership of human tissue: a proposal for federal recognition of human research participants' property rights in their biological material', *Washington and Lee Law Review*, **61**(1), 257–346; and M. Rimmer (2006), 'Miami heat: patent law, informed consent, and benefit-sharing', *Journal of International Biotechnology Law*, **3**, 177–92.
76. The topic of patent law and traditional knowledge is outside the parameters of this study. For further discussion, see Ikechi Mgbeoji (2006), *Global Biopiracy: Patents, Plants, and Indigenous Knowledge*, Vancouver: University of British Columbia Press; Rimmer, M. (2007), 'The genographic project: traditional knowledge and population genetics', *Australian Indigenous Law Review*, **11**(2), 33–55; and *United Nations Declaration on the Rights of Indigenous Peoples* 2007.
77. *Diamond v Chakrabarty* 447 U.S. 303 (1980).
78. *Ranks Hovis McDougall's Application* [1976] 46 AOJP 3915. *American Cyanamid v Berk Pharmaceuticals* [1976] RPC 231; *Re Application for Patent of Abitibi Co.* (1982) 62 CPR (2d) 81.*Re Application for Patent of Connaught Laboratories* (1982) 82 CPR (2d) 32.
79. *Grain Pool Of Western Australia v Commonwealth* (2000) 46 IPR 515.
80. *JEM Ag Supply Inc v Pioneer Hi-Bred International Inc* 534 US 124 (2001).
81. *Monsanto Canada Inc. v Schmeiser* (2004) SCC 34; 2004 SCC 34.
82. Leder, P. and T. Stewart (1988), 'Transgenic non-human mammals', US Patent No: 4,736,866.

83. Leder, P. and T. Stewart (1986), 'Method for producing transgenic animals', European Patent No: EP0169672.
84. *Harvard/Onco-mouse*, 1989 O.J. EPO 451; *Harvard/Onco-mouse*, 1990 O.J. EPO 476; *Harvard/ Onco-mouse* [2003] OJEPO 473; and *Harvard/ Onco-mouse* [2004] T 0315/03–3.3.8.
85. *Harvard College v Canada (The Commissioner of Patents)* [2002] 2 SCR 45.
86. *Harvard College v Canada (The Commissioner of Patents)* [2002] 2 SCR 45.
87. Newman, S. (2003), 'Chimeric embryos and animals containing human cells', US Patent Application 20030079240.
88. *Re Stuart Newman* (2005), United States Patent Office decision, http://patentlaw.typepad.com/patent/files/chimera_final_rejection.pdf.
89. *Laboratory Corp. of America Holdings v Metabolite Laboratories, Inc.* 126 S.Ct. 2921 (2006).
90. *Laboratory Corp. of America Holdings v Metabolite Laboratories, Inc.* 126 S.Ct. 2921 (2006).
91. *Laboratory Corp. of America Holdings v Metabolite Laboratories, Inc.* 126 S.Ct. 2921 (2006).
92. *Laboratory Corp. of America Holdings v Metabolite Laboratories, Inc.* 126 S.Ct. 2921 (2006).
93. *In re Fisher* 421 F.3d 1365 (C.A.Fed., 2005).
94. *In re Fisher* 421 F.3d 1365 (C.A.Fed., 2005).
95. *Madey v Duke University* 307 F.3d 1351 (2002).
96. *Madey v Duke University* 307 F.3d 1351 (2002).
97. *Madey v Duke University* 307 F.3d 1351 (2002).
98. *Merck KGaA v Integra Lifesciences I, Ltd.* 545 U.S. 193 (2005).
99. *Merck KGaA v Integra Lifesciences I, Ltd.* 545 U.S. 193 (2005).
100. Skolnick, M. and D. Goldgar (1995), 'Method for diagnosing a predisposition for breast and ovarian cancer', European Patent No: EP699754; M. Skolnick and D. Goldgar (1995), '17q Linked breast and ovarian cancer susceptibility gene', European Patent No: EP705902; D. Shattuck-Eidens, J. Simard, E. Mitsuru, Y. Nakamura and F. Durocher (1995), 'In vivo mutations in the 17q-linked breast and ovarian cancer susceptibility gene', European Patent No: EP 705903; S. Tavtigian A. Kamb, J. Simard, F. Couch, J. Rommens and B. Weber (1996), 'Chromosome 13-linked breast cancer susceptibility gene BRCA2', European Patent No: EP 785216.
101. *Myriad Genetics Inc. v Cancer Research Campaign Technology Inc.* European Patent Office, Opposition to EP 0858 467 (11 November 2004); *Institut Curie v Myriad Genetics Inc.*, European Patent Office Opposition Division, Division Revoking the European Patent EP0699754 (3 November 2004); *Institut Curie v The University of Utah Research Foundation*, European Patent Office Opposition Division, Interlocutory Decision in Opposition Proceedings Against EP705903 (9 June 2005); *The Belgian Society of Human Genetics and the Institut Curie v The University of Utah Research Foundation*, European Patent Office Opposition Division, Interlocutory Decision in Opposition Proceedings Against EP785216 (29 June 2005); and *Sozialdemokratische Partei der Schweiz and the Institut Curie v The University of Utah Research Foundation*, European Patent Office Opposition Division, Interlocutory Decision in Opposition Proceedings Against EP705902 (19 September 2005).
102. Simons, M. (1992), 'Intron sequence analysis method for detection of adjacent and remote locus alleles as haplotypes', US Patent No: 5,612,179.
103. Simons, M. (1994), 'Genomic mapping method by direct haplotyping using intron sequence analysis', US Patent No: 5,851,762.
104. *Genetic Technologies Limited v Applera Corporation* (2003, United States District Court, Northern District of California), No. c-03-1316 PJH, 2003 WL 23796524.
105. *Auckland District Health Board v Genetic Technologies Ltd.* (2004) (High Court of Auckland).
106. New Zealand, Minister of Health and Associate Minister of Commerce (2003), *Implications of the Granting of Patents Over Genetic Material*, Cabinet Policy

Committee, available at *Internet Archive*, <http://web.archive.org/web/*/http://www.med.govt.nz/buslt/int_prop/genetic-material/cabinet/implications/implications.pdf, November/; and New Zealand, Ministry of Health and Ministry of Commerce (2004), *Memorandum to Cabinet Policy Committee: Report Back with Recommendations and Options for Addressing Genetic Material Patents*, http://www.med.govt.nz/templates/MultipageDocumentTOC____1148.aspx, May, 1.

107. Australian Law Reform Commission (2004), *Genes and Ingenuity: Gene Patenting and Human Health, Report 99*, Sydney: Australian Commonwealth, http://www.austlii.edu.au/au/other/alrc/publications/reports/99/, June, 14; and Advisory Council on Intellectual Property (2005), *Patents and Experimental Use: Final Report*, Canberra: Commonwealth Government, http://www.acip.gov.au/library/ACIP%20Patents%20&%20Experimental%20Use%20final%20report%20FINAL.pdf.

108. *Fertilitescentrum AB and Luminis Pty Ltd* [2004] APO 19; and *Woo-Suk Hwang* [2004] APO 24.

109. Foundation for Taxpayer and Consumer Rights and the Public Patent Foundation (2006), 'Request for re-examination in respect of US patent no: 5,843,780, http://www.pubpat.org/assets/files/warfstemcell/780Request.pdf; Foundation for Taxpayer and Consumer Rights and the Public Patent Foundation (2006), 'Request for re-examination in respect of US patent no: 6,200,806, http://www.pubpat.org/assets/files/warfstemcell/90008139granted.pdf; and Foundation for Taxpayer and Consumer Rights and the Public Patent Foundation (2006), 'Request for re-examination in respect of US patent no: 7,029,913', http://www.pubpat.org/assets/files/warfstemcell/913Request.pdf.

110. *Greenpeace Deutschland e.V. v The University of Edinburgh*, Opposition Division, European Patent Office (24 July 2002); University of Edinburgh (2003), 'Statement of grounds of appeal under Article 108 EPC', European Patent Office, 8 December; Caltech Patent Application, Examining Division, European Patent Office, T522/04–338 (17 October 2003); Wisconsin Alumni Research Foundation Patent Application in Respect of 'Primate embryonic stem cells', Examining Division, June 2004; and Wisconsin Alumni Research Foundation Patent Application in Respect of 'Primate embryonic stem cells', T 1374/04–3.3.08 Interlocutory decision of the Technical Board of Appeal 3.3.08 of 18 November 2005.

111. European Group on Ethics in Science and New Technologies to the European Commission (2002), *Opinion on the Ethical Aspects of Patenting Inventions Involving Human Stem Cells*, Opinion Number 16, 7 May.

112. United Kingdom Patent Office (2003), 'Inventions Involving Human Embryonic Stem Cells', http://www.patent.gov.uk/patent/notices/practice/stemcells.htm, April.

1. Anything under the sun: patent law and micro-organisms

> [U]nder section 101 a person may have invented a machine or a manufacture, which may include anything under the sun that is made by man. (P.J. Federico, Principal draftsman of the *Patent Act* 1952 (US))[1]

A treatise writer, Philip Grubb, comments that biotechnology has a long history, pre-dating the discovery of the double-helix by James Watson and Francis Crick:

> Classical biotechnology may be defined loosely as the production of useful products by living micro-organisms, and as such it has been with us for a long time. The production of ethanol from yeast cells is as old as history, and over 50 years ago the production of various industrial chemicals such as acetic acid and acetone by fermentation processes was well known.[2]

Notably, in 1873, Louis Pasteur was granted a patent by the United States Patent and Trademark Office (USPTO), claiming 'yeast, free from organic germs of disease, as an article of manufacture'.[3] The patent attorney, Grubb, noted: 'In the USA, in spite of the precedent of the Pasteur patent . . . it had become the practice of the Patent Office to refuse claims to living systems as not being patentable subject matter.'[4]

The long-standing practice of the USPTO was to refuse claims to living systems as not being patentable subject matter. In 1889, the Commissioner of Patents rejected a patent application which lay claim to 'cellular tissues of the Pinus australis' tree separated from the 'silicous, resinous, and pulpy parts of the pine needles and subdivided into long, pliant, filaments adapted to be spun and woven'.[5] The Commissioner of Patents ruled that patents could not be granted in respect of 'products of nature':

> It cannot be said that the applicant in this case has made any discovery, or is entitled to patent the idea, or fact, rather, that fiber can be found in the needle of the Pinus australis, or that it is a longer fiber than can be found in other leaves, or that it possesses more or less strength of fineness, because the mere ascertaining of the character or quality of trees that grow in the forest and the construction of the woody fiber and tissue of which they are composed is not a patentable

invention, recognized by the statute, any more than to find a new gem or jewel in the earth would entitle the discoverer to patent all gems which should be subsequently found. The result would be that patents might be obtained upon the trees of the forest and the plants of the earth, which of course would be unreasonable and impossible.[6]

The Commissioner of Patents concluded: '[The product here claimed] is a natural product and can no more be the subject of a patent in its natural state when freed from its surroundings than wheat which has been cut by a reaper or by some new method of reaping can be patented as wheat cut by such a process.'[7]

In 1972, Ananda Chakrabarty of the General Electric Company applied to the USPTO for a patent in respect of 'Microorganisms having multiple compatible degradative energy-generating plasmids and preparation thereof'.[8] Applying the 'products of nature' doctrine, the USPTO rejected the claims in the patent application in respect of the bacteria on the grounds that the claimed micro-organisms were 'products of nature' and that they were drawn to 'live organisms'. It insisted that any exceptions to such a doctrine, such as the *Plant Patent Act* 1930 (US) and the *Plant Variety Protection Act* 1970 (US), had to be explicitly authorized by the United States Congress. The Board of Appeals agreed that patents could not be granted in respect of 'live organisms'. The United States Court of Customs and Patent Appeals ruled that claims were not outside the scope of patentable inventions merely because they were drawn to 'live organisms'.[9] On reconsideration, the United States Court of Customs and Patent Appeals reaffirmed its earlier judgment.[10]

The Supreme Court of the United States granted a writ of certiorari to consider whether a new strain of bacteria produced artificially by bacterial recombination was a patentable invention. The matter provided for a free-wheeling, poly-vocal debate about the merits of biotechnology. Rebecca Eisenberg comments that the case touched upon wider social anxieties about genetic engineering:

> In the anxious rhetoric surrounding genetic engineering in the 1970s, the relationship between nature and human inventors was pictured quite differently. Rather than merely copying from nature, humans seemed to be altering nature's plans in unprecedented ways, making the concerns and intuitions that persuaded previous courts to leave natural products and natural phenomena outside the patent system seem inapposite in this context. By the time the issue was presented to the Supreme Court, the anxiety surrounding genetic engineering had begun to subside, and medically important genes had been cloned in microorganisms. The commercial potential of biotech had become manifest, and a host of *amicus curiae* briefs from the scientific community urged the court to uphold the patentability of genetically engineered microorganisms.[11]

The Court was deluged with amicus curiae briefs from a range of interested stakeholders, including biotechnology companies, pharmaceutical drug manufacturers, universities, researchers and scientists, as well as opponents of genetic engineering.

The Supreme Court of the United States held by a majority of five to four in *Diamond v Chakrabarty* that a new strain of bacteria produced artificially by bacterial recombination was a patentable invention. It found that the bacteria had utility because it could disperse oil slicks. The decision was of wider significance. The Supreme Court of the United States stated that 'anything under the sun made by man' was patentable subject matter. It opened the way for the USPTO to take a broad approach to statutory subject, and grant patents in respect of micro-organisms and other biotechnological inventions.

This chapter charts the history of *Diamond v Chakrabarty*, and considers the significance of the ruling in respect to the patentability of microorganisms, and biotechnological inventions.[12] Jack Wilson has commented:

> The *Chakrabarty* case set a precedent that soon changed how patent law was applied to biotechnology, but curiously did not effect a literal change in the law. Important decisions have all been patent office policy; though several attempts have been made, no relevant legislation has made it through Congress. Although Chakrabarty's bacterium was not created using recombinant DNA techniques, by the time his case was decided in 1980, nearly the complete set of recombinant techniques had been invented/discovered, and some popular reports actually clouded the details of Chakrabarty's case. The first biotechnology companies had been founded, and there were a number of patent applications waiting to be processed.[13]

Section one focuses upon the original patent application by Ananda Chakrabarty of the General Electric Company in respect of 'Microorganisms having multiple compatible degradative energy-generating plasmids and preparation thereof'. It considers the initial rulings of the Patent Examiner, the Board of Appeals and the Court of Customs and Patent Appeals. Section two considers the briefs of the petitioner, the respondent and the amicus curiae in the Supreme Court of the United States case of *Diamond v Chakrabarty*. Section three explores the stark divisions in the five–four decision of the Supreme Court of the United States in *Diamond v Chakrabarty*. It counterpoints the clash between two philosophical positions. The majority judgment of Burger CJ takes a broad, expansive view of patentable subject matter, and expresses disdain for considerations of ethics and morality. The dissenting judgment of Brennan J contends that Congress should be left to determine new eligible subject matter, especially when pressing concerns of public policy are at stake. The conclusion notes that the decision in *Diamond v Chakrabarty*, and the rulings in parallel jurisdictions, opened the way for patent applications in respect of plants, animals and human genes.

I DIAMOND V CHAKRABARTY

Ananda Chakrabarty, a microbiologist then at the General Electric Company, created a novel bacterium that presumably had never before existed in nature.[14] The scientist has recalled the circumstances of the discovery, in which he determined that 'a genetically engineered pseudomonad with various degradative plasmids could have the potential to generate single-cell protein from crude petroleum in significant amounts compared to natural strains'.[15] As Chakrabarty's bacterium showed promise in breaking down crude oil, General Electric decided to apply for a patent because the invention showed a lucrative potential for cleaning up oil spills or remediating toxic waste.

A Application of Chakrabarty

In 1972, respondent Chakrabarty filed a patent application entitled 'Microorganisms having multiple compatible degradative energy-generating plasmids and preparation thereof', assigned to the General Electric Company.[16] The abstract noted:

> This human-made, genetically engineered bacterium is capable of breaking down multiple components of crude oil. Because of this property, which is possessed by no naturally occurring bacteria, Chakrabarty's invention is believed to have significant value for the treatment of oil spills.[17]

The application asserted 36 claims related to Chakrabarty's invention of 'a bacterium from the genus Pseudomonas containing therein at least two stable energy-generating plasmids, each of said plasmids providing a separate hydrocarbon degradative pathway'.[18] The patent claims were of three types: first, there were process claims for the method of producing the bacteria; second, there were claims for an inoculum comprised of a carrier material floating on water, such as straw, and the new bacteria; and third, there were claims to the bacteria themselves.

The Patent Examiner rejected the claims in respect of the bacteria under Title 35 U.S.C. 101 on the grounds that the claimed micro-organisms were 'products of nature' and that they are drawn to 'live organisms'. The Board of Appeals agreed with the appellant that the claimed bacteria were not naturally occurring. However, the Board of Appeals affirmed that live organisms, such as these laboratory-created micro-organisms, were not patentable.

In 1978, the Court of Customs and Patent Appeals reversed the decision, ruling that 'the fact that microorganisms' were alive was 'without legal significance' for purposes of the patent law.[19] Rich J held that the claims were not outside the scope of patentable inventions merely because they were drawn to 'live organisms'.[20] Markey CJ concurred:

> There are but two sources for manufactures and compositions of matter. They are God (or 'nature' if one prefers) and man. As presented to us, the invention is admittedly a 'manufacture' by man. It therefore falls squarely within the language of the statute. The Patent and Trademark Office desires to read into the statute the word 'dead' before 'manufacture' and before 'composition'.[21]

The judge noted that there has long been hostility to frontier technologies: 'As with Fulton's steamboat "folly" and Bell's telephone "toy", new technologies have historically encountered resistance.'[22] His Honour, though, concluded: 'But if our patent laws are to achieve their objective, extra-legal efforts to restrict wholly new technologies to the technological parameters of the past must be eschewed.'[23]

Baldwin J dissented, saying that Chakrabarty had not altered the essential nature of the living subject matter:

> The appellant has not changed this essential nature; he has not created a new life. Rather, he has merely genetically grafted an extra plasmid on to the organism and, thereby, made the organism better at cleaning up oil spills. While this improvement in oil digesting ability does exclude the new organism from classification as a mere product of nature, like the borax-impregnated orange which was a better commercial product because it had a longer shelf life, this improvement in the utility for which the unpatentable starting material was already suited does not change the essential nature of the starting material and does not make the modified thing statutory subject matter.[24]

Miller J also dissented: 'I do not agree that appellant's claimed microorganisms are within the scope of 35 U.S.C. s. 101, and I join in the statement of the board.'[25] The judge added: 'We do not believe that Congress intended 35 U.S.C. 101 to encompass living organisms whether they be plants, modified microorganisms (such as bacteria), or modified multicellular organisms (such as mammals).'[26]

B Application of Bergy

On reconsideration, the United States Court of Customs and Patent Appeals reaffirmed its earlier judgment.[27] The Court of Customs and Patent Appeals referred to its earlier 1977 decision in the *Application of Bergy*.[28] The matter concerned a patent claim in application serial No. 477,766, relating to a biologically pure culture of the microorganism streptomyces vellosus.[29] The USPTO Board of Appeal had affirmed the rejection of the claim. In the Court of Customs and Patent Appeals, Rich J reversed this decision, ruling that it was in the public interest to include micro-organisms within the terms 'manufacture' and 'composition of matter': 'We see no sound reason to refuse patent protection to the microorganisms themselves a kind of tool used by chemists and chemical manufacturers in much the same way as they use chemical elements, compounds, and compositions which are not considered to be alive, notwithstanding their capacities to react and to promote reaction to produce new compounds and compositions by chemical processes in much the same way as do microorganisms.'[30]

Kashiwa J filed a concurring opinion. Miller J filed a dissenting opinion in which Baldwin J joined. The judge asserted that micro-organisms could not be compared to chemicals: 'The nature of organisms, whether microorganisms, plants, or other living things, is fundamentally different from that of inanimate chemical compositions.'[31]

On remand, in 1979, the Court of Customs and Patent Appeals, Rich J held that the claims were within statutory subject matter and should not have been rejected on sole ground that claim was for 'living organism', and the decision was not an extension of *Patent Act* 1952 (US).[32] The judge commented:

> We look at the facts and see things that do not exist in nature and that are manmade, clearly fitting into the plain terms 'manufacture' and 'compositions of matter.' We look at the statute and, plainly, it appears to include them. We look at its legislative history and are confirmed in that belief. We consider what the patent statutes are intended to accomplish and the Constitutional authorization, and it appears to us that protecting these inventions, in the form claimed, by patents will promote progress in very useful arts.[33]

Rich J was scornful of the approach of the USPTO: 'For whatever reason, it decided to reject, first on one ground and then on another, and then set out, lawyer-like, to devise unduly exaggerated justifications spiced with bits and pieces from wholly unrelated plant-patent legislation from nearly half a century ago.'[34]

Baldwin J concurred with the opinion: 'This statute, while not as sweeping as its constitutional basis, is expansive in its scope.'[35] The judge added: 'Indeed, the words of both the Senate and House Reports on the Act indicate that s. 101 is to "include anything under the sun that is made by man." '[36]

Miller J dissented, protesting against the judicial creativity of the majority:

> (T)he patent law is statutory. Our representative form of government requires that the enactments of its Congress must always be, at the very least, the starting

point. There being no common law of patents, we should take care to fill the Holmesian interstices of the statute with judge-made law only under the gravest and most impelling circumstances.[37]

In this context, Sidney Diamond, the Commissioner of the USPTO, sought and won a writ of certiorari from the Supreme Court of the United States to hear an appeal in the case of *Diamond v Chakrabarty*.

II AMICUS CURIAE SUBMISSIONS

After much legal dispute in the lower courts, the Supreme Court of the United States considered whether genetic engineering techniques were patentable. The Bench heard submissions not only from the petitioner and the respondent, but it received amicus curiae submissions from a wide range of interested parties. Academic John Frow has observed of the role of amicus curiae in Supreme Court:

> Each of these is a heterogenous alliance. What they represent is the peculiarly political phenomenon of formations of interest – that is, alliances of quite diverse social groups into a general (but transient) structure of interest. Each side thus represents a massive social pressure, and together they exemplify the social contradictions – the 'calculus of interests' – that the Court must try to reconcile. Part of the juridical ideology within which the Court works, however, is the claim that questions of law are decided on the basis of purely legal criteria.[38]

The Peoples Business Commission led by the redoubtable Jeremy Rifkin argued that patents should not be extended to life forms, such as micro-organisms. By contrast, the fledgling biotechnology company, Genentech Inc., members of the pharmaceutical industry, the University of California and scientists such as Leroy Hood and George Pieczenik urged the Supreme Court of the United States to recognize patents in respect of biotechnological inventions.

A Petitioners' Submission

The petitioner, Sidney Diamond, the Commissioner of the USPTO, submitted that the question of the patentability of living organisms was a matter for Congress, rather than the courts:

> Congress, rather than the judiciary, is empowered and is best able to resolve the complex social, economic, and scientific questions frequently involved in such decisions, and, if an extension is to be made, to tailor the statute to achieve precisely the desired ends. The determination whether living organisms produced by

'genetic engineering' of the kind involved in Chakrabarty's invention should themselves be patentable is just such a decision. It involves social, economic and scientific questions of great complexity. Moreover, if Congress should decide to extend patent protection to such inventions, it might well decide to do so by a specifically tailored statute, similar to those it has provided for certain hybridized plants, rather than by providing generally for the patentability of living organisms under the basic patent law.[39]

The petitioner maintained that, in the absence of a clear congressional intent to afford patent protection to living organisms, the patent statute should not be interpreted to extend coverage to new life forms. Indeed, there seems to be an underlying preference in the submission for sui generis protection of biological inventions. The petitioner argued that Congress did not intend living things to be included within the scope of patent protection. The submission observed that the *Plant Patent Act* 1930 (US) and the *Plant Variety Protection Act* 1970 (US) provided evidence of this intention: 'New legislation was necessary to permit the patenting of such newly created plants.'[40] The submission added: 'That legislation does not encompass Chakrabarty's invention; accordingly, it is not patentable.'[41]

However, underneath this formal position, there were internal divisions within the United States Government over the *Diamond v Chakrabarty* case. Rebecca Eisenberg recounts:

> By this point many people within the [USPTO] favored patent protection for living organisms, including the new Commissioner, Donald Banner, who thought the CCPA decision was correct and was not inclined to seek Supreme Court review. Others within the PTO favored taking the case to the Supreme Court in the hope of getting an affirmance that would give biotechnology investors greater assurance of the validity of their patents, while the Solicitor General of the United States favored reversal.[42]

Thus, in spite of the overt position taken by the petitioner, there lurked a great deal of ambivalence about the merits of allowing patents in respect of biological inventions. Interestingly, after the decision in *Diamond v Chakrabarty*, the United States Government would later become an ardent supporter of intellectual property rights in biotechnology, both at home and in international trade negotiations.

B Respondents' Submission

In a cunningly constructed brief, Chakrabarty's lawyers challenged the government's 'allegation that patenting Chakrabarty's man-made bacterium would amount to extension of the patent laws into new areas'.[43] The lawyers argued that, in fact, the USPTO had a long history of granting patents in relation to biological inventions:

> Patents considered by this and other courts have been issued on living things, including bacteria. In fact, so many have issued that official Patent Office specific subclasses have been established for collection of these patents. Search of these subclasses, and other sources, have located many, many patents to living things. Included are ones in which the Board of Appeals has reversed an Examiner's rejection, thereby causing issuance of the patents. Indeed, a Commissioner of Patents has informed Congress that 'cultures', which are living microorganisms, are patentable.[44]

Chakrabarty's lawyers argued that the policy to grant patents on living things, and specifically on bacteria, could be demonstrated by issuance of patents considered by courts,[45] by the official classifications of issued patents,[46] and by the absence of policy statements to the contrary.[47] General Electric collected over 60 issued patents claiming living subject matter, mostly bacteria, in a limited search of Patent Office records. The submission noted that neither the *Plant Patent Act* 1930 (US) nor the *Plant Variety Protection Act* 1970 (US) had halted the granting of such patents.

The respondents stressed that Chakrabarty's invention did not involve the use of recombinant DNA. The lawyers sought to ward off allegations that genetic technology posed undue risks to human health and the environment:

> Recombinant DNA research, and any controversies concerning it, furnish no basis for denying patents on bacteria because they are alive. Indeed, inhibition to the making and disclosure of recombinant DNA inventions, by outlawing patents on living microorganisms, may adversely affect that research and its great promise for mankind.[48]

General Electric emphasized: 'It is axiomatic that any dangers to the public's health and safety are best prevented by regulation of the source of those dangers, not by an indirect approach that would prevent patenting the results of research.'[49]

To support a broad reading of s. 101, Chakrabarty's lawyers highlighted a passage in the Committee Report on the *Patent Act* 1952 (US): 'A person may have "invented" a machine or a manufacture, which may include anything under the sun that is made by man.'[50] The lawyers maintained that the man-made bacterium fell within the scope of patentable subject matter, because it was a product of 'manufacture'. Alternatively, the counsel submitted that the bacterium was a 'composition of matter'. The lawyers summed up their case:

If the Government wishes to reverse its policy, it should address its desires to the Congress, which can legislate an exclusion, if that is found to be required by the public interest. In the meantime there is no justification for this Court to read the limitation to nonliving subject matter into the patent law.[51]

Citing the authority of *United States v Dubilier Condenser Corp.*, the counsel concluded: 'We should not read into the patent laws limitations and conditions which the legislature has not expressed.'[52]

C The Peoples Business Commission

In an amicus brief, the critic of genetic engineering, Jeremy Rifkin, and the non-government organization, the Peoples Business Commission ('the Commission'), argued that the United States Congress never intended that living organisms should be patentable: 'PBC believes that the ecological, evolutionary, ethical, philosophical, political and economic questions that surround the patenting of living organisms have been given insufficient consideration by the Congress, the country as a whole and the lower court in issuing its ruling in favor of such patents.'[53] The Commission expressed fears that a ruling in favour of patents in respect of life form patents would serve as a precedent in the field of recombinant DNA, and related areas of genetic manipulation: 'Such a ruling would significantly contribute to the profit potential of the genetic industry, thus generating a greater momentum in research and development of genetic engineering technologies.'[54]

First, the Commission submitted: 'The history of the results of the several plant patent acts clearly shows that far from leading to a multiplicity of social benefits, the patenting of plants has in fact cruelly robbed succeeding generations of their own right to a diversified, healthy and vital gene pool.'[55] The group argued that the relatively recent history of granting plant patents illustrated the deleterious genetic and social effects of patenting living organisms, such as the loss of 'the right to a diversified gene pool composed of thousands of varieties of naturally occurring life forms'.[56] The Commission elaborated upon its fears about the patenting of genetic engineering. The group worried that the novel micro-organisms that will be created through various genetic engineering techniques could be, in many cases, 'superbugs', and harm the ecosystem. The Commission feared that the monoculturing of micro-organisms may well prove as deficient as that in food crops. The Commission also expressed concern about biotechnology companies obtaining monopolies in respect of essential genetic resources.

Second, the Commission submitted that the granting of patents on living organisms, and the technology of genetic engineering, taken as a whole, was not in the public interest:

The term 'Biological Revolution' has rightly been used to characterize the astounding and awesome strides being made in the fields of biology and genetics. Because of this Biological Revolution, highly technological societies such as ours are on the threshold of controlling the biological and genetic quality of all living material, from the humblest microorganism to the most proud human. As Dr. George Wald, the Harvard Nobel laureate has said, 'we are moving from the organic design of life to technological specification of living material.' Just as we have manufactured metals and plastics, now there are those who contemplate manufacturing life itself.[57]

The Commission warned: 'If the lower court ruling is upheld, and patents on living organisms are awarded to General Electric and Upjohn, all chance of meaningful public education and participation in the policy decisions surrounding genetic engineering will be lost, for the granting of patents is sure to escalate the drive toward commercial application.'[58] The group argued: 'The genie will be out of the bottle before most Americans have even realized that the bottle was uncorked.'[59]

Third, the Commission noted the submission of the university that 'a microorganism is so close to "the periphery" of life that there should be no obstacle to patenting it, and indeed to proclaiming it non-living'.[60] The group questioned: 'Where and how will we draw the line once we embark on a course of classifying life at "the periphery" as so inconsequential that it is patentable material?'[61] The Commission predicted that the patenting of lower organisms would lead to the patenting of higher forms of life: 'If a ruling in favor of patenting genetically engineered living organisms is forthcoming, then manufactured life – high and low – will have been categorized as less than life, as nothing but common chemicals.'[62] The Commission buttressed its claims with a number of quotations from a range of respectable scientists, philosophers and ethicists. The Commission concluded, citing the cautionary warning of Dr Leon Kass: 'We have paid some high prices for the technological conquest of nature, but none perhaps so high as the intellectual and spiritual costs of seeing nature as mere material for our manipulation, exploitation and transformation.'[63]

The submission of the Peoples Business Commission is worth recalling for a number of reasons. The intervention of the amicus curiae is certainly pioneering in this field. The Peoples Business Commission has been a model for non-government organizations and members of civil society who wish to intervene and participate in patent proceedings, whether through opposing the validity of patent applications, or engaging in wider media campaigning. Furthermore, the group has also been influential in raising wider public policy concerns within the forum of the courts. The submission is a compendium of objections to patenting biological inventions. The brief contains a catalogue of complaints about patenting products of nature, scientific discoveries and life forms. In some ways, the submission was remarkably prescient. The patent granted in *Diamond v Chakrabarty* did indeed open the way for the USPTO to patent both lower and higher life forms. In other respects, the submission is overblown. The fears about the impacts of biotechnology have not necessarily been borne out by developments in the following three decades.

D Amicus Briefs Urging Affirmance

There were a number of amicus briefs from private corporations, industry groups, universities, scientific associations and researchers submitted to the Supreme Court of the United States in support of the General Electric Company. The first established biotechnology company, Genentech Inc, intervened in the *Diamond v Chakrabarty* case as an amicus curiae. The inaugural chief executive of the company, Robert Swanson, reflected:

> This case, which didn't have a lot of commercial relevance, had a great deal of relevance to the biotechnology industry in terms of drug development. We felt that we needed to participate and we did, and fortunately the court made the right decision.[64]

Counsel Thomas Kiley emphasized that the Court's decision would have a profound impact on the question of whether investments in research expenditures and recombinant DNA technology should be made in view of the character of patent protection available.[65] He encouraged the Court to 'confirm the patentability of micro-organisms and both encourage a beneficent science and ensure that broad and forward looking incentives remain for those who would pull the next technology'.[66] Such an intervention demonstrates how Genentech was at the vanguard of legal innovation in respect of biotechnological inventions.

Genentech compared the Peoples Business Commission to the Luddites of early nineteenth-century England who sought to prevent the spread of labour-saving machinery by the simple expedient of destroying it.[67] The biotechnology company observed that the group was unreasonably hostile to new technology

> The attempt to cast this Court in a legislative role is nowhere more evident than in the brief amicus of the Peoples Business Commission (PBC), whose essentially Luddite philosophy would have the Court stand the Patent System on its head, denying patents so as to avoid '. . . generating a greater momentum in research and development of genetic engineering technologies . . . [which] . . . in turn, will lead to the rapid proliferation of genetic techniques in the areas of energy, agriculture, medicine, industrial processes and many other aspects of the nation's economic life'.[68]

Genentech stressed that 'the question before the Court is neither one of ethics, for philosophy, nor politics'.[69] Rather, it submitted: 'It is one of statutory interpretation, of grammar leavened with reason.'[70] The biotechnology company urged the Supreme Court of the United States to 'confirm the patentability of microorganisms' and 'both encourage a beneficent science and ensure that broad and forward-looking incentives remain for those who would pull the next technology, the one now invisible because still down over the horizon of the future, into view and into use'.[71]

The pharmaceutical industry was an enthusiastic promoter of biotechnology, because it was keen to engage in industrial renewal and develop new fields of research and manufacturing. The Pharmaceutical Manufacturers Association argued that patent protection of micro-organisms would be instrumental in encouraging industrial innovation in biotechnology: 'While the Chakrabarty microorganism is economically important for what it consumes, other modified organisms are and will become important for the materials they produce.'[72] In particular, the Association stressed the breakthroughs in medical biotechnology, with the production of human insulin and human growth hormone. The group concluded: 'Given the vast potential of living organism technology for solving many of the ills besetting the country and for contributing significantly to a turnaround of the innovation crisis, it would be tragic indeed if this Court were to accept the "sky-is-falling" arguments of Amicus PBC or the tortured and slavishly technical arguments of Petitioner against its patentability in the abstract.'[73]

The University of California intervened in the case of *Diamond v Chakrabarty*. Counsel Edward Irons and Mary Sears stressed that the case was relevant to academic research on insulin and human growth hormone:

> Whether the University has the right to patent its own newly manufactured microorganisms will depend directly on the disposition that is made in this case. In turn, this will govern whether the University receives income from these inventions, to be significantly shared with its inventors and to use, inter alia, in supporting new research. Indeed, if no patents issue, the health care industry may well elect not to commercialize these important inventions because of its avowed belief that, absent the protection a patent affords, the time and experimental work requisite to obtaining government clearances cannot be justified.[74]

Historian Daniel Kevles stressed that the University of California 'was no more alive than other universities to the hopes of revenues from biotechnology, only more immediately interested, by virtue of the activities on its San Francisco campus'.[75] This statement was echoed and generalized in a single amicus brief filed on behalf of a number of peak organizations in respect of biochemistry and molecular biology. An amicus brief was submitted by the pioneer in biotechnology, Dr Leroy Hood, and several other scientists, along with the American Society of Biological Chemists, the Association of American Medical Colleges, the California Institute of Technology and the American Council on Education.[76] The submission emphasized the need for financial incentives to support commercial funding of public research.

> Though engaged primarily in basic research, Amici have an interest in seeing their work reach commercial development. They fear that adoption of a per se rule excluding all living things from patentability will inhibit commercial development of the advances they are making in recombinant DNA research. Such inhibitions will occur because the incentive to follow through on many scientific advances, so that they will be commercially useful, will be lacking without appropriate financial incentives.[77]

The submission maintained that a per se rule excluding living organisms from patent protection would be improper in light of government action to facilitate the development and application of genetic technologies.

Dr George Pieczenik, a molecular biologist and a computer scientist from Rutgers University, argued that the position of the petitioner lacked a strong scientific basis:

> The distinction between living and non-living matter has no real meaning in relation to this technology. That which is living is typically described in terms of a set of attributes which, when all present, are considered indicia of life. There is no single fundamental property, law of nature, or operating principle, which distinguishes that matter which we call living from that which we do not. To attempt to separate patentable and unpatentable subject matter on the basis of such a concept is to invite confusion in the art, to ignore existing law and to ignore scientific reality.[78]

He recommended 'that the decision of the Court of Customs and Patent Appeals be affirmed but that in any case, the Court's holding provide rational metes and bounds to guide the patenting of subsequent developments in this emerging technology'.[79]

The American Society for Microbiology provided an additional amicus brief, stressing: 'It is particularly important for the Court to be aware that the capability of scientists to make new microorganisms through modification of genetic elements encompasses a variety of scientific techniques and that each of these techniques constitutes the deliberate intervention of man to create a novel microorganism.'[80] The Society contended that the manufacture of novel micro-organisms provided important benefits for the public and for the exchange of scientific information.

The American Patent Association emphasized that the man-made micro-organisms could not be classified as mere products of nature:

Today, however, 'things alive' cannot be equated automatically and invariably to 'products of nature'. Today, the technology or art of genetic engineering is producing 'things alive' which are made by man, not by nature. While these creations demonstrate the properties of 'aliveness', they are not products of nature. They are products of man, products which serve the new and useful ends to which the court referred in *Funk*. The new and useful ends served by respondent's invention were achieved by man. Respondent did not discover a phenomenon of nature but created a phenomenon which does not exist in nature.[81]

The Association maintained that 'there is no justification under our laws known to amicus for a distinction between property rights in living things as opposed to nonliving things'.[82] The submission stressed: 'Valid property rights in living entities have been recognized as long as humans have existed, from the domesticated goat and plots of Indian corn to today's vast herds of sheep, cattle and pigs and vast fields of wheat.'[83] The Association concluded: 'One must ask why the Patent and Trademark Office, through the office of Solicitor General, seeks to have patent rights in living things set apart as some special breed of property right.'[84]

III THE SUPREME COURT OF THE UNITED STATES

In *Diamond v Chakrabarty*, the Supreme Court of United States held by a majority of five to four that Ananda Chakrabarty's patent application in respect of an oil-eating bacterium constituted either a manner of manufacture or a composition of matter and was therefore patentable.[85] There were stark divisions of opinion between the majority and the dissenters. There was a significant argument over methods of hermeneutics and jurisprudence; the rationales and justifications for the patent system; the proper relationship between the judiciary and the Congress; and the relevance or otherwise of questions of philosophy, ethics and politics. For the majority, Burger CJ engaged in a formalistic interpretation of United States patent law. His Honour stated that 'anything under the sun that is made by man' was patentable subject matter. The judge observed that the Court lacked the competence to address arguments about the ethics of genetic engineering and recommends that they be addressed to Congress and the President, as the balancing of competing values and interests was more a matter of high policy for resolution within the legislative process. By contrast, Brennan J dissented that the majority had engaged in considerable judicial creativity by expanding the limits of beyond patentable subject matter, beyond the boundaries explicitly delimited by the United States Congress. His Honour was sensitive to the larger public policy considerations involved in granting patents in respect of life forms. This judgment has been influential amongst sceptics and doubters who question whether it is wise to grant patents in respect of biological inventions.

A Burger CJ

Burger CJ delivered the opinion of the majority of the Supreme Court of the United States. His Honour stressed that the case presented a narrow question of statutory interpretation of s. 101 of the *Patent Act* 1952 (US), which provides: 'Whoever invents or discovers any new and useful process, machine, manufacture, or composition of matter, or any new and useful improvement thereof, may obtain a patent therefor, subject to the conditions and requirements of this title.' His Honour observed: 'Specifically, we must determine whether respondent's micro-organism constitutes a "manufacture" or "composition of matter" within the meaning of the statute.'[86]

Burger CJ emphasized that a broad interpretation should be given to the terms 'manufacture' and 'composition of matter':

> Guided by these canons of construction, this Court has read the term 'manufacture' in §101 in accordance with its dictionary definition to mean 'the production of articles for use from raw or prepared materials by giving to these materials new forms, qualities, properties, or combinations, whether by handlabor or by machinery'.[87] Similarly, 'composition of matter' has been construed consistent with its common usage to include 'all compositions of two or more substances and . . . all composite articles, whether they be the results of chemical union, or of mechanical mixture, or whether they be gases, fluids, powders or solids'.[88] In choosing such expansive terms as 'manufacture' and 'composition of matter', modified by the comprehensive 'any', Congress plainly contemplated that the patent laws would be given wide scope.[89]

Burger CJ emphasized that such a broad construction of patentable subject matter was supported by legislative history. His Honour noted that the original *Patent Act* 1793 (US) embodied Thomas Jefferson's philosophy that 'ingenuity should receive a liberal encouragement'.[90] Moreover, he observed that subsequent patent statutes in 1836, 1870 and 1874 employed this same broad language. Famously, Burger CJ observed that recodification of the patent legislation in 1952 reflected a Congressional intention to allow patent protection on a wide range of technologies and scientific inventions: 'The Committee Reports accompanying the 1952 Act inform us that Congress intended statutory subject matter to "include anything under the sun that is made by man".'[91]

Burger CJ emphasized, though, that a distinction should be drawn between patentable inventions and scientific discoveries:

This is not to suggest that 101 has no limits or that it embraces every discovery. The laws of nature, physical phenomena, and abstract ideas have been held not patentable. Thus, a new mineral discovered in the earth or a new plant found in the wild is not patentable subject matter. Likewise, Einstein could not patent his celebrated law that $E=mc^2$; nor could Newton have patented the law of gravity. Such discoveries are 'manifestations of . . . nature, free to all men and reserved exclusively to none'.[92]

His Honour held: 'Judged in this light, respondent's micro-organism plainly qualifies as patentable subject matter.'[93] In his view, Chakrabarty's 'claim is not to a hitherto unknown natural phenomenon, but to a non-naturally occurring manufacture or composition of matter – a product of human ingenuity "having a distinctive name, character [and] use".'[94] Burger CJ observed that 'Here . . . the patentee has produced a new bacterium with markedly different characteristics from any found in nature and one having the potential for significant utility.'[95] Resonantly, the judge concluded: 'His discovery is not nature's handiwork, but his own; accordingly it is patentable subject matter under 101.'[96]

Burger CJ went on to critique the two main objections put forward by the petitioner. First, he considered the case of the petitioner that the existence of the *Plant Patent Act* 1930 (US) and the *Plant Variety Protection Act* 1970 (US) demonstrated a Congressional understanding that the terms 'manufacture' and 'composition of matter' did not include living things. The judge rejected the argument after due consideration:

> In enacting the *Plant Patent Act* . . . Congress thus recognized that the relevant distinction was not between living and inanimate things, but between products of nature, whether living or not, and human-made inventions. Here, respondent's micro-organism is the result of human ingenuity and research. Hence, the passage of the *Plant Patent Act* affords the Government no support.
>
> Nor does the passage of the 1970 *Plant Variety Protection Act* support the Government's position. As the Government acknowledges, sexually reproduced plants were not included under the 1930 Act because new varieties could not be reproduced true-to-type through seedlings. By 1970, however, it was generally recognized that true-to-type reproduction was possible and that plant patent protection was therefore appropriate. The 1970 Act extended that protection. There is nothing in its language or history to suggest that it was enacted because §101 did not include living things.[97]

In the marginalia, the judge noted that there were examples of patents being granted in respect of micro-organisms by the USPTO. In 1873, the Patent Office granted Louis Pasteur a patent on 'yeast, free from organic germs of disease, as an article of manufacture'.[98] In 1967 and 1968, immediately prior to the passage of the *Plant Variety Protection Act* 1970 (US), the Patent Office granted two patents which stated claims for living micro-organisms.

Second, Burger CJ considered the argument of the petitioner that micro-organisms cannot qualify as patentable subject matter until Congress expressly authorizes such protection. Diamond maintained that genetic technology was unforeseen when Congress enacted section 101. The petitioner contended that the resolution of the patentability of inventions should be left to Congress. Diamond argued that the legislative process was best equipped to weigh the competing economic, social and scientific considerations involved, and to determine whether living organisms produced by genetic engineering should receive patent protection.

Burger CJ was reluctant to consider the policy arguments of the petitioner and the amicus curiae, the Peoples Business Commission, that genetic research and related technological developments pose grave risks to human and animal health, and genetic diversity in the environment. His Honour noted:

> It is argued that this Court should weigh these potential hazards in considering whether respondent's invention is patentable subject matter under 101. We disagree. The grant or denial of patents on micro-organisms is not likely to put an end to genetic research or to its attendant risks. The large amount of research that has already occurred when no researcher had sure knowledge that patent protection would be available suggests that legislative or judicial fiat as to patentability will not deter the scientific mind from probing into the unknown any more than Canute could command the tides. Whether respondent's claims are patentable may determine whether research efforts are accelerated by the hope of reward or slowed by want of incentives, but that is all.[99]

Burger CJ argues that the patentability of genetic research is a question which is divorced from the wider regulation of genetic technology.

In a classic, oft-cited statement, Burger CJ fervently disavows the capacity of the courts to consider wider policy arguments within the context of patent law:

> What is more important is that we are without competence to entertain these arguments – either to brush them aside as fantasies generated by fear of the unknown, or to act on them. The choice we are urged to make is a matter of high policy for resolution within the legislative process after the kind of investigation, examination, and study that legislative bodies can provide and courts cannot. That process involves the balancing of competing values and interests, which in our democratic system is the business of elected representatives. Whatever their validity, the contentions now pressed on us should be addressed to the political branches of the Government, the Congress and the Executive, and not to the courts.[100]

Burger CJ observed that 'Congress is free to amend §101 so as to exclude from patent protection organisms produced by genetic engineering.'[101] He noted that in the past Congress had exempted from patent protection inventions useful solely in the utilization of special nuclear material or atomic energy in an atomic weapon. Burger CJ concluded that the courts should take an inclusive approach to the interpretation of patentable subject matter, in the absence of any particular exclusions. There is a peculiar contradiction to the logic of the Chief Judge: he argues that courts should be involved in higher matters of policy; nonetheless he is willing to second guess the attitudes of Congress as to the patentability of genetic research.

B Brennan J

Brennan J wrote a short, pithy dissent on behalf of White, Marshall and Powell JJ. His Honour agreed that the question before the court was a narrow one:

> Neither the future of scientific research, nor even the ability of respondent *Chakrabarty* to reap some monopoly profits from his pioneering work, is at stake. Patents on the processes by which he has produced and employed the new living organism are not contested. The only question we need decide is whether Congress, exercising its authority under Art. I, 8, of the Constitution, intended that he be able to secure a monopoly on the living organism itself, no matter how produced or how used.[102]

The judge dissented on the basis that the majority had misread the applicable legislation.

Brennan J maintained that the courts should be deferential to the wishes of Congress, and extend patent protection no further than statute provides:

> The patent laws attempt to reconcile this Nation's deep-seated antipathy to monopolies with the need to encourage progress. Given the complexity and legislative nature of this delicate task, we must be careful to extend patent protection no further than Congress has provided . . . In particular, were there an absence of legislative direction, the courts should leave to Congress the decisions whether and how far to extend the patent privilege into areas where the common understanding has been that patents are not available.[103]

His Honour noted that Congress had on two previous occasions passed legislation to provide intellectual property rights to plant varieties: the *Plant Patent Act* 1930 (US) afforded patent protection to developers of certain asexually reproduced plants and the *Plant Variety Protection Act* 1970 (US) extended protection to certain new plant varieties capable of sexual reproduction. Brennan J noted: 'In these two Acts Congress has addressed the general problem of patenting animate inventions and has chosen carefully limited language granting protection to some kinds of discoveries, but specifically excluding others.'[104] In particular, he recognized: 'These Acts strongly evidence a congressional limitation that excludes bacteria from patentability.'[105]

Brennan J complained that the majority had engaged in unwarranted judiciary creativity by broadening the scope of patent protection:

> The Court's decision does not follow the unavoidable implications of the statute. Rather, it extends the patent system to cover living material even though Congress plainly has legislated in the belief that 101 does not encompass living organisms. It is the role of Congress, not this Court, to broaden or narrow the reach of the patent laws. This is especially true where, as here, the composition sought to be patented uniquely implicates matters of public concern.[106]

His last sentiments suggest a willingness to take into account wider policy concerns about genetic research.

CONCLUSION

Reminiscing about the patent litigation, patent applicant Ananda Chakrabarty has reflected upon the legacy of the decision in *Diamond v Chakrabarty*:

> This is both an exciting and a difficult time for a biologist. The technology of animal and human reproduction, as well as the techniques of genetic manipulation, are progressing so rapidly it creates situations that transcend our legal structure and directly affect our social and moral fabrics. It is high time that the United States Congress take a serious look at where the science is going, where it needs to make a positive contribution, and perhaps define the boundaries of our venture into the unknown biological mysteries of nature.[107]

Chakrabarty reflects: 'Very few pharmaceutical companies tried to patent products or cultures before GE applied for the patent on the oil-eating micro-organisms because these companies relied mostly on trade secrets.'[108] The micro-biologist maintains that the majority decision of the Supreme Court of the United States encouraged the dissemination of information and inventions about biotechnology: 'The *Diamond v Chakrabarty* decision has immensely contributed to the growth of the biotechnology industry both by allowing patenting of life forms, as well as facilitating dissemination of scientific ideas, technology, and concepts.'[109]

In the wake of the decision in *Diamond v Chakrabarty*, the USPTO granted a backlog of patent applications in respect of genes and gene sequences. Academic commentator, Rebecca Eisenberg, observed that the decision in *Diamond v Chakrabarty* encouraged the filing of gene patent applications:

> In stark contrast to the public controversy surrounding the patentability of Chakrabarty's invention, the patenting of DNA sequences in the late 1970s and 1980s drew hardly any attention from the media. Following precedents upholding the patentability of purified versions of such naturally occurring products as adrenaline and vitamin B_{12}, the PTO had no trouble allowing patents on 'purified and isolated' DNA sequences and recombinant constructs incorporating such sequences. In the early days of the biotech industry, patenting the genes encoding therapeutic proteins looked like a high-tech variation on the familiar practice of patenting drugs.[110]

Stanley Cohen of Stanford University and Herb Boyer of the University of California, San Francisco obtained a patent for the development of recombinant DNA.[111] This invention made it possible to recombine and clone DNA, thus providing basic scientists with a simple and precise method for studying the structure and function of genes of higher and lower organisms.[112] Sally Smith Hughes comments that the Cohen–Boyer patents were a catalyst for an attitudinal shift among scientists, research institutions and entrepreneurs: 'The patent and its two companions of 1984 and 1988 were instruments in the transformation of perceptions and policy regarding commercial activity in academia.'[113]

The first biotechnology company, Genentech, obtained patent protection in respect of human growth hormone and human insulin.[114] In the wake of its patent applications, Genentech offered its stock to public investors on 14 October 1980. The stock at its initial public offering underwent the most dramatic escalation in value in Wall Street history – the offering of one million shares of stock at $US 35 per share climbed to $US 89 a share within the first 20 minutes of trading.[115] This excitement was reflected in an article in *Time Magazine*, which featured Herb Boyer on the cover, with the headline: 'Shaping Life In The Lab: The Boom In Genetic Engineering'.[116] By the end of the day, the company had raised $US 36 million and was valued at $US 532 million, a sign of investors' enthusiasm for biotechnology at the time.

Rebecca Eisenberg observes that the decision in *Diamond v Chakrabarty* has had larger ramifications for patent jurisprudence and the institutional politics of the courts:

> As predicted by both proponents and opponents of patents on living organisms, investment in biotechnology R & D has flourished in the wake of *Diamond v Chakrabarty*. But the full consequences of the expansive approach to patent eligibility endorsed by the *Chakrabarty* majority continue to be felt far beyond the biotechnology industry . . . A quarter-century ago it was unclear whether the

subject matter boundaries of the patent system were expansive enough to embrace biotechnology and information technology. Today, it is not clear whether the patent system has any subject matter boundaries at all.[117]

Eisenberg comments: 'Over the past quarter-century, following the Supreme Court's broad directive in *Diamond v Chakrabarty*, the Federal Circuit has gradually eviscerated what once appeared to be time-honored categorical exclusions from the patent system for such subject matter as "business methods" and "mathematical algorithms" in favor of a "big tent" approach to patent eligibility.'[118]

It is enlightening to counterpoint the ruling in *Diamond v Chakrabarty* with parallel developments in other jurisdictions. It is worth commenting that such jurisdictions accepted the patentability of micro-organisms, without the larger, divisive policy debate that occurred in the United States. In *American Cyanamid v Berk Pharmaceuticals*, the High Court of Justice Chancery Division held that a patent for a method of producing the antibiotic tetracycline was a manner of manufacture.[119] In *Ranks Hovis McDougall's Application*, the Australian Patent Office held that patents could be granted in respect of living organisms.[120] The Canadian Patent Appeal Board allowed patents in respect of lower life forms in *Re Application for Patent of Abitibi Co*, and *Re Application for Patent of Connaught Laboratories*.[121] However, the Board deferred the larger question of patentability of higher-life forms, as it was not required by the facts of the cases.

Since the contested decision in *Diamond v Chakrabarty* in the Supreme Court of the United States, and the sequence of judgments in other jurisdictions, such as the United Kingdom, Australia and Canada, patent protection of micro-organisms has become an international norm. Article 27(3) of the *TRIPS Agreement* 1994 provides that member States may exclude from patentability 'plants and animals other than micro-organisms, and essentially biological processes for the production of plants and animals'. Members of the World Trade Organization are therefore obliged to provide for patent protection of micro-organisms.

Inventions involving the use of new micro-organisms have presented problems of disclosure in that repeatability often cannot be ensured by means of a written description alone. The *Budapest Treaty* 1977 was developed to provide a uniform international deposit system. The Treaty has three main functions. Under the Treaty, certain culture collections are recognized as 'international depositary authorities'. Any Contracting State which allows or requires the deposit of micro-organisms for the purposes of patent procedure must recognize, for those purposes, a deposit made in any such authorities. The Regulations under the Treaty lay down in detail the procedures which depositors and authorities must follow, the duration of storage of deposited micro-organisms, and the mechanisms for the furnishing of samples. The Treaty and Regulations make various provisions to guard against the loss and consequent non-availability of deposited micro-organisms.

NOTES

1. Federico, P.J. (1951), 'Hearings on H.R. 3760 before Subcommittee No. 3 of the House Committee on the Judiciary', 82d Congress, 1st Session, 37.
2. Grubb, Philip (2004), *Patents for Chemicals, Pharmaceuticals and Biotechnology*, 4th edn, Oxford: Oxford University Press, pp. 224–5.
3. Pasteur, L. (1873), 'Improvement in the manufacture of beer and yeast', US Patent No: 141, 972.
4. Grubb, Philip (2004), *Patents for Chemicals, Pharmaceuticals and Biotechnology*, 4th edn, Oxford: Oxford University Press, p. 227.
5. *Ex parte Latimer* 1889 (CD 46 OG 1638).
6. *Ex parte Latimer* 1889 (CD 46 OG 1638).
7. *Ex parte Latimer* 1889 (CD 46 OG 1638).
8. Chakrabarty, A. (1972), 'Microorganisms having multiple compatible degradative energy-generating plasmids and preparation thereof', US Patent No: 4,259,444.
9. *Application of Chakrabarty* 571 F.2d 40 Cust. & Pat.App (1978).
10. *Application of Bergy* 596 F.2d 952 Cust. & Pat.App. (1979).
11. Eisenberg, R. (2006), 'Biotech patents: looking backward while moving forward', *Nature Biotechnology*, **24**(3), 317–19.
12. Kevles, D. (1994), 'Ananda Chakrabarty wins a patent: biotechnology, law and society, 1972–1980', *Historical Studies in the Physical and Biological Sciences*, **25**, 111–35; Ananda Chakrabarty (2002), 'Patenting of life forms: from a concept to reality', in David Magnus, Arthur Caplan and Glenn McGee (eds), *Who Owns Life?*, Amherst (NY): Prometheus Books, pp. 17–24; and Ananda Chakrabarty (2003), 'Patenting life forms: yesterday, today, and tomorrow', in F. Scott Kieff (ed.), *Perspectives on Properties of the Human Genome Project*, Amsterdam: Elsevier, pp. 3–11.
13. Wilson, Jack (2002), 'Patenting organisms: intellectual property law meets biology', in David Magnus, Arthur Caplan and Glenn McGee (eds), *Who Owns Life?*, Amherst (NY): Prometheus Books, pp. 25–58.
14. For interviews with the inventors, see D. Kevles (2002), 'A history of patenting life in the United States with comparative attention to Canada and Europe', European Group on Ethics in Science And New Technologies To The European Commission, 12 January.
15. Chakrabarty, Ananda (2002), 'Patenting of life forms: from a concept to reality', in David Magnus, Arthur Caplan and Glenn McGee (eds), *Who Owns Life?*, Amherst (NY): Prometheus Books, p. 19.
16. Chakrabarty, A. (1972), 'Microorganisms having multiple compatible degradative energy-generating plasmids and preparation thereof', US Patent No: 4,259,444.
17. Chakrabarty, A. (1972), 'Microorganisms having multiple compatible degradative energy-generating plasmids and preparation thereof', US Patent No: 4,259,444.
18. Chakrabarty, A. (1972), 'Microorganisms having multiple compatible degradative energy-generating plasmids and preparation thereof', US Patent No: 4,259,444.
19. *Application of Chakrabarty* 571 F.2d 40 Cust. & Pat.App (1978).
20. *Application of Chakrabarty* 571 F.2d 40 Cust. & Pat.App (1978).
21. *Application of Chakrabarty* 571 F.2d 40 at 44 Cust. & Pat.App (1978).
22. *Application of Chakrabarty* 571 F.2d 40 Cust. & Pat.App (1978).

23. *Application of Chakrabarty* 571 F.2d 40 Cust. & Pat.App (1978).
24. *Application of Chakrabarty* 571 F.2d 40 Cust. & Pat.App (1978).
25. *Application of Chakrabarty* 571 F.2d 40 Cust. & Pat.App (1978).
26. *Application of Chakrabarty* 571 F.2d 40 Cust. & Pat.App (1978).
27. *Application of Bergy* 563 F.2d 1031 Cust. & Pat.App. (1977).
28. *Application of Bergy* 563 F.2d 1031 Cust. & Pat.App. (1977).
29. Bergy, M., J. Coats and V. Malik (1974), 'Process for preparing lincomycin', US Patent Application No: 477,766.
30. *Application of Bergy* 563 F.2d 1031 at 1038 Cust. & Pat.App. (1977).
31. *Application of Bergy* 563 F.2d 1031 at 1039 Cust. & Pat.App. (1977).
32. *Application of Bergy* 596 F.2d 952 Cust. & Pat.App. (1979).
33. *Application of Bergy* 596 F.2d 952 at 987 Cust. & Pat.App. (1979).
34. *Application of Bergy* 596 F.2d 952 at 987 Cust. & Pat.App. (1979).
35. *Application of Bergy* 596 F.2d 952 at 988 Cust. & Pat.App. (1979).
36. *Application of Bergy* 596 F.2d 952 at 988 Cust. & Pat.App. (1979).
37. *Application of Bergy* 596 F.2d 952 at 1002 Cust. & Pat.App. (1979).
38. Frow, J. (1994), 'Timeshift: technologies of reproduction and intellectual property', *Economy and Society*, **23**, 290.
39. Diamond, S. (1980), 'Brief for the petitioner in *Diamond v Chakrabarty*', 1980 WL 339757, 9–10.
40. Ibid.
41. Ibid.
42. Eisenberg, Rebecca (2006), 'The story of *Diamond v Chakrabarty*: technological change and the subject matter boundaries of the patent system', Jane Ginsburg and Rochelle Cooper Dreyfuss (eds), *Intellectual Property Stories*, New York: Foundation Press, pp. 327–57 at 349.
43. Chakrabarty, A. (1980), 'Brief for the respondent in *Diamond v Chakrabarty*', 1980 WL 339758, 28 January.
44. Ibid., 11.
45. *Funk Bros. Co. v Kalo Co.*, 333 U.S. 127 (1948); and *American Fruit Growers v Brogdex Co.*, 283 U.S. 1 (1931).
46. Class 424 is entitled 'Drug, bio-affecting and body treating compositions'. The official 'Classification Definitions' for Class 424 provide: 'Class 424 provides for compositions containing microorganisms, either alive, dead or attenuated.' Class 424 contains a subclass 93 entitled: 'Whole live microorganism or virus containing.' The January 1979 subclass list of Class 195 'Chemistry, fermentation' contained a Subclass 53 entitled: 'Ferment-containing products . . . living fungi-containing.'
47. 'Manual of patent examining procedure' lists the types of subject matter not considered by the Office to be within the ambit of 35 U.S.C. 101, as 'Printed matter', 'Naturally occurring article', 'Method of doing business' and 'Scientific Principle'. It was silent as to living things.
48. Chakrabarty, A. (1980), 'Brief for the respondent in *Diamond v Chakrabarty*', 1980 WL 339758, 28 January, 31.
49. Ibid., 29.
50. Ibid., 39.
51. Ibid., 54.
52. *United States v Dubilier Condenser Corp.*, 289 U.S. 178, 199 (1933).
53. Peoples Business Commission (1979), 'Appellate brief of peoples business commission as amici curiae in support of the petitioner in *Diamond v Chakrabarty*', 1979 WL 2005, 13 December, 2–3.
54. Ibid., 2–3.
55. Ibid., 10.
56. Ibid., 11.
57. Ibid., 14–15.
58. Ibid., 22.
59. Ibid., 22.

60. Ibid., 28.
61. Ibid., 28.
62. Ibid., 30–31.
63. Ibid., 25.
64. Swanson, R. (1996–97), 'An oral history conducted by Sally Smith Hughes', Bioscience and Biotechnology Archives and Oral Histories, Bancroft Library, http://sunsite.berkeley.edu:2020/dynaweb/teiproj/oh/science/swanson/@Generic__BookView.
65. Genentech Inc. (1980), 'Appellate brief of Genentech Inc. as amicus curiae in support of the respondents in *Diamond v Chakrabarty*', 1980 WL 339766, 28 January.
66. Ibid.
67. Ibid.
68. Ibid., 11.
69. Ibid., 11.
70. Ibid., 11.
71. Ibid., 22.
72. Pharmaceutical Manufacturers Association (1980), 'Appellate brief of the pharmaceutical manufacturers association as amicus curiae in support of the respondents in *Diamond v Chakrabarty*', 1980 WL 339771, 29 January, 3–4.
73. Ibid., 48.
74. The Regents of the University of California (1980), 'Appellate brief of the regents of the University of California as amicus curiae in support of the respondents in *Diamond v Chakrabarty*', 1980 WL 339770, 28 January, 2–3.
75. Kevles, Daniel (1998), '*Diamond v Chakrabarty* and Beyond', in Arnold Thackray (ed.), *Private Science: Biotechnology and the Rise of the Molecular Sciences*, Philadelphia: University of Pennsylvania Press, p. 68.
76. Hood, L. (1980), 'Amicus brief of Dr. Leroy E. Hood, Dr. Thomas P. Maniatis, Dr. David S. Eisenberg, The American Society Of Biological Chemists, The Association Of American Medical Colleges, The California Institute of Technology, and The American Council On Education as amicus curiae in support of the respondents in *Diamond v Chakrabarty*', 1980 WL 339764, 26 January.
77. Ibid., 3.
78. Pieczenik, G. (1980), 'Appellate brief of Dr. George Pieczenik as amicus curiae in support of the respondents in *Diamond v Chakrabarty*', 1980 WL 339773, 29 January, 3–4.
79. Ibid., 15.
80. American Society for Microbiology (1979), 'Appellate brief on behalf of the American Society for Microbiology as amicus curiae in support of the respondents in *Diamond v Chakrabarty*', 1979 WL 200007, 3.
81. American Patent Law Association Inc. (1980), 'Appellate brief on behalf of the American Patent Law Association, Inc., as amicus curiae in support of the respondents in *Diamond v Chakrabarty*', 1980 WL 339772, 29 January, 8.
82. Ibid., 22.
83. Ibid., 22.
84. Ibid., 22.
85. *Diamond v Chakrabarty* 447 US 303 (1980).
86. *Diamond v Chakrabarty* 447 US 303 at 307 (1980).
87. *American Fruit Growers, Inc. v Brogdex Co.*, 283 U.S. 1, 11, 51 S.Ct. 328, 330, 75 L.Ed. 801 (1931).
88. *Shell Development Co. v Watson*, 149 F.Supp. 279, 280 (D.C.1957).
89. *Diamond v Chakrabarty* 447 US 303 at 308 (1980).
90. *Diamond v Chakrabarty* 447 US 303 at 308 (1980).
91. *Diamond v Chakrabarty* 447 US 303 at 309 (1980).
92. *Diamond v Chakrabarty* 447 US 303 at 309 (1980).
93. *Diamond v Chakrabarty* 447 US 303 at 309 (1980).
94. *Diamond v Chakrabarty* 447 US 303 at 309–310 (1980).
95. *Diamond v Chakrabarty* 447 US 303 at 310 (1980).

96. *Diamond v Chakrabarty* 447 US 303 at 310 (1980).
97. *Diamond v Chakrabarty* 447 US 303 at 312 (1980).
98. *Diamond v Chakrabarty* 447 US 303 at 314 (1980); and L. Pasteur (1873), 'Improvement in the Manufacture of Beer and Yeast', US Patent No: 141, 972.
99. *Diamond v Chakrabarty* 447 US 303 at 317 (1980).
100. *Diamond v Chakrabarty* 447 US 303 at 317 (1980).
101. *Diamond v Chakrabarty* 447 US 303 at 318 (1980).
102. *Diamond v Chakrabarty* 447 US 303 at 318 (1980).
103. *Diamond v Chakrabarty* 447 US 303 at 319 (1980).
104. *Diamond v Chakrabarty* 447 US 303 at 319 (1980).
105. *Diamond v Chakrabarty* 447 US 303 at 319 (1980).
106. *Diamond v Chakrabarty* 447 US 303 at 321–322 (1980).
107. Chakrabarty, Ananda (2003), 'Patenting life forms: yesterday, today, and tomorrow'. in F. Scott Kieff (ed.), *Perspectives on Properties of the Human Genome Project*. Amsterdam: Elsevier, pp. 3–11.
108. Ibid.
109. Ibid.
110. Eisenberg, R. (2006), 'Biotech patents: looking backward while moving forward'. *Nature Biotechnology*, **24**(3), 317–19.
111. For a discussion of the patenting of the Boyer–Cohen patent, see S. Smith Hughes (2001), 'Making dollars out of DNA: the first major project in biotechnology and the commercialization of molecular biology, 1974–1980', *Isis*, **92**, 541–78.
112. Cohen, S. and H. Boyer (1979), 'Process for producing biologically functional molecular chimeras', US Patent No: 4,237,224.
113. Smith Hughes, S. (2001), 'Making dollars out of DNA: the first major project in biotechnology and the commercialization of molecular biology, 1974–1980', *Isis*, **92**, 573–4.
114. Rimmer, M. (2002/2003), 'Genentech and the stolen gene: patent law and pioneer inventions', *Bio-Science Law Review*, **5**(6), 198–211.
115. Rifkin, Jeremy (1998), *The Biotech Century: Harnessing the Gene and Remaking the World*, New York: Penguin Putnam Inc, p. 43.
116. Bancroft Library, http://bancroft.berkeley.edu/Exhibits/Biotech/25.html.
117. Eisenberg, Rebecca (2006), 'The story of *Diamond v Chakrabarty*: technological change and the subject matter boundaries of the patent system', Jane Ginsburg and Rochelle Cooper Dreyfuss (eds), *Intellectual Property Stories*, New York: Foundation Press. pp. 327–57 at 357.
118. Eisenberg, R. (2006), 'Biotech patents: looking backward while moving forward'. *Nature Biotechnology*, **24**(3), 317–19.
119. *American Cyanamid v Berk Pharmaceuticals* [1976] RPC 231.
120. *Ranks Hovis McDougall's Application* [1976] 46 AOJP 3915.
121. *Re Application for Patent of Abitibi Co.* (1982) 62 CPR (2d) 81.

2. Franklin barley: patent law and plant breeders' rights

Historically, the patent system was ill-adapted to plant varieties. Plant breeders first sought protection under the industrial patent system. However, a number of technical difficulties were encountered in seeking to apply the rules of a system designed to protect technical inventions to plant varieties, which were thought not to reproduce themselves precisely and whose appearance could vary depending upon the environment in which they are grown. Margaret Llewelyn observes:

> There were two main reasons why the patent system was seen as inappropriate. First, plant material was not regarded as capable of meeting the requirements of novelty, inventive step and disclosure. Secondly, it was not thought to be in the public interest to permit such an extensive monopoly over plant varieties, given their communal importance. Underlying this was the view that it was desirable to retain, in so far as it was possible, the tradition of free exchange of new plant material between plant breeding institutes. This would ensure the widest possible dissemination and use of the new combinations of genetic information.[1]

For these reasons, it was decided to introduce a special form of protection which would be designed to support a specific industry, the plant variety right. The International Convention for New Plant Varieties (the *UPOV Convention* 1961) was adopted in 1961 and an international system for the protection of plant breeders' rights was established.[2]

However, the scope of patentable subject matter expanded, slowly and incrementally, until it covered plants. Bernard Edelman has provided a brief history of intellectual property and biotechnology.[3] The French barrister and philosopher argued that there had been a move away from a strict prohibition against the patenting of nature towards a range of recent decisions allowing the patenting of living matter. Edelman argued that there has been a progressive accommodation of biotechnology within the legal system. He summarized the stages of this passage as follows:

> Life has been integrated into the market as easily as could be imagined because it has been a progressive process. It started with something that was symbolically far removed from mankind, the vegetable domain; from there it passed to the micro-organism, then to the most rudimentary forms of animal life, like the

oyster. The whole of the animal kingdom is now targeted and we are on the verge of the human, weighed down with precedents which ensure the closure of the system and make any resistance difficult. The work of man, which must be remunerated, claims repayment from the whole realm of nature which has traditionally been free of any property claims.[4]

Edelman traced the evolution of the law through key moments in the United States legislation. The *Plant Patent Act* 1930 (US) distinguishes between 'products of nature' and 'human-made inventions'. The *Plant Variety Protection Act* 1970 (US) extends the category of an artificial nature to the reproducibility of plants. The decision of *Diamond v Chakrabarty* determined that genetically engineered organisms are either a manufacture or a composition of matter and are therefore patentable.[5] From single-celled organisms, the line then passes through genetically engineered plants to oysters and transgenic animals – like oncomouse. Bernard Edelman has recently elaborated upon his views on the patenting of genes and gene sequences.[6] He has argued that the contemporary developments over the commercialization of the human genome have raised basic questions as to whether the human species is no more than a product to be used and exploited.

This chapter considers how superior courts in a number of jurisdictions have interpreted the relationship between patent law and plant breeders' rights in light of developments in modern biotechnology. It looks at the range of discourses, including history, constitutional law, intellectual property law, science, economics and international law. It compares and contrasts the approach of three superior courts from Australia, the United States and Canada.[7] Section one considers the High Court of Australia case of *Grain Pool of Western Australia v Commonwealth*.[8] It contrasts the historical methodology of the joint judgment to dealing with plant breeders' rights with the futuristic approach employed by Kirby J in dealing with new scientific and technological developments. Section two examines the significance of the Supreme Court of the United States in *JEM Ag Supply Inc v Pioneer Hi-Bred International Inc*.[9] The majority of the court held that utility patents could be granted in respect of plants in addition to plant patents and plant variety rights. The minority of Breyer and Stevens JJ were concerned about the potential for conflict between the various schemes of intellectual property protection for plants. Section three considers the implications of the decision of the Supreme Court of Canada in *Monsanto Canada Inc. v Schmeiser*.[10] For the majority, McLachlin CJ and Fish J held that a farmer, Percy Schmeiser, had infringed a validly held patent owned by the agricultural biotechnology firm, Monsanto, in respect of glyphosate-resistant canola. For the minority, Arbour J dissented that the scope of the patent was limited to research experiments. In any case, she argued that the patent claims had not been infringed by the farmer and that there should be defences available in respect of farmers' rights and innocent bystanders. The conclusion considers the future interaction between patent law and plant breeders' rights, and discusses the viability of sui generis schemes of protection of biological inventions.

I FEDERATION WHEAT: *GRAIN POOL OF WESTERN AUSTRALIA* v *COMMONWEALTH*

In *Grain Pool of Western Australia v Commonwealth*, the Grain Pool of West Australia challenged the constitutional validity of the *Plant Variety Rights Act* 1987 (Cth) and its successor, the *Plant Breeder's Rights Act* 1994 (Cth).[11]

The case had its origins in an earlier dispute between Cultivaust and the Grain Pool of Western Australia.[12] After obtaining plant breeders' rights in Franklin Barley, the Department of Primary Industry in Tasmania made the South Australian company Cultivaust Pty Ltd the exclusive licensee.[13] In 1991, Cultivaust entered into negotiations with Pool. Cultivaust provided Franklin barley to Grain Pool for the limited purpose of growing trials and malting evaluation. Further negotiations occurred in May 1992, with a view to a permanent licensing arrangement, but no concluded agreement was reached. However, it is said that Pool used the barley provided and other information to exploit the barley in Western Australia. It is said that this was an infringement of the applicants' rights under the plant breeders' rights legislation, a breach of the limited licence granted by Cultivaust, a breach by Grain Pool of a fiduciary duty allegedly owed to Cultivaust arising out of the circumstance of negotiations and a breach by Pool of a duty of utmost good faith owed by Grain Pool to Cultivaust.

In response, the Grain Pool of Western Australia maintained that the legislation was not supported by the intellectual property power under s. 51(xviii) of the *Australian Constitution* 1901, because plant breeders' rights did not fall within the constitutional definition of 'Copyrights, patents of inventions and designs, and trade marks'. Furthermore, the plaintiff argued that the legislation was not supported by the external affairs power under s.51(xxix) of the *Australian Constitution* 1901 because it was not a matter of international concern and the latest iteration of the relevant treaty, the *UPOV Convention* 1991, had not been ratified.

The first defendant, the Commonwealth, in support of the legislation, relied upon s. 51(xviii) and s. 51(xxix) of the *Australian Constitution* 1901. The defendant, Cultivaust, a grain merchant and trader, maintained that it was a licensee from Tasmania, which had the exclusive right to sell and export Franklin barley; and claimed that the plaintiff, by selling within Australia and in exporting Franklin barley, had acted in breach of its rights. The States of Western Australian and Tasmania also intervened.

A Joint Judgment

The joint judgment – undoubtedly written by intellectual property specialist, Gummow J – held that the *Plant Variety Rights Act* 1987 (Cth) and the *Plant Breeder's Rights Act* 1994 (Cth) were valid under the intellectual property power of the *Australian Constitution* 1901.[14] It relied upon a number of sources of authority, including historical studies into the development of intellectual property, constitutional law, and a fine, close reading of the legislation and the case law dealing with plant breeders' rights. The joint judgment concluded that plant variety rights do indeed belong within the ambit of 'patents of invention' in the intellectual property power.

The High Court considered the meaning of the intellectual property power under the *Constitution*, which empowers the Commonwealth to make laws with respect to 'Copyrights, patents of inventions and designs, and trade marks'. The judges reviewed the judicial authorities dealing with intellectual property and constitutional law.[15] The joint judgment endorsed the dissenting judgment of Higgins J in the *Union Label* case:

> These words do not suggest, and what follows in these reasons does not give effect to any notion that the boundaries of the power conferred by s. 51(xviii) are not to be ascertained solely by identifying what in 1900 would have been treated as a copyright, patent, design or trade mark. No doubt some submissions by the plaintiff would fail even upon the application of so limited a criterion. However, other submissions, as will appear, fail, because they give insufficient allowance for the dynamism which, even in 1900, was inherent in any understanding of the terms used in s. 51(xviii).[16]

The judges emphasized that what might answer the description of an invention for the purpose of that section will reflect changes in technology.[17]

The joint judgment of the High Court relied upon a number of historical studies into the development of intellectual property.[18] It emphasized that the formulation of the intellectual property power in the *Australian Constitution* 1901 reflected the crystallization of the legal categories and schema of intellectual property, which had developed in the United Kingdom in the nineteenth century. The joint judgment of the High Court highlighted the recognition of plant variety inventions in 1900. The judges cited with approval the historical overview of Rich J in the United States Court of Appeal case, *Imazio Nursery Inc. v Dania Greenhouses*:

At least as early as 1892, legislation was proposed to grant patent rights for plant-related inventions. Plant patent legislation was supported by such prominent individuals as Thomas Edison who stated that 'nothing that Congress could do to help farming would be of greater value and permanence than to give the plant breeder the same status as the mechanical and chemical inventors now have through the law'. It was also supported by Luther Burbank, a leading plant breeder of the day . . . whose widow stated that her late husband 'said repeatedly that until Government made some such provision [for plant patent protection] the incentive to create work with plants was slight and independent research and breeding would be discouraged to the great detriment of horticulture'.[19]

Callinan J emphasized in the legal proceedings that there was a similar enthusiasm for the protection of plant breeding in Australia.[20] Such comments were incorporated into the final joint judgment, with the note: 'Such views would have been at the time apposite to the position of Australian wheat breeders such as William Farrer, whose Federation cultivar of wheat was named in 1901.'[21]

The High Court considered the evolution of common law and statute law. The joint judgment revisited the watershed Australian case of *NRDC v the Commissioner of Patents*, and noted the concession of the plaintiff that the decision did not present any intrinsic impediment to the patentability of plant varieties.[22] The High Court also cited United States precedents.[23] The joint judgment endorsed the decision of the majority of the Supreme Court of the United States in the case of *Diamond v Chakrabarty*:

> The decision in *Chakrabarty* was that live, human-made, micro-organisms were patentable subject matter within the statutory requirement of an invention or discovery in the *Patents Act* 1952 (US) as being 'any new and useful process, machine, manufacture, or composition of matter, or any new and useful improvement thereof'. However, in the judgment of the Supreme Court of the United States, reference was made to the enactment in 1930 of the *Plant Patent Act* (US), which afforded patent protection to certain asexually produced plants, and to the 1970 *Plant Variety Protection Act* (US), which authorised the grant of patents for certain sexually reproduced plants, but excluded bacteria from its protection.[24]

However, there was no discussion of the dissenting judgment of Brennan J in *Diamond v Chakrabarty*,[25] which took the contrary view that the existence of the *Plant Patent Act* 1930 (US) and the *Plant Variety Protection Act* 1970 (US) suggested that the *Patent Act* 1952 (US) was not intended to cover plant material. This contrary argument was endorsed by a minority of the Supreme Court of the United States in *JEM Ag Supply Inc v Pioneer Hi-Bred International Inc*[26] and the majority of the Supreme Court of Canada in the recent case of *Harvard College v Canada (the Commissioner of Patents)*.[27]

The High Court addressed the argument of the plaintiff that the operation of the intellectual property power under s. 51(xviii) of the *Australian*

Constitution 1901 with respect to patents of invention is limited by what it identifies as certain traditional principles of patent law. In particular, the Grain Pool submitted that there are certain fixed minimum requirements for the 'intellectual effort' required of inventors respecting novelty and inventive step, that there is a crucial distinction between product and process claims, and the term 'patent' involves certain limitations as to exclusivity. The High Court engaged in a close reading of the *Plant Variety Rights Act* 1987 (Cth) and the *Plant Breeder's Rights Act* 1994 (Cth). They considered the threshold criteria for plant breeders' rights – distinctiveness, uniformity and stability – which are known colloquially as the DUS requirements.[28] The High Court ruled that plant variety rights do indeed belong within the ambit of 'patents of invention'. The judges argued that the plant breeders' rights regime featured essential characteristics of the patent regime. The High Court observed: 'A plant variety having those characteristics is an invention in the constitutional sense and the statute secures the benefit of the invention by conferral of particular exclusive rights to control production of other plants with the same essential characteristics.'[29] The High Court observed that a 'plant breeder' is equivalent to an 'inventor' and that a 'plant variety' is like the patent notion of 'an invention'. They commented that the requirement of 'distinctiveness', 'uniformity' and 'stability' under plant breeders' rights is equivalent to 'novelty' and 'inventive step' under patent law. Similarly, the notion of common knowledge was analogous to prior art under patent law. The requirement of 'recent exploitation' was equivalent to the patent rules with respect to secret use.[30]

Finally, the High Court rejected the submission of the Grain Pool of Western Australia that the rights conferred by the *Plant Variety Rights Act* 1987 (Cth) and the *Plant Breeder's Rights Act* 1994 (Cth) amounted to rights 'by way of positive authority to sell and export the protected variety'. The joint judgment held that plant variety rights and plant breeders' rights were negative rights, like those found under patent law, which gave the rights-holder the power to exclude others from using the particular plant material. The Grain Pool of Western Australia had a basic misunderstanding of the nature of intellectual property rights. The organization laboured under the misapprehension that the powers granted under state legislation – the *Grain Marketing Act* 1975 (WA) – trumped federal laws regarding intellectual property.

B Kirby J

Kirby J also held that the *Plant Variety Rights Act* 1987 (Cth) and the *Plant Breeder's Rights Act* 1994 (Cth) were valid. His Honour reached this conclusion not on the basis of the meaning of s. 51(xviii) of the *Australian Constitution* 1901 according, or even by reference, to the accepted understandings of the terms used in 1900. Kirby J instead interpreted the meaning of the phrase 'patents of inventions', in its 'really essential characteristics' as understood in a constitutional context in Australia today.[31] He emphasized the need to be conscious of the future scientific, technological and international developments.

Kirby J considered the debate in constitutional law over the scope of the intellectual property power. His Honour rejected the decision of the majority of the High Court in the *Union Label* case dealing with workers' marks.[32] The judge provided several reasons why this approach should no longer be observed as a criterion for constitutional elaboration of s. 51(xviii) of the *Australian Constitution* 1901. His Honour preferred the decision of Higgins J in that particular case. Kirby J comments:

> Although it is sometimes helpful, in exploring the meaning of the constitutional text, to have regard to the debates in the Constitutional Conventions that led to its adoption and other contemporary historical and legal understandings and presuppositions, these cannot impose unchangeable meanings upon the words. They are set free from the framers' intentions. They are free from the understandings of their meaning in 1900 whose basic relevance is often propounded to throw light on the framers' intentions. The words gain their legitimacy and legal force from the fact that they appear in the Constitution; not from how they were conceived by the framers a century ago.[33]

Kirby J concluded that the court must characterize the limits of the legislative power over 'patents', 'trade marks' and 'copyright law' by identifying the 'really essential characteristics' of the notion referred to. His Honour observed: 'What constitute such 'really essential characteristics' may grow and expand, or may contract over time.'[34] The judge added: 'But the key to finding the meaning is not to be discovered in the statutes and case books before and at 1900 or in the inventions of the framers of the Constitution adopted immediately before and given effect in that year.'[35]

Kirby J took the futuristic view that the legislative powers provided for under the *Australian Constitution* 1901 should be read in such a way as to promote scientific innovation and technological development. He maintains that the objects of the intellectual property power would be destroyed if the notions of 'copyright, patents of inventions and designs and trade marks' were limited to their meaning in 1900. Kirby J rhapsodizes:

> A universal feature of the twentieth century has been the dynamic progress and momentum of science and technology. The principal inventions of the century, which include flight, applied nuclear fission, informatics and biogenetics were all undiscovered, and for the most part unconceived, in 1900. Yet the Constitution

certainly envisaged that the Commonwealth was entering an age of special technological inventiveness. So much can be seen in the specific provision of the post and telecommunications power in such wide terms.[36]

Kirby J refers to Lawrence Lessig's book, *Code and Other Laws Of Cyberspace*, as a general source of authority for a discussion of intellectual property and constitutional law.[37] The joint judgment provides a qualified endorsement of the codified vision of the constitutional power regarding intellectual property. The authors of the joint judgment seem to rely heavily upon historical accounts of intellectual property. By contrast, the judgment of Kirby J seems to adopt a transformative approach. He focuses upon the future developments of technology and science.

Kirby J considered developments in the United States, including the *Diamond v Chakrabarty* decision.[38] His Honour noted that the Supreme Court of the United States initially took a narrow construction of the intellectual property power under the *United States Constitution* in order to promote the development of the public domain and the freedom of competition. However, Kirby J commented that this view was superseded by a broader conception of intellectual property power:

> The advent of biogenetically engineered organisms and of inventions in the field of information technology have stimulated an apparently increased willingness on the part of United States courts to recognise the way in which patents and analogous forms of legal protection can sometimes encourage technological innovation to the economic and social benefit of the United States and beyond. The specific inclusion of s. 51(xviii) in the Australian Constitution affords a further reason for assigning to s. 51(xviii) a meaning that permits the protection of 'products of intellectual effort' in the variety in which such products now manifest themselves and the even greater variety in which they can be expected to appear in the future.[39]

Parenthetically, Kirby J observed that there has been some discussion of copyright protection in relation to the field of biotechnology: 'It is unnecessary now to decide whether copyright law does or could extend to genetically modified organisms.'[40] The debate is most acute in relation to the protection of scientific and genomic databases.[41]

Kirby J also briefly addressed the relationship between intellectual property rights and freedom of speech. His Honour observed in an oblique, cryptic footnote:

> The protection of intellectual property rights must be afforded in a constitutional setting which upholds other values of public good in a representative democracy. In the United States the relevant head of constitutional power has been viewed as containing in-built limitations many of which are derived from

the competing constitutional object of public access to information. In Australia the constitutional setting is different but the existence of competing constitutional objectives, express and implied, is undoubted.[42]

Australian academic Brian Fitzgerald has lauded this statement as a 'landmark footnote'.[43] He speculated upon the implications of this marginalia: 'This reasoning suggests that doctrines such as copyright misuse, which has emerged in the United States in the context of the new technologies, may have relevance in Australia.'[44]

Kirby J was sensitive to the international dimensions of the case. The main problem was that the Federal Government had not ratified the *UPOV Convention* 1991. Kirby J elaborated upon the problems in respect of the external affairs power: 'The position so far as that source of constitutional validity of the federal laws is complicated by revision of the applicable international convention and by the fact that Australia had not, at the time the matter was argued before the court, subscribed to the convention as altered in 1991.'[45] He concluded that 'it would suffice for the Commonwealth and Cultivaust to support the federal laws by reference to the patents power alone. This would leave the question of the ambit of the external affairs power in respect of the subject matter of an international treaty to be elucidated in a future case where such elucidation was essential'.[46] This discussion raises the larger question of the external affairs power and treaty making in relation to intellectual property.

Extra-judicially, Kirby J has expressed support for sui generis protection of biotechnological inventions in a number of forums. His Honour observed in his role as a rapporteur of the UNESCO Committee on Ethics and Intellectual Property:

> Many applications, needs and expectations in this field cannot be accommodated within the framework of intellectual property as it is currently defined. In some cases, responses to such requests for protection could stem from a development of the intellectual property approach. In others, intellectual property could be made to evolve towards the definition of new *sui generis* schemes tailored to the subject matter to be protected, ie genetic resources, along the lines of previous developments aiming to protect plant varieties. One could also contemplate extending intellectual property by adapting existing schemes so as to include, to the largest extent possible, subject matter that is currently not covered.[47]

Kirby J has considered whether there is a need for sui generis protection of biological inventions. His Honour has been willing to contemplate that it might have been better if special legal regimes had been created to deal with the novel intellectual property questions presented by genomics.[48]

The decision of the High Court laid to rest some of the fears that intellectual property legislation would be vulnerable to constitutional challenges.

Jill McKeough and Andrew Stewart, for instance, complained: 'This formula has the disadvantage of being limited to those forms of protection which were familiar at the turn of the century, preventing expansions in traditional areas, certainly precluding the adoption of entirely new regimes.'[49] Such a pall has been lifted by the High Court decision. It seems that there will be no constitutional obstacles to the introduction of legislation dealing with subject matter on the outer limits of intellectual property, such as certification trade marks, databases, publicity rights and the so-called 'neighbouring rights', 'performers' rights'.[50] In light of this decision, there does not seem to be quite the same urgency to implement the recommendation of the Australian Constitutional Commission that s. 51(xviii) be amended to enable the Commonwealth to legislate for: 'Copyright, patents of inventions and designs, trade marks, and other like protection for the products of intellectual activity in industry, science, literature, and the arts'.[51] The High Court has given a clear signal that it will interpret the intellectual property power in a broad and flexible fashion.

In spite of its failed constitutional challenge, the Grain Pool of Western Australia was successful in defending itself from allegations of plant breeders' rights infringement in the Federal Court and Full Federal Court of Australia.[52]

II FIRST THE SEED: *JEM AG SUPPLY v PIONEER HI-BRED INTERNATIONAL INC*

In the case of *JEM Ag Supply Inc v Pioneer Hi-Bred International Inc*, the Supreme Court of the United States considered whether utility patents could be granted in respect of plants.[53]

Pioneer Hi-Bred International Inc had obtained 17 utility patents for its inbred and hybrid corn seed products. It sold the patented hybrid seed to merchants and growers under a limited licence, the terms of which only permitted the production of grain and forage from that seed and prohibited re-sale and use of that seed for propagation, seed multiplication or the production or development of a new hybrid or variety. Pioneer's hybrid corn plant 3394 was 'characterized by superior yield for maturity, excellent seedling vigor, very good roots and stalks, and exceptional stay green'.[54]

JEM Ag Supply Inc., trading as Farm Advantage, bought patented seed from Pioneer under such a licence and resold it. Pioneer brought proceedings against Farm Advantage alleging patent infringement. In reply, Farm Advantage counter-claimed that Pioneer's patents were invalid, because sexually reproducing patents were not patentable subject matter.

The District Court granted summary judgment to Pioneer, relying on a broad construction of the decision in *Diamond v Chakrabarty* in finding that utility patents covered plant life.[55] It found that, in enacting the *Plant Patent Act* 1930 (US) and the *Plant Variety Protection Act* 1970 (US), Congress had not expressly or impliedly removed plants from the scope of patent protection. The United States Court of Appeals for the Federal Circuit affirmed this decision.[56] JEM Ag Supply appealed to the Supreme Court of the United States.

For the petitioners, the Corn Growers Association and the National Farmers Union expressed their concerns about the potential impacts of utility patents upon agriculture, in particular upon genetic erosion, plant uniformity and the exchange of information and germplasm.[57] They were also alarmed that the expansion of intellectual property rights would result in a consolidation of the seed industry, and undermine traditional farming practices of saving seed. Malla Pollack and other law professors also supported the case of JEM Ag Supply.[58]

For the respondents, a number of amicus curiae supported the submission of Pioneer Hi-Bred International. Corporate firms such as Monsanto and Delta and Pine Land Company argued that utility patents should be granted in respect of plants.[59] Trade organizations like the American Crop Protection Association, the American Seed Trade Association and the Biotechnology Industry Organization also stressed the importance of general patent protection in respect of agriculture and biotechnology.[60] Furthermore law groups such as the American Intellectual Property Law Association and the American Bar Association supported the case of the respondent.[61] Finally, the United States Government lent its support to Pioneer Hi-Bred International.[62] By this time, the Solicitor-General of the United States Government was a firm supporter of the majority decision of the Supreme Court of the United States in *Diamond v Chakrabarty*.[63]

A   Thomas J

Thomas J delivered the opinion of the majority of the Supreme Court of the United States, in which Rehnquist CJ and Kennedy, Souter and Ginsburg JJ joined. His Honour engaged in a historical review of the *Plant Patent Act* 1930 (US), the *Plant Variety Protection Act* 1970 (US) and *Diamond v Chakrabarty*,[64] and concluded that utility patents could be granted in respect of plant subject matter. Scalia J concurred with this position in a separate judgment.

Thomas J cites Jack Kloppenburg's groundbreaking book *First The Seed*, a social history of plant breeding and agricultural biotechnology.[65] The judge discussed the historical origins of intellectual property protection of plants. There were a number of legislative models proposed for protecting plants, which were modelled on trade mark law, unfair competition, patent law and sui generis systems.[66] Kloppenburg comments upon the impetus for this legislation:

> The *Morrill Act* of 1862 was intended, in the words of the legislation, to 'assure agriculture a position in research equal to that of industry.' Seedsmen were painfully aware that this was not the case. Private cereal and fruit breeders began calling for establishing of a plant patent system as early as 1885. A proposal that a committee of experts should be empowered to recommend new varieties of appropriate quality for patent registration was rejected in 1901 by the American Pomological Society as 'socialistic' . . . Legislators were not ready to countenances proprietary rights to genetic information.[67]

Thomas J noted: 'Furthermore, like other laws protecting intellectual property, the plant patent provision must be understood in its proper context. Until 1924, farmers received seed from the Government's extensive free seed program that distributed millions of packages of seed annually.'[68] His Honour observed, citing Kloppenburg, 'In 1930, seed companies were not primarily concerned with varietal protection, but were still trying to successfully commodify seeds. There was no need to protect seed breeding because there were few markets for seeds.'[69]

Thomas J noted the significance of the United States Congress passing the *Plant Patent Act* 1930 (US).[70] This legislation provided a special form of protection, which was limited to asexually reproduced varieties of plants which did precisely reproduce themselves and called a plant patent. Thomas J maintained that the *Plant Patent Act* 1930 (US) does not limit the scope of utility patents. His Honour noted: 'Whatever Congress may have believed about the state of patent law and the science of plant breeding in 1930, plants have always had the potential to fall within the general subject matter of s. 101, which is a dynamic provision designed to encompass new and unforeseen inventions'.[71]

Thomas J considered how Congress passed the *Plant Variety Protection Act* 1970 (US) in an effort to harmonize with a number of European countries which protected plant breeders' rights under sui generis legislation. This legislation provided protection to developers of novel, sexually reproduced plants. Thomas J held that the *Plant Variety Protection Act* 1970 (US) did overlap with utility patents, but such conflicts were not irreconcilable. His Honour observed: 'It is much more difficult to obtain a utility patent for a plant than to obtain a plant variety certificate because a utility patentable plant must be new, useful, and non-obvious.'[72] Thomas J therefore deduced: 'Because of the more stringent requirements, utility patent holders receive greater rights of exclusion than the holders of a PVP certificate. Most notably, there are no exceptions for research or saving seed under a utility patent.'[73]

Thomas J denied that granting utility patents in respect of plants would render the exceptions under plant breeders' rights obsolete. He acknowledged that the *Plant Variety Protection Act* 1970 (US) also contained exemptions for saving seed and for research. A farmer who legally purchases and plants a protected variety can save the seed from these plants for replanting on his own farm.[74] In addition, a protected variety may be used for research.[75] The utility patent statute did not contain similar exemptions. In footnote number 12, Thomas J denied that utility patents would undercut farmers' rights and the breeders' exception: 'Since 1985 the PTO has interpreted §101 to include utility patents for plants, and there is no evidence that the availability of such patents has rendered the PVPA and its specific exemptions obsolete.'[76] His Honour maintains that the *Plant Variety Protection Act* 1970 (US) continues to co-exist happily alongside the system of utility patents.

Thomas J stressed that the language in *Diamond v Chakrabarty* was extremely broad and noted that the Court explicitly rejected the argument in that case that Congress must expressly authorize protection for new patentable subject matter.[77] The judge referred to the opinion of Burger CJ in *Diamond v Chakrabarty* that there was nothing in the language or the history of the *Plant Patent Act* 1930 (US) and the *Plant Variety Protection Act* 1970 (US) to suggest that the *Patent Act* 1952 (US) does not include living organisms.[78] Sympathetic to such logic, Thomas J expressed the view that the Government, the Congress, and the Executive are better suited to the balancing of competing policy interests raised by the friends of the court. His Honour gave short shrift to the arguments made in the amicus curiae submissions, such as by the Corn Growers Association, the National Farmers Union and the coalition of law professors.

B Breyer and Stevens JJ

Dissenting, Breyer and Stevens JJ held that the two specific plant statutes – namely the *Plant Patent Act* 1930 (US) and the *Plant Variety Protection Act* 1970 (US) – embodied a legislative intent to deny coverage under the Utility Patent Statute to those plants covered in existing legislation.

Breyer and Stevens JJ sought to divine the original intent of the *Plant Patent Act* 1930 (US). The judges observed that the legislation provides patent protection for any person 'who has invented or discovered and asexually reproduced any distinct and new variety of plant, other than a tuber-propagated plant'. It is particularly helpful to those breeders who reproduced plants through grafts – such as, say, apple trees. Breyer and Stevens JJ commented:

Given these characteristics, the PPA is incompatible with the claim that the Utility Patent Statute's language ('manufacture, or composition of matter') also covers plants. To see why that is so, simply imagine a plant breeder who, in 1931, sought to patent a new, distinct variety of plant that he invented but which he has never been able to reproduce through grafting, i.e. asexually. Because he could not reproduce it through grafting, he could not patent it under the more specific terms of the PPA.[79]

Breyer and Stevens JJ considered whether such a breeder could nonetheless patent the plant under the more general Utility Patent Statute language 'manufacture, or composition of matter'. The judges concluded: 'Even a prescient court would have had to say, as of 1931, that the 1930 *Plant Patent Act* had, in amending the Utility Patent Statute, placed the subject matter of the PPA – namely, plants – outside the scope of the words "manufacture, or composition of matter".'[80]

Breyer and Stevens JJ argued that nothing in the history, language or purposes of the *Plant Variety Protection Act* 1970 (US) suggested an intention to enlarge and expand the scope of patentable subject matter:

> The PVPA proved necessary because plant breeders became capable of creating new and distinct varieties of certain crops, corn, for example, that were valuable only when reproduced through seeds – a form of reproduction that the earlier Act freely permitted. Just prior to its enactment a special Presidential Commission, noting the special problems that plant protection raised and favoring the development of a totally new plant protection scheme, had recommended that '[a]ll provisions in the patent statute for plant patents be deleted . . .'.[81] Instead Congress kept the PPA while adding the PVPA.[82]

The judges noted that it is an interesting quirk of history that the United States should have both plant patents and plant variety protection. They observed that the *Plant Variety Protection Act* 1970 (US) gave protection to plants reproduced by seed, and it excluded the requirement that a breeder have 'asexually reproduced' the plant. It imposed certain specific requirements, notably, that the variety must be new, distinct, uniform and stable. Furthermore, the *Plant Variety Protection Act* 1970 (US) also created two important exceptions: the farmer's right and the breeder's research exception.

The two judges were concerned that the expansion of utility patents to include plant subject matter would undermine the exceptions provided for under the *Plant Variety Protection Act* 1970 (US):

> Why would anyone want to limit the exemptions – related to seed-planting and research – only to those new plant varieties that are slightly less original? Indeed, the research exemption would seem to be more useful in respect to more original, not less original, innovation. The Court has advanced no sound reason why

Congress would want to destroy the exemptions in the *Plant Variety Protection Act* that Congress created. And the Court's reading would destroy those exemptions.[83]

The judges were conscious that the defence of farmers' privilege had been read down and limited in a previous Supreme Court of the United States decision. In *Asgrow Seed Company v Winterboer*, the respondents contended that they were entitled to a statutory exemption from liability under s. 2543, which provides that a farmer may save seed and use such saved seed in the production of a crop for use on his farm, or for sale for reproductive purposes.[84] The majority of the Supreme Court of the United States held that a farmer who meets the requirements set forth in s. 2543's proviso may sell for reproductive purposes only such seed as he has saved for the purpose of replanting his own acreage. It found that the respondents were not eligible for the exception because their planting and harvesting were conducted as 'a step in marketing'. However, Stevens J dissented that Congress intended to preserve the farmer's right to engage in so called 'brown-bag' sales of seed to neighbouring farmers. His Honour believed that Congress would have used a term such as 'sale' if they intended the farmer's privilege exemption to have a narrow operation.

Breyer and Stevens JJ argued that the decision in *Diamond v Chakrabarty* does not control the outcome in the case, because its impact is limited to micro-organisms.[85] They champion the dissenting judgment of Brennan J in that case. In *Diamond v Chakrabarty*, Brennan J maintained that the scope of patentable inventions did not include living organisms.[86] His Honour commented that the *Patent Act* 1952 (US) should be read in light of the *Plant Patent Act* 1930 (US) and the *Plant Variety Protection Act* 1970 (US).[87] Brennan J draws two findings from the existence of such legislation. First, he infers that the legislation is evidence that Congress was of the understanding that the *Patent Act* 1952 (US) did not include living organisms. Second, he notes that Congress had specifically addressed bacteria in the *Plant Variety Protection Act* 1970 (US), saying that it was excluded from the scope of protection. Brennan J concludes: 'It is the role of Congress, not this Court, to broaden or narrow the reach of the patent laws.'[88]

Finally, Breyer and Stevens JJ emphasized that the majority wrongly relied upon the canon of implied repeal:

> Those who write statutes seek to solve human problems. Fidelity to their aims requires use to approach an interpretive problem not as if it were a purely logical game, like a Rubik's Cube, but as an effort to divine human intent that underlies the statute. Here that effort calls not for an appeal to canons, but for an analysis of language, structure, history, and purpose. Those factors make clear that the Utility Patent Statute does not apply to plants. Nothing in *Chakrabarty* holds to the contrary.[89]

The judges resolved that the United States Congress would have to expressly amend United States patent law, if it wanted utility patents to apply to plants.

Mark Janis and Jay Kesan commented in *Nature Biotechnology*[90] that the decision of the Supreme Court of the United States in *JEM Ag Supply v Pioneer Hi-Bred International* leaves a number of issues unresolved: 'It will now fall to the lower courts to work out how numerous other issues of patent law doctrine apply to patents, and to Congress to consider broader policy issues concerning the relationship among IP regimes for plants.'[91] Janis and Kesan raise a number of outstanding questions, such as the protection of non-obvious plants; patent infringement via pollen drift, plant breeding research and seed saving; and the enforceability of technology user agreements. A number of lower court decisions have started to address such matters.[92]

III SEEDS OF HOPE: *MONSANTO CANADA INC. v PERCY SCHMEISER*

In Canada, Percy Schmeiser, an elderly canola farmer from Bruno, Saskatchewan, was sued for patent infringement by the biotechnology company Monsanto. The petty infringement matter blossomed into a symbolic dispute about genetically modified crops, which attracted international attention. Bruce Ziff commented:

> The dispute at the bottom of *Monsanto Canada Inc. v. Schmeiser* became a *cause célèbre* of the highest order. It has served as a flashpoint in the ongoing political conflict concerning genetically modified foods, the propriety of patenting living organisms, the impact of genetically modified organisms (GMOs) on the environment (the rise of superweeds, etc.), and the plight of small farmers living under the shadow of the giants of agribusiness.[93]

The litigation between Monsanto Inc. and the doughty Percy Schmeiser passed through the Federal Court and the Full Federal Court of Canada and ended up being resolved in the Supreme Court of Canada. The case attracted widespread media attention, and a great deal of academic commentary.[94]

The case concerned a Canadian patent granted to Monsanto in 1993 for an invention named 'Glyphosate-Resistant Plants'.[95] The patent was for 'man-made genetically engineered genes, and cells containing those genes which, when inserted in plants, in this canola, make those plants resistant to glyphosate herbicides' such as Monsanto's product Roundup Ready.

In 1997, Monsanto sent private investigators, ex-Mounties, to take samples from the canola farm of Percy Schmeiser. It claimed that Schmeiser planted glyphosate-resistant seeds to grow a crop of genetically modified canola, for harvest. It contended that the farmer used, reproduced and created genes, cells, plants and seeds containing the genes and cells claimed in the patent.

The Federal Court[96] and the Full Federal Court[97] held that Percy Schmeiser knew or should have known that those plants were glyphosate-resistant when he saved their seeds in 1997 and planted those seeds the following year. It was the cultivation, harvest and sale of the 1998 crop that made Percy Schmeiser vulnerable to Monsanto's infringement claim.

On 20 January 2004, the Supreme Court of Canada heard an appeal against the judgment of the Federal Court in *Monsanto Canada Inc. v Schmeiser*.[98] The matter raised important questions about gene patents, innocent infringement and farmers' rights. Are biological inventions patentable – like mechanical inventions? Can an innocent bystander be held liable for infringing a patent? Should farmers' privileges to save seed trump patent rights and technology user agreements?

A number of other parties intervened in the case. Percy Schmeiser was supported by a consortium of six non-government organizations, including farmers' unions, environmental organizations and anti-biotechnology ginger groups. The National Farmers Union argued that the lower court decision would have an adverse impact upon production costs, farm profitability and biodiversity.[99] The Council of Canadians expressed concerns about the commercialization of the field of agriculture.[100] The International Center for Technology Assessment observed that 'patenting is the principal tool used by corporations such as Monsanto to concentrate their power over agricultural and natural resources'.[101] The ETC Group emphasized that the decision of the Federal Court of Canada had far-reaching and adverse impacts upon the practice of saving seed.[102] The Sierra Club of Canada also supported the position of Schmeiser. Dr Vandana Shiva of the India-based Research Foundation for Science, Technology and Ecology, expressed apocalyptic concerns about the wider ramifications of the ruling for farm-saved seed in other jurisdictions.[103]

For its part, Monsanto was supported by the Canadian Seed Trade Association and the Canadian Canola Growers Association, and the industry group BIOTECanada. The Canadian Canola Growers Association highlighted the positive benefits of genetically modified crops.[104] BIOTECanada expressed concerns that, if the arguments of Schmeiser and his supporters were accepted, the 'development of new and promising technologies for the benefit not only of Canadians but for all citizens of the world will be inhibited by reducing the level of research and development for biotechnological inventions in Canada'.[105]

The provincial Government of Ontario was also a party to the case because of its concerns about the impact of gene patents on biomedical research and health care. The province was sensitive to the issue because it had been threatened in the past with legal action for patent infringement by the Utah biotechnology firm, Myriad Genetics.[106] In the course of oral argument in *Monsanto Canada Inc. v Schmeiser*, Sara Blake, on behalf of the Attorney-General of Ontario, sought to bring the Supreme Court of Canada's attention to the impact of gene patents in the health care field. She expressed concern that a ruling in *Monsanto Canada Inc. v Schmeiser* could have an inadvertent and unforeseen impact upon health care.[107]

The Supreme Court of Canada ruled in favour of Monsanto against Percy Schmeiser by a majority of five to four.[108] McLachlin CJ and a new appointment to the bench, Fish J, wrote the leading majority judgment. First of all, the court held that Monsanto's patent on 'Glyphosate-Resistant Plants' was valid. Second, the Supreme Court held that Percy Schmeiser had infringed the patent of Monsanto by using the genetically modified canola crop on his land. It ruled: 'Mr Schmeiser was not an innocent bystander; rather, he actively cultivated Roundup Ready Canola.'[109] Third, the court denied that Schmeiser was permitted under patent law to save and reuse seed. Finally, the court held that Monsanto was not entitled to an account of profits because Percy Schmeiser earned no profit from the invention. It ordered each party to bear its own costs.

Arbour J, the next United Nations Human Rights Commissioner, wrote the dissenting judgment. She maintained that the patent claims could not be extended over whole plants and there was no infringing use.

A Biological Inventions

First of all, the Supreme Court of Canada considered whether biological inventions, such as Monsanto's patented genetically modified canola crop, were patentable subject matter.[110] Professor Brad Sherman comments that the case posed fundamental conceptual problems for the operation of patent law:

> One of the recurring themes in patent law has been the instability of biological inventions. Many of the problems they have posed for patent law can be traced to the fact that, unlike mechanical inventions which are inert and stable, biological inventions are volatile, unstable and dynamic.[111]

The modern patent system was a product of the industrial revolution, and designed to protect mechanical and chemical inventions. The regime was not designed to provide protection for living organisms. Indeed, life forms were considered to be discoveries of nature, rather than scientific inventions. However, there has been a progressive accommodation within patent law of biotechnological inventions, including plants, animals, and even human beings.

There has been a backlash in Canada against the dramatic expansion of the scope of patent law. In the case of *Harvard College v Canada (The Commissioner of Patents)*, the Supreme Court of Canada considered whether the transgenic animal, the Harvard oncomouse, could be the subject of a patent.[112] The Court decided by a majority of five to four that higher life forms were not patentable subject matter. Trained in a civil, French legal tradition, the majority judges emphasized the ethics of patenting. By contrast, the minority judges from a common law, British tradition stressed the commercial goals of patenting.

In the leading judgment, Bastarache J emphasized that Parliament must give an express legislative direction to authorize the patenting of higher life forms:

> Patenting higher life forms would involve a radical departure from the traditional patent regime. Moreover, the patentability of such life forms is a highly contentious matter that raises a number of extremely complex issues. If higher life forms are to be patentable, it must be under the clear and unequivocal direction of Parliament.[113]

Bastarache J indicates that there are also a number of reasons why Parliament might want to be cautious about encouraging the patenting of higher life forms, such as plants, seeds, animals and human beings. In his view, whether higher life forms such as Oncomouse ought to be patentable is a matter for Parliament to determine.

In the case of *Monsanto Canada Inc. v Schmeiser*, the majority of the Supreme Court held that Monsanto's patent was valid.[114] In its view, the invention did not offend the prohibition against the patenting of higher life forms because it related to a component of the plant, not the whole of the plant. McLachlin CJ and Fish J distinguished the case from the precedent of *Harvard College v Canada (The Commissioner of Patents)*:

> This case is different from Harvard Mouse, where the patent refused was for a mammal. The Patent Commissioner, moreover, had allowed other claims, which were not at issue before the Court in that case, notably a plasmid and a somatic cell culture. The claims at issue in this case, for a gene and a cell, are somewhat analogous, suggesting that to find a gene and a cell to be patentable is in fact consistent with both the majority and the minority holdings in Harvard Mouse.[115]

The judges emphasized: 'Under the present Act, an invention in the domain of agriculture is as deserving of protection as an invention in the domain of mechanical science.'[116] Given such sentiments, it would have been more sensible for the court to overturn the precedent of *Harvard College v Canada (The Commissioner of Patents)* completely.

In dissent, Arbour J held that Monsanto's patent claims did not extend to plants, seeds and crops. She stressed that the gene claim did not extend patent protection to the plant. Her Honour relied upon the precedent of *Harvard College v Canada (The Commissioner of Patents)*:

> The trial judge interpreted the scope of the Monsanto patent without the benefit of the holding in Harvard College that higher life forms, including plants, are not patentable. Both lower court decisions 'allo[w] Monsanto to do indirectly what Canadian patent law has not allowed them to do directly: namely, to acquire patent protection over whole plants'. Such a result is hard to reconcile with the majority decision in Harvard College. It would also invalidate the Patent Office's long-standing policy of not granting exclusive rights, expressed in a patent grant, over higher life forms, that was upheld in Harvard College.[117]

Her Honour noted: 'Monsanto is on the horns of a dilemma; a narrow construction of its claims renders the claims valid but not infringed, the broader construction renders the claims invalid.'[118] In light of the decision in *Harvard College v Canada (The Commissioner of Patents)*, she concluded that the patent claims here cannot be interpreted to extend patent protection over whole plants and that there was no infringing use.

In 2002, the Canadian Biotechnology Advisory Committee released an advisory memorandum on 'Higher Life Forms and The Patent Act'. It maintains: 'If the Government of Canada wishes higher life forms to be patentable, it must propose amendments to the *Patent Act* and gain Parliament's agreement.'[119] It stressed that Canada has an unprecedented opportunity to define the special characteristics of biological inventions at the legislative level. In 2006, the Canadian Biotechnology Advisory Committee released its report, *Human Genetic Materials, Intellectual Property and the Health Sector*.[120] The Chair, Arnold Naimark, advised that 'we believe that action is needed now to enhance Canada's intellectual property regime so that it is better prepared, in the context of rapid advances in genetic technologies, to meet the dual objectives of encouraging innovation and making the benefits of such innovation readily accessible to Canadians.'[121] Academic Tim Caulfield has commented that the Canadian Federal Government has been slow to respond to the policy issues raised by gene patents: 'Parliament has yet to consider any of the suggested reforms of the Canadian patent system and there are no formal proposals pending. Altering the existing patent system could take many years.'[122]

B Innocent Infringement

Second, there was much argument in the case as to whether an 'innocent bystander' could infringe a patent. Percy Schmeiser argued that the GM crops on his land were the result of accidental contamination, such as crossfield breeding by wind or insects, or by seed being blown off neighbours' trucks, which did not have their tarpaulin firmly secured. In any case, he maintained that he did not derive any benefit from the GM canola because he did not spray it with Roundup Ready.

Monsanto maintained that the presence of the GM canola on the farm was not accidental. It conducted a number of tests on canola taken from the field of Percy Schmeiser. The results of these tests showed the presence of the patented gene in a range of 95–98 per cent of the canola sampled. The majority of the Supreme Court ruled that Percy Schmeiser was not a mere innocent bystander, but an active user of the patented GM canola. McLachlin CJ and Fish J stressed:

> This case concerns a large scale, commercial farming operation that grew canola containing a patented cell and gene without obtaining licence or permission. The main issue is whether it thereby breached the Patent Act. We believe that it did.[123]

The majority of the Supreme Court of Canada noted: 'Had he been a mere "innocent bystander", he could have refuted the presumption of use arising from his possession of the patented gene and cell.'[124] It stressed that such matters were better addressed by politicians: 'If Parliament wishes to respond legislatively to biotechnology inventions concerning plants, it is free to do so.'[125]

Dissenting, Arbour J held that the cultivation of plants containing the patented gene and cell did not constitute an infringement. She maintained that the plants containing the patented gene could have no stand-by value or utility. Arbour J was concerned by the possibility that an innocent bystander could infringe a patent. She observed: 'The complexities and nuances of innocent bystander protection in the context of agricultural biotechnology should be expressly considered by Parliament because it can only be inadequately accommodated by the law on use.'[126] Indeed, the Canadian Biotechnology Advisory Committee has suggested the creation of such a defence.[127]

The University of Ottawa academic, Jeremy deBeer, has commented that the term 'innocent infringer' is an oxymoron, and unhelpful in legal analysis:

> It must be emphasized, however, that classification of 'innocent bystanders' is inherently misleading. Whether a classic property owner who exercises the normal freedom of ownership, such as possessing and using the property, is himself innocent, is precisely the question. The fact that the IP right is statutory, while the classical property right is not, says nothing about the justice of favouring one over the other.[128]

The case has ramifications for the regulation of genetic technology. There is a potential conflict between the patent regime and the Gene Technology Regulator, for instance, in Australia.[129] In the case of a farmer who was an 'innocent bystander', the Gene Technology Regulator would place legal responsibility upon a biotechnology company to clean up any environmental contamination. However, such a biotechnology company could sue a farmer who had patented genetically modified plants on their land. There is a need to resolve such potential disharmony between the two regimes.

C Farmers' Rights

Finally, there was much debate in the Supreme Court of Canada about the status of farmers' rights in modern agricultural economies. As was common practice for a number of canola farmers in the Bruno area, Percy Schmeiser routinely saved a portion of the canola harvested on his property to serve as seed for the next generation of crops. He observed: 'My wife and I are known on the Prairies as seed developers in canola and as seed savers.'[130] The farmer maintained that he was entitled to save and reuse seed under the ancient notion of farmers' rights.

Agricultural chemical and biotechnology companies have sought to erode farmers' rights. They have attempted to limit the capacity of farmers to save and reuse seed through the means of patent law, contract law and genetic use restriction technologies.

First of all, there is some limited recognition for the protection of farmers' rights at a national and international level. The *Plant Breeders' Rights Act* 1990 (Can) protects the right of farmers to save and reuse seed. However, unlike the sui generis regime of plant breeders' rights, the *Patent Act* 1985 (Can) provides no farm-saved seed exception. Therefore, when farmers use patented seed, they do not have the right to save the seed from a crop and reuse that seed in the next year. Accordingly, Monsanto has sued Percy Schmeiser under the general regime of patent law, rather than the specific system of plant breeders' rights. As such, it maintains that Percy Schmeiser should not have been allowed to save and reuse patented seed.

Second, agricultural and biotechnology companies have become increasingly reliant upon contract law and technology user agreements in their commercial dealings with farmers and growers. The terms and conditions of such agreements are quite restrictive. Under the standard Monsanto agreement for canola, the growers are required to pay a technology fee and a premium rate for the GM seed. They are required to relinquish the right to save and reuse seed. Furthermore, the grower could only use Monsanto's Roundup Ready brand of glyphosate herbicide (there are other brands). The company has the right to inspect and test their fields for up to three years. If any of these conditions were breached, Monsanto could seek liquidated damages. There remains legal debate as to whether such private contacts are valid and enforceable, and can override the public defence to save and reuse seed.

Third, agricultural chemical and biotechnology companies are investing in genetic use restriction technologies, known as 'GURTs' for short. Upset by a patent in respect of the control of plant gene expression granted to the Delta and Pine Land Co. and the United States Government,[131] the Rural Advancement Foundation International famously dubbed GURTs 'terminator technologies'.[132] Such technologies render seed sterile, so that growers are forced to buy new seeds each year from a biotechnology company. They are designed to protect the seed producer against multiplication of the seed by a third party. Thus the GURTs technology can be used to prevent infringement of plant breeders' rights and patent law. Analogies could be drawn with copyright law and the provisions banning circumvention devices and other technological protection measures. However, there have been concerns that the use of GURTs technology provides excessive protection for the holders of intellectual property rights.[133]

The majority of the Supreme Court of Canada held that the patent regime did not contain a defence in relation to farm-saved seed. McLachlin CJ and Fish J observed:

> The appellants argue, finally, that Monsanto's activities tread on the ancient common law property rights of farmers to keep that which comes onto their land. Just as a farmer owns the progeny of a 'stray bull' which wanders onto his land, so Mr. Schmeiser argues he owns the progeny of the Roundup Ready Canola that came onto his field. However, the issue is not property rights, but patent protection. Ownership is no defence to a breach of the Patent Act.[134]

The judges emphasized that saving seed could amount to patent infringement: 'Saving and planting seed, then harvesting and selling the resultant plants containing the patented cells and genes appears, on a common sense view, to constitute "utilization" of the patented material for production and advantage.'[135]

Dissenting, Arbour J held that it was inappropriate that plants could gain dual protection under plant breeders' rights and patent law. She observed: 'Patents should not necessarily be available when other, more tailored intellectual property protection exists.'[136] Her Honour seems to suggest that the plant breeders' rights regime may be better adapted to dealing with agricultural biotechnology.

The Canadian Biotechnology Advisory Committee recommended that a farmers' privilege provision be included in the *Patent Act* 1985 (Can).[137] It should specify that farmers are permitted to save and sow seeds from patented plants or to breed patented animals, as long as these progeny are not sold as commercial propagating material or in a manner that undermines the commercial value to its creator of a genetically engineered animal, respectively. Further action would be necessary to ensure that farmers' rights could not be overridden by contract law or technological measures.

D Remedies

The Supreme Court of Canada found that Monsanto held a valid patent, and that Percy Schmeiser had infringed that patent. However, it refused to grant Monsanto an account of profits because the farmer had not benefited from the invention:

> Their profits were precisely what they would have been had they planted and harvested ordinary canola. They sold the Roundup Ready Canola they grew in 1998 for feed, and thus obtained no premium for the fact that it was Roundup Ready Canola. Nor did they gain any agricultural advantage from the herbicide resistant nature of the canola, since no finding was made that they sprayed with Roundup herbicide to reduce weeds. The appellants' profits arose solely from qualities of their crop that cannot be attributed to the invention.[138]

The Supreme Court of Canada ordered that each party should bear its own costs. Obviously it was reluctant to rule against Percy Schmeiser in respect of costs, lest he be forced to sell his farm.

In spite of being found to have infringed Monsanto's patent, Percy Schmeiser claims that the Supreme Court decision is a 'moral and personal victory'. He has taken some small comfort in the ruling in his favour on damages: 'I have said all along that I didn't take advantage or profit from Monsanto's technology in my fields.'[139] Schmeiser has exhausted all legal avenues in relation to this particular case. Nonetheless he vowed that he will continue to be an activist for farmers' rights: 'I will continue to support any efforts to strengthen the rights of a farmer to save and re-use his own seed.'[140] However, he is conscious that he will need to pay for outstanding legal costs: 'I still have legal bills to pay and I am grateful to all for any past and future contributions.'[141]

Monsanto welcomed the decision of the Supreme Court of Canada that the subject matter claimed within its patent for Roundup Ready canola fell within the *Patent Act* 1985 (RSC) and that Percy Schmeiser and Schmeiser Enterprises Ltd infringed that patent. Carl Casale, the executive vice-president of Monsanto, commented:

> We are gratified the Supreme Court of Canada found that Monsanto's patent pertaining to the Roundup Ready gene is valid and enforceable. The Supreme

Court has set a world standard in intellectual property protection and this ruling maintains Canada as an attractive investment opportunity. Patent protection encourages innovations that will lead to the next generation of value-added products for Canadian farmers.[142]

Monsanto must also feel vindicated that the Supreme Court largely agreed with its interpretation of the facts. No doubt it hopes that Percy Schmeiser will not have the same credibility as an advocate against genetically modified crops after this decision.

In a sequel to the Schmeiser litigation, in *Hoffman v Monsanto*, Percy Schmeiser's lawyer Terry Zareski has been unsuccessful in his efforts to bring a class action on behalf of all organic grain farmers in Saskatchewan against the biotechnology companies Monsanto Canada Inc. and Bayer Cropscience Inc.[143]

CONCLUSION

The superior courts have been required to consider the historical development of intellectual property. They have been required to determine the significance of such landmarks as the *Plant Patent Act* 1930 (US), the *Plant Variety Protection Act* 1970 (US), and the case of *Diamond v Chakrabarty*. Keith Aoki comments in a survey of the recent skirmishes in the 'seed wars':

> *Chakrabarty* left a lacuna: if living organisms transformed by human agency were patentable subject matter under the Patent Statute, 35 USC 191, what was the relation of the *Plant Patent Act* and the *Plant Variety Protection Act*?[144]

There has been a noted divergence in the approach of superior courts to this lacuna. The High Court of Australia, the majority of the Supreme Court of the United States, and the minority of the Supreme Court of Canada have taken a broad reading of *Diamond v Chakrabarty*,[145] and concluded that patents can be granted in respect of plant subject matter. They support the co-existence of a number of overlapping regimes of protection: plant breeders' rights, plant patents and standard patents. By contrast, the majority of the Supreme Court of Canada and a vocal minority of the Supreme Court of the United States conclude that plants are exclusively protected by the *Plant Patent Act* 1930 (US) and the *Plant Variety Protection Act* 1970 (US). They are reluctant to draw the implication from *Diamond v Chakrabarty* that plants could be additionally protected under patent law. They would prefer that the legislatures provide express direction to the courts.

There have been reservations expressed in the superior courts that the unchecked expansion of patent law would render the exceptions provided under plant breeders' rights obsolete. Margaret Llewelyn comments upon the resistance within rural and regional communities to the imposition of patent law to plant subject matter:

> To impose the strict patent ideal of an absolute monopoly is likely in this instance to have the effect of alienating a farming community already suspicious of the motives lying behind the need to obtain patent protection over crops, fodder material and farm animals. It is important to remember that the farming community is not experienced in dealing with patent law principles, nor does it automatically see how the patent system has a direct application in the context of farming. Simply to state that the rights which a patent holder has will be enforced regardless of the wishes or traditional practices of the farmers would, it is submitted, be both arrogant and foolish.[146]

In the case of *Grain Pool of Western Australia v Commonwealth*, Kirby J highlighted the disparities between the range of exceptions under patent law and plant breeders' rights.[147] In the case of *JEM Ag Supply Inc v Pioneer Hi-Bred International Inc*, Thomas J denied that the farmers' privilege and the research exemption were under threat.[148] Breyer and Stevens JJ were concerned that the patent system will override the exceptions granted under plant breeders' rights. In *Monsanto Canada Inc v Schmeiser*, McLachlin CJ and Fish J insisted that it was possible to have dual protection of agricultural biotechnology inventions under both patent law and plant breeders' rights.[149] For the minority, Arbour J submitted that agricultural biotechnology was better dealt with under a sui generis system, such as plant breeders' rights.

The superior courts have considered the relationship between the intellectual property regimes of patent law and plant breeders' rights. They have examined the role of a sui generis system of protection alongside a general regime of intellectual property protection. Graham Dutfield poses the question: are plant breeders' rights obsolete in light of developments in patent law and the science of biotechnology? He observes:

> It is tempting to assume that a system that is dear to the hearts of many plant breeders but not to those of corporate patent lawyers or to the businesses they all work for is doomed to wither away and be replaced by patents, which provide stronger and broader protection. After all, so many seed companies have been taken over by the life science and other corporations that now dominate this industrial sector. Why should the views of breeders and the no longer independent seed companies carry any weight within the corporations they are now part of when they contribute such a small share of the profits of these giants?[150]

However, Graham Dutfield maintains that the plant breeders' rights scheme remains a viable scheme. He notes that the advantages of the plant breeders rights system are better understood by the patent lawyers and the life science corporations. Alternatively, in his view, 'these corporations are happy to let their seed subsidiaries do what they think is right with respect to IP protection without interfering'.[151] He concludes: 'But wherever the truth lies, it seems that, as long as an IP system has corporate users who believe they benefit from its existence, its future is secure.'[152] Indeed, a number of judges believe that the plant breeders' rights system provides an ideal model for the development of sui generis protection of biological inventions. Far from being redundant, the regime of plant breeders' rights may show the way forward for the future development and evolution of intellectual property.

NOTES

1. Llewelyn, M. (1997), 'The legal protection of biotechnological inventions: an alternative approach', *European Intellectual Property Review*, **19**(3), 115–27 at 117.
2. International Union for the Protection of Varieties of Plants, http://www.upov.int/.
3. Edelman, Bernard (1998), 'Vers une approche juridique du vivant', in Bernard Edelman and Marie-Angèle Hermitte (eds), *L'Homme, La Nature et le Droit*, Paris: Christian Bourgois, pp. 28–9, trans. John Frow (1997), *Time and Commodity Culture: Essays in Cultural Theory and Postmodernity*, Oxford: Oxford University Press, pp. 195–7.
4. Edelman, Bernard (1998), 'Vers une approche juridique du vivant', in Bernard Edelman and Marie-Angèle Hermitte (eds), *L'Homme, La Nature et le Droit*, Paris: Christian Bourgois, p. 142, trans. by John Frow (1997), *Time and Commodity Culture: Essays in Cultural Theory and Postmodernity*, Oxford: Oxford University Press, p. 197.
5. *Diamond v Chakrabarty* (1980) 447 US 303.
6. Edelman, B. (2001), 'International symposium on ethics, intellectual property, and genomics', February, rapporteur, M. Kirby, 'Intellectual property and the human genome', *Australian Intellectual Property Journal*, **12**, 61–81 at 77.
7. For accounts of patent law and plant breeders' rights under the European Community Biotechnology Directive, see M. Llewelyn (2006), 'European bio-protection laws: rebels with a cause', in M. Rimmer (ed.), *Patent Law and Biological Inventions, Law in Context*, **24**(1), 11–33; Margaret Llewelyn and Mike Adcock (2006), *European Plant Intellectual Property*, Oxford: Hart Publishing, p. 578; and J. Sanderson (2006), 'Essential derivation, law, and the limits of science', in M. Rimmer (ed.), *Patent Law and Biological Inventions, Law in Context*, **24**(1), 34–53.
8. *Grain Pool of Western Australia v Commonwealth* (2000) 46 IPR 515.
9. *JEM Ag Supply v Pioneer Hi-Bred International Inc* 534 US 124 (2001).
10. *Monsanto Canada Inc. v Schmeiser*, [2004] 1 S.C.R. 902, 2004 SCC 34.
11. *Grain Pool of Western Australia v Commonwealth* (2000) 46 IPR 515.
12. *Cultivaust Pty Ltd v Grain Pool of WA* [2000] FCA 974.
13. Department of Primary Industry Tasmania (1989), 'Franklin Barley', *Plant Varieties Journal*, **2**(2), Application No: 1989/018.
14. The joint judgment consisted of Gleeson CJ, Gaudron, McHugh, Gummow, Hayne and Callinan JJ.
15. *Attorney-General (NSW) v Brewery Employees Union of NSW (the Union Label Case)* (1908) 6 CLR 469; *Australian Tape Manufacturers Association Ltd v Commonwealth* (1993) 176 CLR 480; and *Nintendo v Centronics Systems* (1994) 181 CLR 134.
16. *Grain Pool of Western Australia v Commonwealth* (2000) 46 IPR 515 at 522.
17. *Grain Pool of Western Australia v Commonwealth* (2000) 46 IPR 515 at 520.

18. Sherman, Brad and Lionel Bently (1999), *The Making of Modern Intellectual Property Law: The British Experience, 1760–1911*, Cambridge: Cambridge University Press, pp. 138–9.
19. *Imazio Nursery Inc. v Dania Greenhouses* 69 F 3d 1560 (1995).
20. *Grain Pool of Western Australia v Commonwealth* (5 October 1999) transcript.
21. *Grain Pool of Western Australia v Commonwealth* (2000) 46 IPR 515 at 523.
22. *NRDC v the Commissioner of Patents* (1959) 102 CLR 252.
23. There has been a similar enthusiasm for following United States patent law in the field of information technology. Heerey J in *Welcome Real Time SA v Catuity Inc* [2001] 51 IPR 327 comments: 'It may be true, as the respondents argue, that US patent law has a different historical source owing little or nothing to the Statute of Monopolies . . . But the social needs the law has to serve in that country are the same as in ours. In both countries, in similar commercial and technological environments, the law has to strike a balance between, on the one hand, the encouragement of true innovation by the grant of monopoly and, on the other, freedom of competition.' By contrast, the Australian courts display a great reluctance to adopt United States copyright law.
24. *Grain Pool of Western Australia v Commonwealth* (2000) 46 IPR 515 at 527–8.
25. *Diamond v Chakrabarty* 447 US 303 (1980).
26. *JEM Ag Supply v Pioneer Hi-Bred International Inc* 534 US 124 (2001).
27. *Harvard College v Canada (The Commissioner of Patents)* [2002] SCC 76.
28. S. 43 of the *Plant Breeders' Rights Act* 1994 (Cth).
29. *Grain Pool of Western Australia v Commonwealth* (2000) 46 IPR 515 at 534.
30. *Sun World International Inc v Registrar, Plant Breeders' Rights* (1995) 33 IPR 106; (1997) 39 IPR 161; (1998) 42 IPR 321 [1998].
31. *Grain Pool Of Western Australia v Commonwealth* (2000) 46 IPR 515.
32. *Attorney-General (NSW) v Brewery Employees Union of NSW (the Union Label Case)* (1908) 6 CLR 469.
33. *Grain Pool of Western Australia v Commonwealth* (2000) 46 IPR 515 at 544.
34. *Grain Pool of Western Australia v Commonwealth* (2000) 46 IPR 515 at 549.
35. *Grain Pool of Western Australia v Commonwealth* (2000) 46 IPR 515 at 549.
36. *Grain Pool of Western Australia v Commonwealth* (2000) 46 IPR 515 at 550.
37. Lessig, Lawrence (1999), *Code and Other Laws Of Cyberspace*, New York: Basic Books.
38. *Diamond v Chakrabarty* (1980) 447 US 303.
39. *Grain Pool of Western Australia v Commonwealth* (2000) 46 IPR 515 at 551.
40. *Grain Pool of Western Australia v Commonwealth* (2000) 46 IPR 515 at 550.
41. Rimmer, M. (2003), 'Beyond blue gene: intellectual property and bioinformatics', *International Review of Industrial Property and Copyright Law*, **34**(1), 31–49.
42. *Grain Pool of Western Australia v Commonwealth* (2000) 46 IPR 515 at 551.
43. Fitzgerald, B. (2001), 'Case comment: *Grain Pool of WA v The Commonwealth*: Australian constitutional limits of intellectual property rights', *European Intellectual Property Review*, **23**(2), 103.
44. Ibid.
45. *Grain Pool of Western Australia v Commonwealth* (2000) 46 IPR 515 at 539.
46. *Grain Pool of Western Australia v Commonwealth* (2000) 46 IPR 515 at 539.
47. Kirby, M. (2002), 'Report of the International Bioethics Committee on Ethics, Intellectual Property and Genomics', International Bioethics Committee, 10 January 2002, p. 6, http://unesdoc.unesco.org/images/0013/001306/130646e.pdf#xml=http://unesdoc.unesco.org/ulis/cgi-bin/ulis.pl?database=ged&set=3F0E5B42_2_12&hits_rec=1&hits_lng=eng.
48. Kirby, M. (2000), 'The human genome and patent law', *Reform*, **79**, 10–13.
49. McKeough, Jill and Andrew Stewart (1997), *Intellectual Property in Australia*, 2nd edn, Sydney: Butterworths, p. 3.
50. *Grain Pool of Western Australia v Commonwealth* (5 October 1999) transcripts.
51. Australian Constitutional Commission (1988), *Final Report of the Constitutional Commission*, Canberra: Australian Government Publishing Service, **2**, paras 10.140–10.153.

52. *Cultivaust v Grain Pool Pty Ltd* [2004] FCA 638; and *Cultivaust v Grain Pool Pty Ltd* [2005] FCAFC 223.
53. *JEM Ag Supply Inc v Pioneer Hi-Bred International Inc* 534 US 124 (2001).
54. Niebur, W., R. Riley and S. Noble (1996), 'Hybrid corn plant and seed', US Patent No: 5,491,295.
55. *Pioneer Hi-Bred International Inc v JEM Ag Supply Inc* 1998 WL 1120829.
56. *Pioneer Hi-Bred International Inc v JEM Ag Supply Inc* 200 F.3d 1374 C.A.Fed. (Iowa), (2000).
57. American Corn Growers Association and National Farmers Union (2001), 'Brief for Amici Curiae American Corn Growers Association and National Farmers Union in Support of the Petitioners in *JEM Ag Supply Inc v Pioneer Hi-Bred International Inc*', 2001 WL 490944, 4 May.
58. Pollock, M. (2001) 'Brief for Amici Malla Pollack and other Law Professors Supporting Reversal in *JEM Ag Supply Inc v Pioneer Hi-Bred International Inc*', 2001 WL 476088, 2 May.
59. Monsanto Inc. (2001), 'Brief of Amicus Curiae Monsanto Company in Support of Respondent in *JEM Ag Supply Inc v Pioneer Hi-Bred International Inc*', 2001 WL 674207, 15 June; and Delta and Pine Land Company (2001), 'Brief of Amicus Curiae Delta and Pine Land Company in Support of Respondent in *JEM Ag Supply Inc v Pioneer Hi-Bred International Inc*', 2001 WL 689283, 15 June.
60. American Crop Protection Association (2001), 'Brief of Amicus Curiae American Crop Protection Association in Support of Affirmance in *JEM Ag Supply Inc v Pioneer Hi-Bred International Inc*', 2001 WL 674199, 15 June; American Seed Trade Association (2001), 'Brief of American Seed Trade Association in Support of Respondent in *JEM Ag Supply Inc v Pioneer Hi-Bred International Inc*', 2001 WL 670055, 13 June; and Biotechnology Industry Organization (2001), 'Brief of Amicus Curiae Biotechnology Industry Organization in *JEM Ag Supply Inc v Pioneer Hi-Bred International Inc*', 2001 WL 689273, 15 June.
61. American Intellectual Property Law Association (2001), 'Brief for Amicus Curiae American Intellectual Property Law Association in Support of Respondent Supporting Affirmance in *JEM Ag Supply Inc v Pioneer Hi-Bred International Inc*', 2001 WL 649829, 11 June; American Bar Association (2001),'Brief for Amicus Curiae American Bar Association in Support of Respondent in *JEM Ag Supply Inc v Pioneer Hi-Bred International Inc*', 2001 WL 674189, 15 June.
62. United States Government (2001), 'Brief for the United States Government as Amicus Curiae Supporting Respondent in *JEM Ag Supply Inc v Pioneer Hi-Bred International Inc*', 2001 WL 689516, 15 June.
63. *Diamond v Chakrabarty* 447 US 303 (1980).
64. *Diamond v Chakrabarty* 447 US 303 (1980).
65. Kloppenburg, Jack (1988), *First The Seed: The Political Economy of Plant Biotechnology 1492–2000*, Cambridge: Cambridge University Press. The University of Wisconsin academic must be one of the few Marxists to be cited with approval by the Supreme Court of the United States.
66. Janis, M. and J. Kesan (2002), 'U.S. plant variety protection: sound and fury . . .?', *Houston Law Review*, **39**, 727–78 at 730–45.
67. Kloppenburg, Jack (1988), *First The Seed: The Political Economy of Plant Biotechnology 1492–2000*, Cambridge: Cambridge University Press, p. 132.
68. *JEM Ag Supply Inc v Pioneer Hi-Bred International Inc* 534 US 124 at 127 (2001).
69. Kloppenburg, Jack (1988), *First The Seed: The Political Economy of Plant Biotechnology 1492–2000*, Cambridge: Cambridge University Press, p. 132.
70. Fowler, C. (2000), 'The Plant Patent Act of 1930: a sociological history of its creation', *Journal of the Patent and Trademark Office Society*, **82**(9), 621–44.
71. *JEM Ag Supply v Pioneer Hi-Bred International Inc* 534 US 124 at 135 (2001).
72. *JEM Ag Supply v Pioneer Hi-Bred International Inc* (2001) 534 US 124 at 142 (2001).
73. *JEM Ag Supply v Pioneer Hi-Bred International Inc* (2001) 534 US 124 at 143 (2001).

74. S. 2543 of the *Plant Variety Protection Act* 1970 (US) provides: '[I]t shall not infringe any right hereunder for a person to save seed produced by the person from seed obtained, or descended from seed obtained, by authority of the owner of the variety for seeding purposes and use such saved seed in the production of a crop for use on the farm of the person'; see also *Asgrow Seed Company v Winterboer et al.* 513 US 179 (1995).
75. S. 2544 of the *Plant Variety Protection Act* 1970 (US) provides: 'The use and reproduction of a protected variety for plant breeding or other bona fide research shall not constitute an infringement of the protection provided under this chapter.'
76. *JEM Ag Supply v Pioneer Hi-Bred International Inc* 534 US 124 at 140 (2001).
77. *JEM Ag Supply v Pioneer Hi-Bred International Inc* 534 US 124 at 130 (2001).
78. *Diamond v Chakrabarty* 447 US 303 (1980).
79. *JEM Ag Supply v Pioneer Hi-Bred International Inc* 534 US 124 at 152 (2001).
80. *JEM Ag Supply v Pioneer Hi-Bred International Inc* 534 US 124 at 152 (2001).
81. President's Commission on the Patent System (1967), *To Promote the Progress of Useful Arts*, S. Doc. No. 5, 90th Cong., 1st Sess., 20–21.
82. *JEM Ag Supply v Pioneer Hi-Bred International Inc* 534 US 124 at 153 (2001).
83. *JEM Ag Supply v Pioneer Hi-Bred International Inc* 534 US 124 at 155 (2001).
84. *Asgrow Seed Company v Winterboer et al.* 513 US 179 (1995); and Seth Shulman (1999), *Owning The Future*, New York: Houghton Mifflin, pp. 83–106.
85. *Diamond v Chakrabarty* 447 US 303 (1980).
86. The judgment of Brennan J was joined by White, Marshall and Powell JJ.
87. *Diamond v Chakrabarty* 447 US 303 at 319 (1980).
88. *Diamond v Chakrabarty* 447 US 303 at 322 (1980).
89. *JEM Ag Supply v Pioneer Hi-Bred International Inc* 534 US 124 at 156 (2001).
90. Janis, M. and J. Kesan (2002), 'Intellectual property protection for plant innovation: unresolved issues after JEM v Pioneer', *Nature Biotechnology*, **20**, 1161–65; see also M. Janis (2006), 'Rules v standards for patent law in the plant sciences', in M. Rimmer (ed.), *Patent Law and Biological Inventions, Law in Context*, **24**(1), 54–66.
91. Ibid.
92. *Pioneer Hi-Bred International, Inc. v Ottawa Plant Food, Inc.* 283 F.Supp.2d 1018 N.D. Iowa (2003); and 219 F.R.D. 135 N.D. Iowa (2003); *Monsanto Co. v McFarling* 2002 WL 32069634 (2002); *Monsanto Co. v McFarling*, 363 F.3d 1336 (2004); *McFarling v Monsanto Co.* 125 S.Ct. 2956 (Mem) U.S (2005); *Monsanto Co. v Good*, 2004 WL 1664013 (2003); and *Sample v Monsanto Co.* 218 F.R.D. 644 (2003).
93. Ziff, B. (2005) 'Travels with my plant: *Monsanto v Schmeiser* revisited', *University of Ottawa Law and Technology Journal*, **2**(2), 493–509 at 493.
94. Burrell, R. and S. Hubicki (2005), 'Patent law and genetic drift: *Schmeiser v Monsanto Canada Inc*', *Environmental Law Review*, **7**(3), 278–98.
95. Shah, D., S. Rogers, R. Fraley and R. Horsch (1986), 'Glysophate-resistant plants', Canadian Patent No: 1,313,830.
96. *Monsanto Canada Inc. v Schmeiser* 2001 FCT 256 (CanLII), (2001), 202 F.T.R. 78, 12 C.P.R. (4th) 204.
97. *Schmeiser v Monsanto Canada Inc.*, 2002 FCA 448 (CanLII), [2003] 2 F.C. 165, 218 D.L.R. (4th) 31.
98. *Monsanto Canada Inc. v Schmeiser*, [2004] 1 S.C.R. 902, 2004 SCC 34.
99. National Farmers Union (2003), 'Amicus Brief of National Farmers Union in *Monsanto Canada Inc. v Schmeiser*', http://www.canadians.org/food/documents/NFU_Affidavit_final.pdf, 26 September.
100. The Council of Canadians (2002), 'Amicus Brief of the Council of Canadians in *Monsanto Canada Inc. v Schmeiser*', http://www.canadians.org/food/documents/COC_Affidavit.pdf, October.
101. International Center for Technology Assessment (2003), 'Amicus brief of the International Center for Technology Assessment in *Monsanto Canada Inc. v Schmeiser*', http://www.canadians.org/food/documents/ICTA_Affidavit.pdf, 26 September.

102. The ETC Group (2003), 'Amicus brief of the ETC group in *Monsanto Canada Inc. v Schmeiser*', http://www.canadians.org/food/documents/ETC_Affidavit.pdf, 24 September.
103. Research Foundation for Science, Technology and Ecology (2003), 'Amicus brief of the Research Foundation for Science, Technology and Ecology in *Monsanto Canada Inc. v Schmeiser*', http://www.canadians.org/food/documents/RSTE_Affidavit.pdf, 19 September.
104. Canadian Canola Growers Association (2003), 'Canadian Canola Growers Association standing up for the interests of Canadian canola growers', *Seed Quest*, http://www.seedquest.com/News/releases/2003/december/7224.htm, 8 December.
105. BIOTECanada (2003), 'Amicus brief of BIOTECanada in *Monsanto Canada Inc. v Schmeiser*', http://www.biotech.ca/media.php?mid=838, December.
106. Garforth, K. (2005), 'Health care and access to patented technologies', *Health Law Journal*, **13**, 77–97 at 80.
107. *Monsanto Canada Inc. v Schmeiser*, [2004] 1 S.C.R. 902, 2004 SCC 34.
108. *Monsanto Canada Inc. v Schmeiser*, [2004] 1 S.C.R. 902, 2004 SCC 34.
109. *Monsanto Canada Inc. v Schmeiser*, [2004] 1 S.C.R. 902, 2004 SCC 34.
110. *Monsanto Canada Inc. v Schmeiser*, [2004] 1 S.C.R. 902, 2004 SCC 34.
111. Sherman, B. (2002), 'Biological inventions and the problem of passive infringement', *Australian Intellectual Property Journal*, **13**, 146–54.
112. *Harvard College v Canada (The Commissioner of Patents)* [2002] SCC 76.
113. *Harvard College v Canada (The Commissioner of Patents)* [2002] SCC 76.
114. *Monsanto Canada Inc. v Schmeiser*, [2004] 1 S.C.R. 902, 2004 SCC 34.
115. *Monsanto Canada Inc. v Schmeiser*, [2004] 1 S.C.R. 902, 2004 SCC 34.
116. *Monsanto Canada Inc. v Schmeiser*, [2004] 1 S.C.R. 902, 2004 SCC 34.
117. *Monsanto Canada Inc. v Schmeiser*, [2004] 1 S.C.R. 902, 2004 SCC 34.
118. *Monsanto Canada Inc. v Schmeiser*, [2004] 1 S.C.R. 902, 2004 SCC 34.
119. Canadian Biotechnology Advisory Committee (2002), *Patenting of Higher Life Forms*, Ottawa: Canadian Biotechnology Advisory Committee, June.
120. Canadian Biotechnology Advisory Committee (2006), *Human Genetic Materials, Intellectual Property, and the Health Sector*, Ottawa: Canadian Biotechnology Advisory Committee, http://cbac-cccb.ca/epic/internet/incbac-cccb.nsf/en/ah00578e.html.
121. Ibid.
122. Caulfield, T. (2005), 'Policy conflicts: gene patents and health care in Canada', *Community Genetics*, **8**, 223–7.
123. *Monsanto Canada Inc. v Schmeiser*, [2004] 1 S.C.R. 902, 2004 SCC 34.
124. *Monsanto Canada Inc. v Schmeiser*, [2004] 1 S.C.R. 902, 2004 SCC 34.
125. *Monsanto Canada Inc. v Schmeiser*, [2004] 1 S.C.R. 902, 2004 SCC 34.
126. *Monsanto Canada Inc. v Schmeiser*, [2004] 1 S.C.R. 902, 2004 SCC 34.
127. Canadian Biotechnology Advisory Committee (2002), *Patenting of Higher Life Forms*, Ottawa: Canadian Biotechnology Advisory Committee, June.
128. J. deBeer (2005), 'Reconciling property rights in plants', *Journal of World Intellectual Property*, **8**(1), 5–31 at 13.
129. Gene Technology Regulator, http://www.ogtr.gov.au/, and the *Gene Technology Act 2000* (Cth).
130. Schmeiser, P. (2003), 'In his own words', Vancouver Central Library, http://www.gmofreemendo.com/press_releases/percyschmeiser_to_mendo.html, 10 December.
131. Oliver, M., J. Quisenberry, N. Trolinder and D. Keim (1995), 'Control of plant gene expression', US Patent No: 5,723,765.
132. Rural Advancement Foundation International (1998), 'US patent on new technology will prevent farmers saving seed', http://www.etcgroup.org/en/materials/publications.html?id=420, 11 March; and Rural Advancement Foundation International (1998), 'Biotech activists oppose terminator technology', http://www.etcgroup.org/en/materials/publications.html?id=418, 13 March.
133. Dutfield, G. (2003) 'Should we terminate terminator technology?', *European Intellectual Property Review*, **25**(11), 491–5; and B. Sherman and S. Hubicki. (2005),

'The killing fields: intellectual property and genetic use restriction technologies'. *University of New South Wales Law Journal*, **28**(3), 740–57.
134. *Monsanto Canada Inc. v Schmeiser*, [2004] 1 S.C.R. 902, 2004 SCC 34.
135. *Monsanto Canada Inc. v Schmeiser*, [2004] 1 S.C.R. 902, 2004 SCC 34.
136. *Monsanto Canada Inc. v Schmeiser*, [2004] 1 S.C.R. 902, 2004 SCC 34.
137. Canadian Biotechnology Advisory Committee (2002), *Patenting of Higher Life Forms*. Ottawa: Canadian Biotechnology Advisory Committee, June.
138. *Monsanto Canada Inc. v Schmeiser*, [2004] 1 S.C.R. 902, 2004 SCC 34.
139. Schmeiser, P. (2004), 'Percy Schmeiser claims personal and moral victory in Supreme Court decision', *Grain*, http://www.grain.org/bio-ipr/?id=397, 22 May.
140. Ibid.
141. Ibid.
142. Monsanto (2004), 'Supreme Court finds in favor of Monsanto in *Schmeiser v Monsanto* patent infringement case', Press Release, http://www.monsanto.com/monsanto/layout/media/04/05-21-04.asp, 21 May.
143. *Hoffman v Monsanto* (2005) 2005 SKQB 225, http://www.canlii.org/sk/cas/skqb/2005/2005skqb225.html; and J. deBeer (2007), 'The rights *and* responsibilities of ag-biotech patent owners', *The University of British Columbia Law Review*, **40**(1), 343–73.
144. Aoki, K. (2003), 'Weeds, seeds and deeds: recent skirmishes In the Seed Wars', *Cardozo Journal of International and Comparative Law*, **11**, 247–311 at 302.
145. *Diamond v Chakrabarty* 447 US 303 (1980).
146. Llewelyn, M. (1997), 'The legal protection of biotechnological inventions: an alternative approach', *European Intellectual Property Review*, **19**(3), 115–27 at 125.
147. *Grain Pool of Western Australia v Commonwealth* (2000) 46 IPR 515.
148. *JEM Ag Supply v Pioneer Hi-Bred International Inc* 534 US 124 (2001).
149. *Monsanto Canada Inc. v Schmeiser*, [2004] 1 S.C.R. 902, 2004 SCC 34.
150. Dutfield, Graham (2003), *Intellectual Property Rights and the Life Science Industries*. Aldershot: Ashgate Publishing Limited, p. 193.
151. Ibid, p. 193.
152. Ibid, p. 193.

3. The human chimera patent initiative: patent law and animals

There is much reliance upon animal models, including drosophila, mice, zebra fish, as well as pigs, chimpanzees and monkeys, in both genetics and stem cell research.[1] There has been much legal, ethical and commercial debate over the patenting of animals, in relation to polyploid oysters, transgenic animals such as the Harvard oncomouse, animal models, cloned animals and human–animal chimera.

In the past decade, biotechnology developers, public research institutions and industry groups have strongly supported the extension of patent law to include animals. Canadian researchers, Vincent Amanor-Boadu, Morris Freeman and Larry Martin, sum up the positive arguments advanced for broadening intellectual property protection to include the animal kingdom:

> The increased investment in animal biotechnology research and development implies an increased likelihood of finding solutions to some of the human and animal diseases that currently defy treatments. In this way, consumers may benefit from improved food and health care products coming from farm animals.[2]

The commentators conclude: 'Under a stronger IP system, processors and retailers would benefit from innovation leading to cheaper and/or more improved food products and/or new products such as pharmaceutical products and chemicals from animals.'[3]

Some policy makers have been willing to adapt the patent system so that it can better accommodate animals into its framework, so as to better balance the interests of technology developers against those of agricultural producers. United States House of Representatives member, Robert Kastenmeier, introduced the *Transgenic Animal Patent Reform Act* 1989 (US) into the United States Congress. The legislation provided defences in respect of the reproduction of a patented transgenic farm animal through breeding; the use of such an animal in a farming operation; and the sale of such an animal and its offspring. The Canadian Biotechnology Advisory Committee recommended that higher life forms, such as plants and animals, should be patentable, as long as the other patent criteria were met, such as novelty, inventive step, and utility.[4] The Committee advised the Canadian Government that a farmers' privilege and an innocent bystanders' defence should be adopted, and made available both in respect of plants and animals. In Australia, the Advisory Council on Intellectual Property flirted with the idea of including animal varieties within the second tier system of innovation patents.[5]

By contrast, some commentators have argued that animal varieties should be protected under a sui generis scheme, much like the way plant varieties are dealt with under plant breeders' rights.[6] Andrew Christie and Nicholas Peace have argued that there should be a sui generis scheme of protection for animal breeding as an incentive for animal breeders: '[A] Plant Variety Rights scheme may provide a useful model for the development of a new intellectual property system to address the current inadequacies of legal protection for the products of animal breeding.'[7] However, there has been little enthusiasm by policy makers for minting sui generis schemes of intellectual property protection, especially for animal breeders. The producers of traditional livestock, such as dogs, horses and cattle, have been concerned that the introduction of such a scheme could interfere with traditional agricultural practices.

However, animals' rights activists, environmental groups and some farmers' unions have adamantly argued that animals should be the subject of patent protection – or any other similar form of intellectual property protection. As an illustrative example, the American Anti-Vivisection Society is currently running a campaign, entitled 'Stop Animal Patents'.[8] The group maintains that animals should not be considered to be patentable subject matter because they are sentient beings, nor mere mechanical inventions:

> Just like toasters, clocks, and other inanimate object inventions, animals are being patented in the United States. Private companies, universities, and individual 'bioentrepreneurs,' have been granted over 470 patents on animals such as monkeys, mice, dogs, cats, sheep, and chimpanzees . . . It is our position that it is an inappropriate use of the patent system and unethical to issue patents for sentient beings.[9]

The Society's concerns embrace a number of inter-related themes. The organization takes the view that higher life forms such as animals should not be the subject of patent protection. The group expresses a particular concern about animal experimentation. As the Society notes elsewhere, 'Inherent in the patenting of animals is animal suffering, as the limits of what suffering and pain they can tolerate are explored and violated.'[10] There is an underlying anxiety about biotechnology and genetic engineering. There is also a wider concern about the commercialization and commodification of life forms, hence the remarks about 'bioentrepreneurs'. The Society claims that there is widespread ethical and moral opposition amongst the community to animal patents.[11]

This chapter explores the legal, commercial and ethical debate over the patenting of animals. It considers the key conflicts that have pitted technology developers and researchers against animal rights' groups, farmers' unions and environmental groups. Section one examines the ruling in *Ex Parte Allen* that the United States Patent and Trademark Office (USPTO) could grant in respect of a polyploid oyster.[12] It also explores the unsuccessful challenge in *Animal Legal Defense Fund v Quigg* by farmers and animal rights organizations to a notice issued by the USPTO that recognized that animals could be patentable subject matter.[13] Section two considers the litigation in various jurisdictions over the Harvard oncomouse – a transgenic animal designed to be genetically predisposed to develop cancerous tumours. It compares and contrasts the approach of the USPTO, the European Patent Office and the Canadian Intellectual Property Office. Close attention is paid to the split decision by the Supreme Court of Canada in *Harvard College v Canada (The Commissioner of Patents)* that the Harvard oncomouse was a higher life form, and therefore not eligible subject matter for patent protection.[14] Section three considers the unsuccessful application by Stuart Newman and Jeremy Rifkin for a United States patent in respect of 'Chimeric embryos and animals containing human cells'.[15]

I THE POLYPLOID OYSTER

In the wake of the decision of *Diamond v Chakrabarty*, there was some speculation and conjecture amongst patent attorneys, researchers and industry that the USPTO would allow patents in respect of higher life forms, such as plants and animals. In the case of *Ex Parte Allen*, the Board of Patent Interferences and Appeals held that it was indeed possible that polyploid oysters could constitute patentable subject matter.[16] The USPTO issued a notice, announcing that animals could be patentable subject matter. This direction was the subject of an unsuccessful challenge by animal rights groups and farmers' groups. The United States Congress considered whether there was a need for law reform, with debate over Robert Kastenmeier's reform bill for transgenic animals.

A *Ex Parte Allen*

Testing this hypothesis, Standish Allen and Sandra Downing of the University of Washington and Jonathan Chaiton of the Coast Oyster Company applied for a patent in September 1984 on the production of the triploid-sterile Pacific oyster. The inventor's lawyer, David Maki, sought to extend the claim to include the triploid oyster itself. He observed: 'I wanted to provide maximum protection for my client. Besides, there was a transition occurring in case law on living organisms, and there were rumors around that the scope of patentability might be enlarged to include living animals.'[17] The inventors maintained that the polyploid oysters were novel, inventive and useful because they were sterile and did not devote significant portions of their body weight to reproduction, thereby remaining edible year around.

The examiner rejected a number of claims in the patent application on the grounds that polyploid oysters were living entities and therefore not patentable subject matter. Relying upon a number of precedents,[18] the examiner held that the animal produced by the method claimed was 'controlled by laws of nature and not a manufacture by man that is patentable'.[19] The examiner distinguished the decision in *Diamond v Chakrabarty* on the basis that the opinion categorized the claimed micro-organisms as 'more akin to inanimate chemical compositions such as reactants, reagents, and catalysts than they are to horses and honeybees or raspberries and roses'.[20] The examiner also held that a number of the claims were obvious in light of a previous publication, which recommended polyploidy as a way to increase growth in cultured oysters.[21]

In *Ex Parte Allen*, the Board of Patent Appeals and Interferences ruled that the claims in the patent application were indeed directed to patentable subject matter.[22] The Board observed:

[T]he Supreme Court made it clear in its decision in *Diamond* v. *Chakrabarty*, supra, that Section 101 includes man-made life forms. The issue, in our view, in determining whether the claimed subject matter is patentable under Section 101 is simply whether that subject matter is made by man. If the claimed subject matter occurs naturally, it is not patentable subject matter under Section 101. The fact, as urged by the examiner, that the oysters produced by the claimed method are 'controlled by the laws of nature' does not address the issue of whether the subject matter is a nonnaturally occurring manufacture or composition of matter. The examiner has presented no evidence that the claimed polyploid oysters occur naturally without the intervention of man, nor has the examiner urged that polyploid oysters occur naturally.[23]

The Board concluded that 'the claimed polyploid oysters are non-naturally occurring manufactures or compositions of matter within the confines of patentable subject matter under 35 USC 101'.[24] However, the Board agreed with the examiner 'that in view of the express recommendation by Stanley et al., experts in the art who have successfully induced polyploidy in one species of oysters, it would have been obvious to one of ordinary skill in the art to induce polyploidy in Pacific Crassostrea gigas oysters'.[25]

The Court of Appeals for the Federal Circuit affirmed the decision of the Board of Patent Appeals and Interferences. Friedman J of the Federal Circuit observed that the polyploid oysters were obvious:

> To state that the Pacific oyster is genetically and phenotypically distinct from the Atlantic oyster does not rebut the prima facie case. Both are a species of oyster which are sterile and, therefore, larger when polyploid. The Board found that the Allen declaration did not rebut the prima facie obviousness determination. We have no reason to disagree with that conclusion.[26]

The Federal Circuit did not disturb the finding that animals could be patentable subject matter.

B *Animal Legal Defense Fund v Quigg*

In response to the decision of the Supreme Court of the United States in *Diamond v Chakrabarty*[27] and the ruling of the Board of Patent Appeals and Interferences in *Ex Parte Allen*,[28] the USPTO released a notice, announcing that animals could constitute patentable subject matter:

> The Patent and Trademark Office now considers nonnaturally occurring non-human multicellular living organisms, including animals, to be patentable subject matter within the scope of 35 U.S.C. 101.
>
> The Board's decision does not affect the principle and practice that products found in nature will not be considered to be patentable subject matter under 35 U.S.C. 101 and/or 102. An article of manufacture or composition of matter occurring in nature will not be considered patentable unless given a new form, quality, properties or combination not present in the original article existing in nature in accordance with existing law.
>
> A claim directed to or including within its scope a human being will not be considered to be patentable subject matter under 35 U.S.C. 101.[29]

The notice concluded: 'Accordingly, the Patent and Trademark Office is now examining claims directed to multicellular living organisms, including animals.'[30] The notice added the rider: 'To the extent that the claimed subject matter is directed to a non-human 'nonnaturally occurring manufacture or composition of matter – a product of human ingenuity' (*Diamond v Chakrabarty*), such claims will not be rejected under 35 U.S.C. 101 as being directed to nonstatutory subject matter.'[31]

In *Animal Legal Defense Fund v Quigg*, a number of farmers and animal rights organizations filed a lawsuit challenging the notice issued by the USPTO that recognized that animals could be patentable subject matter.[32] The opponents included farmers, animal husbanders and animal protection organizations, including the Animal Legal Defense Fund, The American Society for the Prevention of Cruelty to Animals, The Marin Humane Society, the Wisconsin Family Farm Defense Fund, John Kinsman, Michael Cannell, Humane Farming Association, Association of Veterinarians for Animal Rights, and People for the Ethical Treatment of

Animals. This coalition of organizations sought to stop the issuance of patents for animals.

The United States District Court for the Northern District of California dismissed the challenge at first instance.[33] Smith J observed that the notice was supported by relevant legal and administrative case law: 'these decisions hold precisely what the Rule states: that non-naturally occurring, non-human multicellular living organisms, including animals, are patentable subject matter under 35 U.S.C. Section 101'.[34] The judge held that the USPTO did not exceed its statutory authority in promulgating that rule. Smith J, though, noted: 'Whether *Allen* itself or any actual "animal" patents issued to applicants under *Allen* and *Chakrabarty* exceed the PTO's authority under 35 U.S.C. Section 101 is a different question and one that is not raised by this action.'[35]

The United States Court of Appeals for the Federal Circuit also dismissed the challenge from farming and animal rights groups.[36] First, Nies CJ held that the notice was an 'interpretative rule', which was exempt from the notice and comment requirements of the *Administrative Procedure Act 1966* (US): 'The Notice clearly corresponds with the interpretations of section 101 set out by the Board in *Allen* and *Hibberd*, in reliance on *Chakrabarty*, with the only caveat being the statement that section 101 does not extend to humans.'[37] The judge ruled that the notice was consistent with the Supreme Court of the United States decision in *Diamond v Chakrabarty*, and the USPTO rulings in *Ex parte Allen* and *Hibberd*.

Second, Nies CJ ruled that the farmers, husbandry groups and animal rights organizations did not have standing to seek a declaration that animals are not patentable subject matter and an injunction against the issuance of animal patents:

> Appellants are in effect attempting to intervene as third parties in the prosecution of *all* animal patent applications. By analogy, merely because appellants make a broadside attack gives no greater right of intervention against all than against one.[38]

The court dismissed the allegations of farmers and agricultural groups that they would suffer economic loss because of the higher fees associated with the purchase of patented, genetically altered animals, and increased competition from more productive non-naturally occurring animals. The judge observed that 'their allegation that their costs of operation will *increase* by reason of "royalties" is at best speculative'.[39] His Honour added: 'Similarly, the farmers' alleged injury from increased competition can only result from the development and commercialization of genetically improved animals – not from the grant of a patent.'[40]

Citing with approval the remarks of Burger CJ in *Diamond v Chakrabarty*, the judge observed that research into genetically altered animals could not be prevented by patent law alone.[41] The judge noted that 'were we to enjoin issuance of patents for non-naturally occurring animals, the requested relief would not prevent the development of such animals'.[42] He added: 'It should hardly need saying that the issuance of a patent gives no right to make, use or sell a patented invention, or that the absence of a patent creates no legal prohibition against continued research or development.'[43]

C Transgenic Animal Patent Reform Act

In 1987 and 1989, a number of House of Representatives committees, chaired by Wisconsin Democrat Robert Kastenmeier, considered the patentability of animals.[44] The chair observed:

> We've approved the patenting of plants in 1930, seeds in 1970, microbes in 1980, and now we've moved to considering patenting animals. Next is human beings. We need to look at this and consider some ground rules for proceeding down this path.[45]

Interestingly, Kastenmeier was an independent thinker on matters of patent law. He had observed: 'The proprietors always seem to get what they want: once in a while, the public ought to get its way.'[46] Kastenmeier even expressed reservations about the Supreme Court of the United States decision in *Diamond v Chakrabarty*: 'I'm not one of those people who feel everything under the sun should be patentable.'[47]

The United States Congress heard a number of policy objections to animal patents from a range of interest groups and stakeholders. A coalition of environmental groups, animal rights activists, religious organizations and farmers' unions took issue with animal patents. Jeremy Rifkin, the campaigner against biotechnology, argued:

> Patenting higher forms of life will essentially allow multinational corporations to literally own and control entire animal gene pools, from apes to insects. It will reduce all animals to matter for manipulation. And it enables companies to define the genetic blueprints of all living things as their own property. It is a profound change in how we regard animals.[48]

The president of the Animal Legal Defense Fund, Steven M. Wise, observed: 'Congress never intended that animals should be viewed as compositions of matter under the patent law.'[49] He argued: 'Animals are sentient, have consciousness, and are far above mere compositions of matter.'[50] A consultant for the National Farmers Union, Howard Lyman, supported a moratorium on animal patenting, declaiming that 'we are moving much too quickly into an uncharted area.'[51]

In rejoinder, government agencies, biotechnology developers and some agricultural groups supported patents being granted in respect of animals. Donald J. Quigg, assistant Commerce secretary and Commissioner of the USPTO, told the subcommittee, 'Neither the imposition of compulsory licensing nor a moratorium on the patenting of inventions pertaining to the field of transgenic animals is in the national interest.'[52] William H. Duffey, general patent counsel to the Monsanto Corporation, argued, 'It would be unthinkable for the United States to selectively exclude animal patents from patentable subject matter because of the outburst from a minority who are attacking the patent system simply because it is the only forum in which they can currently obtain a hearing.'[53] Speaking for the American Farm Bureau Federation, Donald Haldeman said that the organization favoured 'strong patent support to encourage' biotechnology research and development.[54]

The House Judiciary Subcommittee concluded that a moratorium on animal patents would be unwise and unnecessary. The Committee held that farmers should be exempted from paying royalties on the offspring of the patented livestock they bought. Furthermore, it submitted that humans should not be patented. The chairman, Kastenmeier, observed:

> There is broad apprehension in the farm community, in the religious community and in other communities about developments in biotechnology. In some cases it is fear over what kinds of organisms will be produced. In other cases it is suspicion about who really benefits.[55]

Kastenmeier introduced the *Transgenic Animal Patent Reform Act* 1989 (US) into the United States Congress. The legislation would have provided patent defences for farmers in respect of the reproduction, use and sale of a patented transgenic farm animal and its offspring. The Kastenmeier bill passed the House of Representatives; however, it was not debated in the Senate before the end of Congress. After Kastenmeier lost his seat in the 1990 Congressional elections, the legislation was never reintroduced into the United States Congress.[56]

II   THE HARVARD ONCOMOUSE

The Harvard oncomouse has an active oncogene in order to give it a genetic disposition to develop cancerous tumours and hence be a better laboratory animal for testing new anti-cancer drugs and therapies.[57] The transgenic animal has been the subject of great public controversy in a number of jurisdictions. There has been much legal and political controversy over the Harvard oncomouse in the United States, the European Union and Canada.[58] The litigation surrounding the Harvard oncomouse has attracted much academic debate.[59]

In April 1988, a month after the United States Court of Appeals for the Federal Circuit affirmed the decision in *Ex Parte Allen*, the USPTO issued a patent on the Harvard oncomouse to two genetics researchers, Philip Leder of Harvard Medical School and Timothy Stewart of San Francisco, who assigned it to the president and trustees of Harvard College.[60] Du Pont then made arrangements with Charles River Laboratories to market the Harvard oncomouse.

By contrast, in the European Union, there was an epic, two-decade legal battle over the validity of the equivalent patent in respect of the Harvard oncomouse. In 1989, the Examining Division of the European Patent Office refused a patent application on the Harvard oncomouse on the basis that Article 53(b) of the *European Patent Convention* 1973 (EU) which establishes an exception to patentability for 'plant or animal varieties or essentially biological processes for the production of plants or animals', excluded patent protection for all animals per se.[61] However, on appeal, in 1990, the European Patent Office Technical Board of Appeal held that Article 53(b) did not exclude, per se, the patenting of animals.[62] The Board held that the test to be applied under Article 53(a) was one of 'unacceptability', based on the weighing up of potentially detrimental effects of the grant of a patent on the one hand and the invention's usefulness to humankind. Applying this test, in 1992, the Examination Division concluded that the potential medical benefits of the mouse outweighed any concerns of animal suffering and risks of escape into the environment. It was considered that the invention did not offend the provisions of Article 53(a) of the *European Patent Convention* 1973 (EU) and thus a patent on the Harvard oncomouse was granted.[63]

In 2000, the European Patent Office informed the relevant parties that the Harvard oncomouse patent application would be considered anew in light of the *European Union Directive on the Legal Protection of Biotechnological Inventions* 1998 (the Directive).[64] The Opposition Division in Munich invited various parties – Harvard University as the patent proprietor and 16 different groups, individuals, political parties and organizations wanting the patent revoked – to oral proceedings in 2001. In 2003, the European Patent Office Opposition Division affirmed that there was no doubt that living matter and in particular plants and animals could be patented.[65] It also added that the exclusion in respect of animals was limited to animal varieties only and that it could not be extended to animals in general. The European Patent Office Opposition Division decided to maintain Harvard

University's patent application in an amended form. The Opposition Division ruled that the patent must be limited to 'transgenic rodents' containing an additional cancer gene, rather than 'transgenic non-human mammals'. In 2004, the European Patent Office Technical Board of Appeal further reduced the scope of the patent claims from transgenic rodents to transgenic mice.[66] The Board lamented the long delays involved in the case. The patent in respect of the Harvard oncomouse expired in 2005.

Harvard College's patent application in respect of the Harvard oncomouse fared even worse in Canada. In *Harvard College v Canada (The Commissioner of Patents)*, the Supreme Court of Canada considered an appeal against the decision of the Full Federal Court that the Harvard oncomouse was patentable subject matter under Canadian law.[67]

Harvard College sought to protect the process by which oncomice are produced and the end product of that process: the founder mice and the offspring whose cells are affected by the oncogene.[68] The patent examiner refused to accept the claims that pertained to transgenic mammals as the products of the invention. The Commissioner of Patents in Canada refused to grant a patent for the product claims in 1995. The Federal Court of Canada dismissed an appeal by Harvard College on 21 April 1998.[69] The judge decided that a transgenic mammal is not truly reproducible because too much is left to chance, including the chromosomal location of the transgene and the degree of transgene expression. Consequently, the judge concluded that the transgenic mammal was not sufficiently reproducible to be a 'composition of matter' or an 'article of manufacture' under the *Patent Act* 1985 (RSC). Harvard then appealed its case to the Canadian Federal Court of Appeal.

On 3 August 2000, the majority of the appellate court determined that the oncomouse was a composition of matter and sent the case back to the Commissioner of Patents with the direction to grant a patent on the transgenic animal claims.[70] In the name of the Commissioner of Patents, the Attorney General of Canada filed an application to seek appeal to the Supreme Court of Canada. On 14 June 2001, the Supreme Court of Canada granted the application for appeal.

A number of submissions were made to the Supreme Court of Canada from friends of the court. The amicus curiae included religious groups such as the Canadian Council of Churches and the Evangelical Fellowship of Canada, environmental organizations like Greenpeace Canada, the ETC Group and the Canadian Institute for Environmental Law and Policy, and animals' rights activists such as the Animal Alliance of Canada, the International Fund for Animal Welfare, and Zoocheck Canada.

The Supreme Court of Canada ruled by a five to four majority that the Harvard oncomouse was not patentable subject matter.[71] The majority consisted of judges who expressed legal and ethical concerns about the patenting of higher life forms, without explicit legislative direction from the Canadian Parliament, including Bastarache, Gonthier, Iacobucci, L' Heureux-Dube and Le Bel JJ. The minority was composed of judges who believed that the patenting of biological inventions was necessary to encourage research and development in new technologies, including McLachlin CJ and Binnie, Major and Arbour JJ dissenting. The division between the judges represented major ideological differences as to the patenting of biotechnological inventions. As Sean Robertson has observed, the Harvard College case raised larger policy issues about 'farmers' rights, the public domain, indigenous traditional knowledge frameworks, public trust doctrine, the ethics of patenting higher life, and concern for biosafety and food security'.[72]

A Bastarache J

In his leading judgment, Bastarache J emphasized that Parliament must give an express legislative direction to authorize the patenting of higher life forms:

> Patenting higher life forms would involve a radical departure from the traditional patent regime. Moreover, the patentability of such life forms is a highly contentious matter that raises a number of extremely complex issues. If higher life forms are to be patentable, it must be under the clear and unequivocal direction of Parliament. For the reasons discussed above, I conclude that the current Act does not clearly indicate that higher life forms are patentable. Far from it. Rather, I believe that the best reading of the words of the Act supports the opposite conclusion – that higher life forms such as the oncomouse are not currently patentable in Canada.[73]

Bastarache J indicated that there were also a number of reasons why Parliament might want to be cautious about encouraging the patenting of higher life forms, such as plants, seeds, animals and human beings. In his view, whether higher life forms such as oncomouse ought to be patentable was a matter for Parliament to determine. However, Bastarache J affirmed that it is acceptable to engage in the patenting of lower life forms, like bacteria, yeast and moulds. His Honour observed that 'it is far easier to analogize a micro-organism to a chemical compound or other inanimate object than it is to analogize a plant or an animal to an inanimate object'.[74]

In the course of his decision, Bastarache J assumed that the distinction between lower and higher life forms is defensible on the basis of 'common sense' differences between the two. However, this judgment has been criticized for its vagueness and arbitrariness.[75] Scientist William Leiss, for instance, says acerbically: 'There is no place in the book of DNA for such brittle categories as "higher" and "lower" life forms.'[76] He suggests that modern scientific developments would render the distinction between higher and lower life forms utterly redundant and meaningless.

Engaging in statutory interpretation, Bastarache J considered whether the words 'manufacture' and 'composition of matter', within the context of the *Patent Act* 1985 (RSC), were sufficiently broad to include higher life forms such as 'inventions'. His Honour explicitly rejected the approach of the majority of the Supreme Court of the United States in *Diamond v Chakrabarty*, which presumed that 'anything under the sun that is made by man' was patentable:

> I cannot however agree with the suggestion that the definition is unlimited in the sense that it includes 'anything under the sun that is made by man'. In drafting the *Patent Act*, Parliament chose to adopt an exhaustive definition that limits invention to any 'art, process, machine, manufacture or composition of matter'. Parliament did not define 'invention' as 'anything new and useful made by man'. By choosing to define invention in this way, Parliament signalled a clear intention to include certain subject matter as patentable and to exclude other subject matter as being outside the confines of the Act. This should be kept in mind when determining whether the words 'manufacture' and 'composition of matter' include higher life forms.[77]

Bastarache J observes that biological inventions cannot be analogized with mechanical works: 'With respect to the meaning of the word "manufacture" (*fabrication*), although it may be attributed a very broad meaning, I am of the opinion that the word would commonly be understood to denote a non-living mechanistic product or process.'[78] His Honour adds that 'composition of matter' does not include a higher life form such as oncomouse.

Bastarache J maintained that such a literal interpretation of the *Patent Act* 1985 (RSC) is supported by the higher policy objectives of the legislation:

> The patenting of higher life forms raises unique concerns which do not arise in respect of non-living inventions and which are not addressed by the scheme of the Act. Even if a higher life form could, scientifically, be regarded as a 'composition of matter', the scheme of the Act indicates that the patentability of higher life forms was not contemplated by Parliament.[79]

The judge concluded: 'Owing to the fact that the patenting of higher life forms is a highly contentious and complex matter that raises serious practical, ethical and environmental concerns that the Act does not contemplate, I conclude that the Commissioner was correct to reject the patent application.'[80] The decision reflects concerns about the patenting of gene therapy, germline treatments, stem cell research and human cloning.[81]

Considering adjacent legislative regimes, Bastarache J maintained that the existence of the *Plant Breeders' Rights Act* 1990 (RSC) was relevant to the issue of whether Parliament intended higher life forms to be patentable under the *Patent Act* 1985 (RSC):

> Far more significant, in my view, is that the passage of the *Plant Breeders' Rights Act* demonstrates that mechanisms other than the *Patent Act* may be used to encourage inventors to undertake innovative activity in the field of biotechnology. As discussed above, the *Plant Breeders' Rights Act* is better tailored than the *Patent Act* to the particular characteristics of plants, a factor which makes it easier to obtain protection. The *quid pro quo* is that a narrower monopoly right is granted.[82]

Bastarache J cited the opinion of the Minister of Agriculture Honourable Donald Mazankowski that the *Plant Breeders' Rights Act* 1990 (RSC) was passed to accommodate the special characteristics of crossbred plants as self-reproducing higher life forms while at the same time striking an appropriate balance between the holder of the monopoly right and others: 'The legislation is designed to deal with the complexities of the issue and that is why we have chosen this route rather than to amend the *Patent Act*.'[83] His Honour concluded that the special regime for plant breeders' rights provided a model for sui generis protection of biological inventions.

Bastarache J commented that there is a need to reform the patent system to include defences in respect of agricultural biotechnology: 'Because higher life forms reproduce by themselves, the grant of a patent covers not only the particular plant, seed or animal sold, but also all of its progeny containing the patented invention.'[84] The judge emphasized that there is a need for farmers' privilege provision to be included within the scope of the patent legislation. He envisaged that the privilege would permit farmers to collect and reuse seeds harvested from patented plants and to breed patented animals for their own use, so long as these were not sold for commercial breeding purposes.[85] Bastarache J also stressed the need for a defence of innocent infringement in respect of agricultural biotechnology patents. He recommended that the *Patent Act* 1985 (RSC) be reformed to include a provision that would allow the so-called 'innocent bystander' to rebut the usual presumption concerning knowledge of infringement in respect of inventions capable of reproducing, such as plants, seeds and animals.[86]

Finally, Bastarache J commented that the special regime for plant breeders' rights provided a model for sui generis protection of biological inventions: 'If a special legislative scheme were needed to protect plant varieties, a subset of higher life forms, a similar scheme may also be necessary to deal with the patenting of higher life forms in general.'[87]

B Binnie J

Binnie J wrote the minority opinion on behalf of the dissenting judges. It is a mixture of tenacious argument and eloquent exasperation. In a rebuttal of the arguments of Bastarache J, Binnie J contends that there is no prohibition on the patenting of higher life forms under the *Patent Act* 1985 (RSC).

Emphasizing the commercial and scientific context of intellectual property and biotechnology, Binnie J argued that 'the massive investment of the private sector in biotechnical research is exactly the sort of research and innovation that the *Patent Act* was intended to promote'.[88] His Honour observed that intellectual property rights are an important contributor to financing research and development:

> Nevertheless it is indisputable that vast amounts of money must be found to finance biomedical research. It is necessary to feed the goose if it is to continue to lay the golden eggs. The *Patent Act* embodies the public policy that those who directly benefit from an invention should be asked, through, the patent system, to pay for it, at least in part.[89]

Binnie J emphasized: 'One would think it in the public interest to shorten the time and reduce the cost of research designed to minimize human suffering, and to reward those who develop research tools that might make this possible.'[90] His Honour feared that Canada would be deprived of the benefits of biotechnology if the patenting of higher life forms was banned.

Binnie J engaged in a broad statutory interpretation of the definition of 'invention' under the *Patent Act* 1985 (RSC). The key provision was section 2 of the *Patent Act* 1985 (RSC), which provides that an ' "invention' means any new and useful art, process, machine, manufacture, or composition of matter, or any new and useful improvement in any art, process, machine, manufacture, or composition of matter'. Binnie J facetiously ridiculed the majority decision for being too narrow in its interpretation of 'composition of matter' and 'manner of manufacture':

> 'Matter' is a most chameleon-like word. The expression 'grey matter' refers in everyday use to 'intelligence' – which is about as incorporeal as 'spirit' or 'mind' . . . If the oncomouse is not composed of matter, what, one might ask, are such things as oncomouse 'minds' composed of? The Court's mandate is to approach this issue as a matter (that slippery word in yet another context!) of law, not murine metaphysics.[91]

Binnie J maintained that 'manufacture' and 'composition of matter' should necessarily include biological inventions. He noted that the tradition of patent jurisprudence has been expansive, not restrictive, citing the opinion of the 1851 text *Godson on Patents* that the possible objects of 'manner of manufacture' were 'almost infinite'.[92]

In a systematic fashion, Binnie J argued that the distinction between lower and higher life forms is not axiomatic, counting at least ten possible positions.[93] After cataloguing this array of distinctions, Binnie J concludes: 'With respect, there seems to be as many versions of "common sense" as there are commentators.'[94]

Taking an international perspective, Binnie J emphasizes that patents have been granted on higher-life forms in comparable jurisdictions. He commented: 'We were not told of any country with a patent system comparable to Canada's (or otherwise) in which a patent on the oncomouse had been applied for and been refused.'[95] Binnie J contended that Canada was out of step with comparable jurisdictions with similar intellectual property legislation. He observes that there is nothing unique about the definition of 'invention' in Canadian legislation: 'The truth is that our legislation is not unique. The Canadian definition of what constitutes an invention, initially adopted in pre-Confederation statutes, was essentially taken from the United States *Patent Act of 1793*, a definition generally attributed to Thomas Jefferson.'[96] Binnie J dismisses the objections of anti-globalization groups that the patenting of life forms will disadvantage the interests of developing countries. His Honour concluded that the mobility of capital and technology make it desirable for there to be international harmonization in relation to intellectual property and biotechnology.

Binnie J denied that the court should take from the passage in the *Plant Breeders' Rights Act* 1990 (RSC), the negative inference that plants were not intended by Parliament to be patentable under the *Patent Act* 1985 (RSC). Firstly, he argued that there was nothing in the *Plant Breeders' Rights Act* 1990 (RSC) that expressly barred an application under the *Patent Act* 1985 (RSC) which confers much more exclusive and valuable rights. The *Plant Breeders' Rights Act* 1990 (RSC) merely granted protection for 18 years on the sale and propagation for sale of enumerated new plant varieties: cultivars, clones, breeding lines or hybrids that can be cultivated. Secondly, he maintained that the use of specific terms such as 'strain' or 'hybrid' would undermine the generality that section 2 seeks to achieve by use of the term 'composition of matter'. Thirdly, the judge stressed that plant breeders' rights and patent rights can co-exist happily. He noted that similar arguments about inconsistency were rightly rejected by the Supreme Court of the United States in *JEM Ag Supply Inc. v Pioneer Hi-Bred International Inc.*[97] In any case, Binnie J argued that the Canadian Commissioner of Patents was inconsistent in opposing the oncomouse patent in respect of a transgenic animal, when supporting a Monsanto patent in relation to round-up ready canola.[98] His Honour suggested that there was a fundamental inconsistency in the case of the Canadian Commissioner of Patents.[99]

Acknowledging that there has been much scholarly controversy in Canada over the role of intellectual property in biotechnology,[100] Binnie J noted that, there, some thoughtful critics suggest that patents in this field may in fact deter rather than promote innovation.[101] His Honour noted that there have been advocates in Canada of the 'farmers' privilege' to avoid farmers being subject to patent enforcement in the case of the progeny of patented plants and animals. Binnie J also observed that others had advocated protection for 'innocent bystanders' who inadvertently made use of a genetically engineered plant or animal, unaware of its being patented. His Honour argued, though, that such proposals for legislative reform had not been adopted by the Court or Parliament to date, and neither the Commissioner of Patents nor the courts had the authority to declare, in effect, a moratorium on life (or 'higher' life) patents until Parliament chose to act: 'The respondent is entitled to have the benefit of the *Patent Act* as it stands.'[102]

Finally, Binnie J was unwilling to entertain the policy submissions from amicus curiae who were concerned about the impact of the decision upon animals' rights, the environment and the sanctity of life. His Honour stressed:

> In this appeal, however, we are only dealing with a small corner of the biotechnology controversy. The legal issue is a narrow one and does not provide a proper platform on which to engage in a debate over animal rights, or religion, or the arrogance of the human race.[103]

Binnie J noted that Parliament may instead wish to regulate the creation and use of higher life forms outside the framework of the *Patent Act* 1985 (RSC). He observes: 'Even a partial listing of the possibilities demonstrates why it should occasion no surprise that such regulatory structures are not crammed into the *Patent Act*, which has always had the more modest and focussed objective of simply encouraging the disclosure of the fruit of human inventiveness in exchange for the statutory rewards.'[104] Such comments echo the jeremiad of Burger CJ in *Diamond v Chakrabarty* against judicial dabbling in matters of politics and ethics.[105]

C   The Canadian Biotechnology Advisory Committee

The decision of the Supreme Court of Canada in *Harvard College v Canada (The Commissioner of Patents)* has had wider implications for the patenting of plants, animals and human genes in the jurisdiction of Canada.[106] The judgment alarmed many in the biotechnology industry. The lawyers for Harvard College, David Morrow and Colin Ingram of the Ottawa firm Smart and Biggar, said: 'There is no rational basis for interpreting the definition of "invention" in a manner which excludes higher life forms from patentability.'[107] The patent holder Harvard College was understandably disappointed by the outcome of the case: 'The Court's disappointing narrow decision leaned on technical aspects of a 19th century patent law and is counter to the recommendations made earlier this year by the Canadian government's own biotech committee.'[108] The president of BIOTECanada, Janet Lambert, was livid at the decision, contending that it was bad news for the Canadian biotechnology community and consumers: 'This decision stops our pursuit of knowledge and innovation dead in its tracks. It is a great loss to Canada at both the social and economic level.'[109]

However, other commentators have been pleasantly surprised by the decision of the Supreme Court of Canada. Professor Martin Phillipson of the University of Saskatchewan welcomed the judgment: 'I am not anti-biotech or some sort of Neo-Luddite. I just think that the decision will force the government to engage in widespread consultation on what is a hugely significant question.'[110] Similarly, Montreal lawyer, Helen D'Iorio of Gowling Lafleur Henderson suggested that any adverse impact on the biotechnology industry had been exaggerated: 'The decision will not, in all likelihood, have a major impact on the intellectual property and research and development communities.'[111]

In response to the decision, the Canadian Biotechnology Advisory Committee has released an advisory memorandum on 'Higher Life Forms and The Patent Act'.[112] It sought to allay fears that the decision of the Supreme Court of Canada spelt the ruin of the Canadian biotechnology industry: 'Sorting out the implications of the special characteristics of higher life forms for the patent regime will not be accomplished overnight.'[113] The Committee concluded: 'If the Government of Canada wishes higher life forms to be patentable, it must propose amendments to the *Patent Act* and gain Parliament's agreement.'[114] It stressed that Canada has an unprecedented opportunity to define the special characteristics of biological inventions at the legislative level. Ryan Atkinson suggests that Canadian legislators need to come up with a more flexible formula for distinguishing between products of nature and synthetics: 'Life forms, DNA, protein and other subject matter could overcome the prohibition on patenting products of nature if the inventor demonstrates uniform mass production.'[115]

III   THE HUMAN CHIMERA PATENT INITIATIVE

In 1997, the biologist Professor Stuart Newman of the New York Medical College and the long-standing opponent of biotechnology, Jeremy Rifkin, announced that they would seek a patent on methods to create a chimera, a hybrid human and animal. The project was entitled the Human Chimera Patent Initiative. Newman explains the nature of the project:

> I had no intention of producing such creatures, nor does US patent law require that an actual prototype for an invention be supplied, only that feasibility be demonstrated, as well as novelty and utility. While a decision as to patentability by the [USPTO] would not control whether or not it would be legal to produce human-animal chimeras, or other types of biologically manipulated humans, we considered that applying for a chimera patent would raise these issues before the public and the legal system in a particularly dramatic fashion.[116]

The patent application was an attempt to force the USPTO to grapple with the ethical dimensions of patent law. It was also a 'thought experiment' to promote public debate about the ethics of biotechnology, animal research and human cloning.

A Patent Application

In 1998, a patent application was lodged with the USPTO for a human–animal chimera, an 'invention' that involves the fusing, at an early stage of development, a human and a non-human embryo to form a single, chimeric, embryo potentially capable of developing into a chimeric adult.[117] The abstract said that the invention was 'a mammalian embryo developed from a mixture of embryo cells, embryo cells and embryonic stem cells, or embryonic stem cells exclusively, in which at least one of the cells is derived from a human embryo, a human embryonic stem cell line, or any other type of human cell, and any cell line, developed embryo, or animal derived from such an embryo'.[118] The first claim related to 'a chimeric embryo comprising cells from a first and one or more second animal species, wherein said first animal species is human, wherein said second animal species is non-human, and wherein said second animal species is non-primate'.[119]

Valerie Phillips noted that much care and thought had gone into the patent application: '[The patent application] was carefully crafted to avoid being summarily rejected on the grounds that it was outside the novelty and non-obviousness requirements that all patents are supposed to meet.'[120] The intent of the patent application was to force the USPTO to engage in a consideration of the ethics and morality of patent applications in the field of biotechnology.

In response, the USPTO issued a media advisory on the patent application, observing:

> The Patent and Trademark Office is required by law to keep all patent applications in confidence until such time as a patent may be granted. However, the existence

of a patent application directed to human/non-human chimera has recently been discussed in the news media. It is the position of the PTO that inventions directed to human/non-human chimera could, under certain circumstances, not be patentable because, among other things, they would fail to meet the public policy and morality aspects of the utility requirement.[121]

The USPTO emphasized that it would not issue a patent for an invention of incredible or specious utility or for inventions whose utilization was not adequately disclosed in the application. The USPTO referred to case law, in which the courts had interpreted the utility requirement to exclude inventions deemed to be 'injurious to the well being, good policy, or good morals of society'.[122]

B United States Patent and Trademark Office Rulings

In March 1999, the USPTO rejected the first application by Stuart Newman and Jeremy Rifkin. The examiner stated that 'the PTO believes that Congress did not intend 35 USC to include the patenting of human beings'.[123] The examiner added: 'Since applicant's claimed invention embraces a human being, it is not considered to be patentable subject matter.'[124] Stuart Newman said of this rejection:

> As it attempted with the *Chakrabarty* patent application, the PTO rejected our chimera patent in its initial reviews. Of course, the major difference between the *Chakrabarty* case and ours is that the PTO no longer opposes patents on organisms. Instead, it would like to draw a line between obviously troublesome inventions of the sort we propose and other life forms they have allowed to be patented, such as human bone-marrow cells and pigs containing human genes ... Concealed within the patent issue is the deeper one of how far we as a society will go in permitting technology to blur the lines between human and non-human, person and artifact.[125]

The USPTO determined that claimed techniques in the application were not sufficiently distinguished from the prior art to allow claims for the creation of human–animal chimeras. Finally, the application did not include an enabling detailed description, including a best mode for practising the invention.

In 2002, Newman and Rifkin refiled their patent application. The pair argued that the patent claims were not directed to a human being or human embryo, but rather a man-made chimeric animal developed from a chimeric embryo. Even if the claims cover human beings, the statute does not restrict patentability based on whether the claims embrace a human being.

In 2004, the USPTO examiner, Deborah Crouch, rejected the patent application on the grounds that the claimed invention was directed to non-statutory subject matter.

Applicant argues that the statute does not restrict patentability based on whether the claims cover a human being, and that the Director lacks authority to impose a limitation on patenting a human. For reasons already stated on the record, the Office does not agree that humans are patentable subject matter.[126]

The examiner noted that, with regard to the allowance of claims encompassing humans, section 634 of the *Consolidated Appropriations Act* 2004 (US) provided that 'none of the funds appropriated or otherwise made available under this Act may be used to issue patents on claims directed to or encompassing a human organism'.[127]

The examiner held that the claims were not described in the specifications in such a way as to enable one skilled in the art to make or use the invention so that it will operate as intended without undue experimentation:

> Using either applicant's definition of chimeric animal or the broader definition provided by the art at the time of filing, the art cited by applicant in their response does not enable any of the possible chimeric animals encompassed by the claims. The specification fails to enable the production of a chimeric human/nonhuman animal that contains contributions from both parental cell types in all its organs and tissues. Likewise, the art fails to enable the production of a human/nonhuman primate chimeric animal that contains contributions from both parental cell types in all its organs and tissues.[128]

The examiner ruled that the specifications failed to demonstrate possession of the invention by actual reduction to practice, clear depiction of the invention in a detailed drawing, or description with sufficient relevant identifying characteristics of the invention.

The examiner also ruled that Newman was not entitled to a patent invention because the application had been anticipated, or was otherwise obvious. Crouch commented in particular upon claim 50, which was directed to a descendant of a chimeric animal:

> If the germ cell subsequently used in reproduction was human, and a human was used as a mate, then, the descendant would be totally human. If the germ cell, which subsequently was used in reproduction was a nonhuman primate and the same species animal was used as a mate, then, the descendant would be totally a nonhuman primate. Therefore, the descendants would not be any different from humans, or nonhuman primates found in nature. Therefore, the descendant would be anticipated by, or made obvious over known humans and nonhuman primates.[129]

The examiner was not convinced that the patent application would result in any invention which was a novel or inventive improvement upon the prior art.

The examiner also rejected the patent application on the basis that there was a lack of specific and substantial utilities. Newman had maintained that the application would be useful in toxicology assays and development studies. Crouch was unpersuaded by such arguments: 'Even assuming toxicology studies are a critical step in the development of new drugs, there is no specific explanation showing that observing developmental disorders in chimeras would have any practical utility.'[130] The examiner concluded that the claimed invention had not been brought to the point where specific benefit exists in an available form. Such an approach is consistent with the USPTO utility guidelines, which require a specific, substantial utility.

C Policy Debate

Despite the decision of the USPTO, Stuart Newman claimed a 'moral victory' in the legal proceedings. He observed that the action had demonstrated that the USPTO lacked a criterion for determining the relative humanity of a genetically engineered organism:

> But if you could genetically engineer the chimera so that the human component will be a known percentage of the organism then the USPTO might be better satisfied. I don't think that the rejection of this patent will impede research in the field. I do hope, however, that it stimulates legislative guidelines. With commercial incentive alone it is only a matter of time before such an organism is made.[131]

Newman and Rifkin hoped that the 'thought experiment' had demonstrated that ethical judgments were inescapable in patent determinations by the USPTO.

An editorial in *Nature Biotechnology* noted that 'no country's patent system has yet found a way of extricating itself from the philosophical and political morass associated with patent applications that encroach on definitions of humanness'.[132] The piece suggested that the USPTO rejection of the Newman and Rifkin application was based on the notion that 'there will be no patents on monsters'.[133] The editorial noted that 'moral standards are clearly an unsatisfactory benchmark for establishing patentability: morality (like obscenity) is one of those things that arbiters (more specifically, patent examiners) are likely to have a hard time defining'.[134] The piece called for better definitions of human beings: 'One potential criterion, for example, could be to reject patent applications on any product that requires the use or inclusion of human embryos over 14-days old (the point at which development of the nervous system and potentially human sentience begins).'[135]

In the spirit of such an inquiry, Sander Rabin observed that the dispute over the Human Chimera Patent Initiative raised a number of larger policy questions.[136] He observed: 'The issues raised by the Newman–Rifkin applications require legislation establishing the determinants of what is "human" in a way that respects human rights and protects the financial incentives of the biotech industry in creating "human" products of value.'[137] First, Rabin suggests that the USPTO should adopt a rebuttable presumption against patents on genes or cells known to endow sentience or to affect human intellect, emotion or behaviour. He maintains that the presumption against using such material in a chimera or other transgenic system should be rebuttable.[138] Second, he maintains that the USPTO should refuse to issue patents involving genes with unknown functions. He maintains that 'patents involving transfer of genes with unknown function should also be considered illegitimate subject matter'.[139] Rabin concludes: 'These proposals provide guidelines to ban "humanization" of animals without imposing undue burdens on the biotech industry.'[140]

In a piece entitled 'Patenting the Minotaur', Stanković also calls for a reformation of patent law as it applies to human–animal chimeras.[141] He submits that the United States Congress should create a standard-based statute to regulate the patenting of life forms: 'A standard-based statutory provision is flexible and suitable for the rapidly-advancing field of biotechnological innovation; it might allow for better accommodation of the legal (patentability) treatment of future chimaeric creatures, such as biomechanic hybrids and other types of organisms that we cannot now envision.'[142] Stanković concludes: 'Instead of becoming mired in moral and ethical controversies, critics of patenting the Minotaur should call for more comprehensive regulations on genetic engineering, not a ban on patenting of chimeras.'[143]

Barry Edwards has argued that Congress should act to enact a ban or a moratorium on transgenic animal patents until they are better understood: 'As the realities of science surpass what only recently passed for science fiction, Congress should reconsider legislation regulating the issuing of patents for transgenic animals.'[144] At the very least, Edwards argues that the USPTO should revive the moral utility requirement in respect of transgenic animals.

CONCLUSION

In the United States, there has been a clear trend towards the USPTO and the courts expanding the limits of patentable subject matter to include the animal kingdom, from simple organisms such as the polyploid oyster to more complex organisms, such as the Harvard oncomouse. Professor Margo Bagley from the University of Virginia School of Law comments that it is nigh impossible to reverse such a trend, at least in the United States:

> In conclusion, patents on some categories of morally controversial biotech subject matter are here to stay. It is very hard for Congress to retrench and remove subject matter from patent eligibility after patents covering such subject matter have issued. I do not think there is any going back in the area of stem cell patents or transgenic animal patents. However, the patent eligibility of other subject matter, such as humans, is still in flux, and is a topic that Congress needs to study and address.[145]

Bagley questions whether patent applicants should decide whether morally controversial inventions are subject to patent protection: 'Patent applicants are no better equipped to make that determination than the average person, and yet they are the ones making these high policy decisions by virtue of the content of the applications they file with the USPTO.'[146] She wonders whether it is appropriate for such ethical questions to be deferred by default to Patent Offices and the courts. Bagley argues that Congress needs to play a more active role in determining the boundaries of patentable subject matter: 'Specific legislation, detailing exceptions to patent eligibility or at least its outer limits, would provide greater guidance to the USPTO and courts in making patentability determinations.'[147] Similarly, it would be worthwhile exploring policy options such as exceptions to patent infringement, for example the farmers' privilege and an innocent bystanders' defence.

At an international level, there remains flexibility in respect of the eligibility of animals for patent protection. The decision of the Supreme Court of Canada may provide some inspiration for developing countries who wish to exclude animals from the scope of patentable subject matter. Richard Gold of McGill University comments:

> The Harvard College decision is of particular relevance to developing countries whose courts have yet to make determinations as to the patent eligibility of the products of biotechnology. By following the example of the Supreme Court of Canada, courts in these countries may avoid the institutional difficulties engendered from an overly ambitious enlargement of judicial jurisdiction.[148]

Article 27(2) of the *TRIPS Agreement* 1994 provides that member states may exclude inventions from patentability if prevention of the commercial exploitation of an invention is necessary to protect '*ordre public* or morality', including 'to protect human, animal or plant life or health or to avoid serious prejudice to the environment'. Article 27(3) of the *TRIPS Agreement* 1994 declares that member states may exclude from patentability 'diagnostic, therapeutic and surgical methods for the treatment of humans or animals' and 'plants and animals other than micro-organisms, and essentially biological processes for the production of plants and animals'.

NOTES

1. Brookes, Martin (2002), *Fly: The Unsung Hero of 20th Century Science*, New York: Ecco; Trans National Institutes of Health Mouse Initiatives, http://www.nih.gov/science/models/mouse/; Nolch, G. (2003), 'Wallaby genome a short hop away: Professor Jenny Graves and the Centre for Kangaroo Genomics', *Australasian Science*, **24**(7), 5; and National Institutes of Health, Animal Models, Zebra Fish: http://www.nih.gov/science/models/zebrafish.
2. Amanor-Boadu, V., M. Freeman and L. Martin (1995), 'The potential impacts of patenting biotechnology on the animal and agri-food sector', Intellectual Property Policy Directorate, Industry Canada, http://strategis.ic.gc.ca/pics/ip/martinef.pdf, p. xii.
3. Ibid.
4. Canadian Biotechnology Advisory Committee (2002), *Patenting of Higher Life Forms*, Ottawa: Canadian Biotechnology Advisory Committee, June.
5. Advisory Council on Intellectual Property (2002), *Innovation Patent – Exclusion of Plant and Animal Subject Matter*, Canberra: Department of Industry, http://www.acip.gov.au/library/Innovation%20Patent%20Issues%20Paper.PDF.
6. Lesser, W. (1993), 'Animal variety protection: a proposal for a US model law', *Journal of the Patent and Trademark Office Society*, **75**, 398.
7. Christie, A. and N. Peace (1996), 'Intellectual property protection for the products of animal breeding', *European Intellectual Property Review*, **18**(4), 213–33 at 230, 233.
8. Stop Animal Patents, http://www.stopanimalpatents.org/.
9. Stop Animal Patents, http://www.stopanimalpatents.org/.
10. Stop Animal Patents, http://www.stopanimalpatents.org/faq.php.
11. A survey commissioned by the American Anti-vivisection Society of 1008 American adults found that 68 per cent believed that it was unethical to issue animal patents, whereas only 25 per cent agreed that it was ethical to grant animal patents; http://www.stopanimalpatents.org/images/surveyresultswebsite.ppt#4.
12. *Ex Parte Allen* 2 USPQ 2d, p. 1425 (1987).
13. *Animal Legal Defense Fund v Quigg* 710 F.Supp. 728, 9 USPQ2d 1816 (N.D.Cal.1989); and *Animal Legal Defense Fund v Quigg* 932 F.2d 920 C.A.Fed. (Cal.), (1991).
14. *Harvard College v Canada (The Commissioner of Patents)* [2002] SCC 76.
15. *Re Stuart Newman* (2005), US Patent Office decision, http://patentlaw.typepad.com/patent/files/chimera_final_rejection.pdf.
16. *Ex Parte Allen* 2 USPQ 2d, p. 1425 (1987).
17. Kevles, Daniel (1998), '*Diamond* v. *Chakrabarty* and beyond: the political economy of patenting life', in Arnold Thackray (ed.), *Private Science: Biotechnology and the Rise of the Molecular Sciences*, Philadelphia: University of Pennsylvania Press, p. 73.
18. *In re Merat*, 519 F.2d 1390, 186 USPQ 471 (CCPA 1975) and *In re Bergy*, 563 F.2d 1031, 195 USPQ 344 (CCPA 1977).
19. Quoted in *Ex Parte Allen* 2 USPQ 2d, p. 1425, 2 (1987).
20. Quoted in *Ex Parte Allen* 2 USPQ 2d, p. 1425, 2 (1987).
21. Stanley, J., H. Hidu and S. Allen (1984), 'Growth of American oysters increased by polyploidy induced by blocking meiosis I but not Meiosis II', *Aquaculture*, **37**, 147–55; quoted in *Ex Parte Allen* 2 USPQ 2d, p. 1425, 2 (1987).
22. *Ex Parte Allen* 2 USPQ 2d, p. 1425 (1987).
23. *Ex Parte Allen* 2 USPQ 2d, p. 1425, 2 (1987).
24. *Ex Parte Allen* 2 USPQ 2d, p. 1425, 2 (1987).
25. *Ex Parte Allen* 2 USPQ 2d, p. 1425, 3 (1987).

26. *In re Allen* 846 F.2d 77 (1988), 1988 WL 23321.
27. *Diamond v Chakrabarty* 447 US 303 (1980).
28. *Ex Parte Allen* 2 USPQ 2d, p. 1425 (1987).
29. United States Patent and Trademark Office (1987), 'Notice: animals – patentability', Official Gazette, United States Patent and Trademark Office, **1077**, 8, 21 April.
30. Ibid.
31. Ibid.
32. *Animal Legal Defense Fund v Quigg* 710 F.Supp. 728, 9 USPQ2d 1816 (N.D.Cal.1989); and *Animal Legal Defense Fund v Quigg* 932 F.2d 920 C.A.Fed. (Cal.),1991.
33. *Animal Legal Defense Fund v Quigg* 710 F.Supp. 728, 9 USPQ2d 1816 (N.D.Cal.1989).
34. *Animal Legal Defense Fund v Quigg* 710 F.Supp. 728 at 732 USPQ2d 1816 (N.D.Cal.1989).
35. *Animal Legal Defense Fund v Quigg* 710 F.Supp. 728 at 732, 9 USPQ2d 1816 (N.D.Cal.1989).
36. *Animal Legal Defense Fund v Quigg* 932 F.2d 920 C.A.Fed. (Cal.), (1991).
37. *Animal Legal Defense Fund v Quigg* 932 F.2d 920 C.A.Fed. (Cal.), (1991).
38. *Animal Legal Defense Fund v Quigg* 932 F.2d 920 C.A.Fed. (Cal.), (1991).
39. *Animal Legal Defense Fund v Quigg* 932 F.2d 920 C.A.Fed. (Cal.), (1991).
40. *Animal Legal Defense Fund v Quigg* 932 F.2d 920 C.A.Fed. (Cal.), (1991).
41. *Animal Legal Defense Fund v Quigg* 932 F.2d 920 C.A.Fed. (Cal.), (1991).
42. *Animal Legal Defense Fund v Quigg* 932 F.2d 920 C.A.Fed. (Cal.), (1991).
43. *Animal Legal Defense Fund v Quigg* 932 F.2d 920 C.A.Fed. (Cal.), (1991).
44. House Committee on the Judiciary (1987), *Patents and the Constitution: Transgenic Animals*, Hearings before the Subcommittee on Courts, Civil Liberties, and the Administration of Justice, 100th Congress, 1st Session; and the House Subcommittee on the Judiciary, Subcommittee on Courts, Intellectual Property, and the Administration of Justice (1989), *Transgenic Animal Patent Reform Act* of 1989, HR 1556, 101st Congress, 1st Session.
45. Kastenmeier, R. (1987), 'Patenting life', *The New York Times*, July 26.
46. Andrews, E. (1990), 'Kastenmeier's loss will have a major impact on inventors', *The New York Times*, 18 November, H3.
47. Ibid.
48. Schneider, K. (1988), 'Life patents: doubts are registering', *The New York Times*, 7 August.
49. Ibid.
50. Ibid.
51. Weiss, R. (1988), 'Animal patent report lacks support', *Science News*, 9 April.
52. Greene, R. (1989), 'Administration opposes farmer exemption', The Associated Press, 14 September.
53. Schenider, K. (1987), 'Agency and Congress face clash over patenting of animals', *The New York Times*, 23 July.
54. Biotechnology Newswatch (1989), 'Author of Animal Patents Bill hears dire warnings pro, con', *Biotechnology Newswatch*, 2 October, **9**(19), 4.
55. Schneider, K. (1988), 'Life patents: doubts are registering', *The New York Times*, 7 August.
56. Andrews, E. (1990), 'Kastenmeier's loss will have a major impact on inventors', *The New York Times*, 18 November, H3.
57. Leder, P. and T. Stewart (1988), 'Transgenic non-human mammals', US Patent No: 4,736,866.
58. Kevles, D. (2002), 'Of mice and money: the story of the world's first animal patent', *Daedalus*, **131**(2), 78–88.
59. Leese, Mark (1996), 'Is an American mouse a European mouse? Towards a sociology of patents', in Andrew Webster and Kathryn Packer (eds), *Innovation and the Intellectual Property System*, London: Kluwer Law International Ltd, p. 171; Mandelker, B. (1998–99), 'Commentary: Harvard College *v* Canada', *Intellectual Property Journal*, **13**, 87–106; Haraway, Donna (1997), *Modest Witness, Second*

Millennium: FemaleMan Meets OncoMouse. Feminism And Technoscience, New York: Routledge; Myerson, George (2000), *Donna Haraway and GM Foods*, Cambridge: Icon Books; Gold, E.R. (2003–4), 'The reach of patent Law and institutional competence', *The University of Ottawa Law and Technology Journal*, **1**, 263–84; Mohammed, E.A.C. (2004), 'Cat in the hat, a mouse in the house: comparative perspectives on Harvard mouse', *Intellectual Property Journal*, **18**, 169–85; Atkinson, R. (2005), 'Mixed messages: Canada's stance on patentable subject matter in biotechnology', *Intellectual Property Journal*, **19**, 1–27; and Robertson, S. (2005), 'Re-imagining economic alterity: a feminist critique of the juridical expansion of biopropety in the Monsanto decision at the Supreme Court of Canada', *University of Ottawa Law and Technology Journal*, **2**, 227–53.

60. Leder, P. and T. Stewart (1988), 'Transgenic non-human mammals', US Patent No: 4,736,866.
61. *Harvard/Onco-Mouse*, 1989 O.J. EPO 451.
62. *Harvard/Onco-Mouse*, 1990 O.J. EPO 476.
63. Leder, P and T. Stewart (1986), 'Method for producing transgenic animals', European Patent No: EP0169672.
64. *Directive 98/44/EC of the European Parliament and of the Council* of 6 July 1998 on the Legal Protection of Biotechnological Inventions, <http://europa.eu.int/eur-lex/pri/en/oj/dat/1998/l_213/l_21319980730en00130021.pdf.>
65. *Harvard/ Onco-mouse* [2003] OJEPO 473.
66. *Harvard/ Onco-mouse* [2004] T 0315/03–3.3.8.
67. *Harvard College v Canada (The Commissioner of Patents)* [2002] SCC 76.
68. Leder, P. and T. Stewart (1985), 'Transgenic animals', Canadian Patent No: CA 1341442.
69. *Harvard College v Canada (The Commissioner of Patents)* [1998] 3 FC 510 (Fed T.D.).
70. *Harvard College v Canada (The Commissioner of Patents)* [2000] 4 FC 528 (Fed C.A.)
71. *Harvard College v Canada (The Commissioner of Patents)* [2002] SCC 76.
72. Robertson, S. (2005), 'Re-imagining economic alterity: a feminist critique of the juridical expansion of biopropety in the Monsanto decision at the Supreme Court of Canada', *University of Ottawa Law and Technology Journal*, **2**, 227–53 at 230.
73. *Harvard College v Canada (The Commissioner of Patents)* [2002] SCC 76 [166].
74. *Harvard College v Canada (The Commissioner of Patents)* [2002] SCC 76 [203].
75. Editorial (2003), 'Patenting pieces of people', *Nature Biotechnology*, **21**(4), 341, http://www.nature.com/nbt/journal/v21/n4/full/nbt0403-341.html.
76. Leiss, W. (2002), 'Higher life forms before the law', http://www.leiss.ca/index.php?option=com_content&task=view&id=41&Itemid=43.
77. *Harvard College v Canada (The Commissioner of Patents)* [2002] SCC 76 [158].
78. *Harvard College v Canada (The Commissioner of Patents)* [2002] SCC 76 [159].
79. *Harvard College v Canada (The Commissioner of Patents)* [2002] SCC 76 [155].
80. *Harvard College v Canada (The Commissioner of Patents)* [2002] SCC 76 [155].
81. Shulman, S. (2002), 'Of oncomice and men, owning the future', *Technology Review*, September; and Rimmer, M. (2003), 'The attack of the clones: patent law and stem cell research', *Journal of Law and Medicine*, **10**(4), 488–505.
82. *Harvard College v Canada (The Commissioner of Patents)* [2002] SCC 76 [188].
83. House of Commons (1989), *Minutes of Proceedings and Evidence of the Legislative Committee on Bill C-15, An Act Respecting Plant Breeders' Rights*, **1**, 11 October, p. 1115.
84. *Harvard College v Canada (The Commissioner of Patents)* [2002] SCC 76 [170].
85. *Harvard College v Canada (The Commissioner of Patents)* [2002] SCC 76 [171]; and Canadian Biotechnology Advisory Committee (2002), *Patenting of Higher Life Forms*, Ottawa: Canadian Biotechnology Advisory Committee, June, p. 12.
86. *Harvard College v Canada (The Commissioner of Patents)* [2002] SCC 76 [172]; and Canadian Biotechnology Advisory Committee (2002), *Patenting of Higher Life Forms*, Ottawa: Canadian Biotechnology Advisory Committee, June, pp. 13–14.
87. *Harvard College v Canada (The Commissioner of Patents)* [2002] SCC 76 [196].

88. *Harvard College v Canada (The Commissioner of Patents)* [2002] SCC 76 [18].
89. *Harvard College v Canada (The Commissioner of Patents)* [2002] SCC 76 [25].
90. *Harvard College v Canada (The Commissioner of Patents)* [2002] SCC 76 [20].
91. *Harvard College v Canada (The Commissioner of Patents)* [2002] SCC 76 [45].
92. *Harvard College v Canada (The Commissioner of Patents)* [2002] SCC 76 [57].
93. *Harvard College v Canada (The Commissioner of Patents)* [2002] SCC 76 [52].
94. *Harvard College v Canada (The Commissioner of Patents)* [2002] SCC 76 [52].
95. *Harvard College v Canada (The Commissioner of Patents)* [2002] SCC 76 [2].
96. *Harvard College v Canada (The Commissioner of Patents)* [2002] SCC 76 [3].
97. *JEM Ag Supply v Pioneer Hi-Bred International Inc* 534 US 124 (2001).
98. *Harvard College v Canada (The Commissioner of Patents)* [2002] SCC 76 [48].
99. *Monsanto Canada Inc. v Schmeiser* 2001 FCT 256 (CanLII), (2001), 202 F.T.R. 78, 12 C.P.R. (4th) 204; *Schmeiser v Monsanto Canada Inc.*, 2002 FCA 448 (CanLII), [2003] 2 F.C. 165, 218 D.L.R. (4th) 31; and *Monsanto Canada Inc. v Schmeiser*, [2004] 1 S.C.R. 902, 2004 SCC 34.
100. Gold, E. Richard (1996), *Body Parts: Property Rights and the Ownership of Human Biological Materials*, Washington DC: Georgetown University Press; Gold, E. Richard (1999), 'Making room: reintegrating basic research, health policy, and ethics into patent law', in Tim Caulfield and Bryn Williams-Jones (eds), *The Commercialization of Genetic Research: Ethical, Legal, and Policy Issues*, New York: Kluwer Academic, p. 63; Caulfield, T. (2000), 'Underwhelmed: hyperbole, regulatory policy, and the genetic revolution', *McGill Law Journal*, **45**, 437–60; Knoppers, B. (2000), 'Reflections: the challenge of biotechnology and public policy', *McGill Law Journal*, **45**, 559–65; Mooney, P.R. (2001), *The Impetus for and Potential of Alternative Mechanisms for the Protection of Biotechnological Innovations*, Ottawa: Canadian Biotechnology Advisory Committee, March, p. 13.
101. Heller, M. and R. Eisenberg (1998), 'Can patents deter innovation? The anticommons in biomedical research', *Science*, **280**, 698–701; and Gold, E.R. (2000), 'Biomedical patents and ethics: a Canadian solution', *McGill Law Journal*, **45**, 413–35.
102. *Harvard College v Canada (The Commissioner of Patents)* [2002] SCC 76 [114].
103. *Harvard College v Canada (The Commissioner of Patents)* [2002] SCC 76 [1].
104. *Harvard College v Canada (The Commissioner of Patents)* [2002] SCC 76 [109].
105. *Diamond v Chakrabarty* (1980) 447 US 303.
106. *Harvard College v Canada (The Commissioner of Patents)* [2002] SCC 76.
107. Scott Alexander, J. (2003), 'Mouse trap', Canadian Bar Association, October, p. 29.
108. Harvard College (2002), 'Statement regarding Canadian Supreme Court 5–4 Decision Dec. 5, 2002 denying the Patentability of "Oncomouse" in Canada', Boston, 5 December.
109. BIOTECanada (2002), 'BIOTECanada responds to Supreme Court decision on Harvard Mouse Case', Ottawa, 5 December.
110. Scott Alexander, J. (2003), 'Mouse trap', Canadian Bar Association, October, p. 30.
111. Ibid., p. 32.
112. Canadian Biotechnology Advisory Committee (2003), *Advisory Memorandum: Higher Life Forms and the Patent Act*, Ottawa: Canadian Biotechnology Advisory Committee, 24 February.
113. Ibid., p. 5.
114. Ibid., p. 5.
115. Atkinson, R. (2005), 'Mixed messages: Canada's stance on patentable subject matter in biotechnology', *Intellectual Property Journal*, **19**, 1–27 at 26.
116. Newman, S. (2002), 'The human chimera patent initiative', New York Medical College, http://www.nymc.edu/sanewman/PDFs/Lahey_Winter_2002.pdf.
117. Newman, S. (2003), 'Chimeric embryos and animals containing human cells', US Patent Application: 20030079240.
118. Ibid.
119. Ibid.
120. Phillips, V. (2005), 'Half-human creatures, plants and indigenous peoples: musings on ramifications of Western notions of intellectual property and the attempt to patent a

theoretical half-human creature', *Santa Clara Computer & High Technology Law Journal*, **21**, 383–450 at 388–9.
121. United States Patent and Trademark Office (1998) 'Media advisory: facts on patenting life forms having a relationship to humans', http://www.uspto.gov/web/offices/com/speeches/98-06.htm, 1 April.
122. *Lowell v Lewis*, Fed. Cas. No. 8568 (C.C. Mass. 1817) (Story, J.).
123. Quoted in Rabin, S. (2006), 'The human use of humanoid beings: chimeras and patent law', *Nature Biotechnology*, **24**, 517–19, 1 May.
124. Ibid.
125. Newman, S. (2002), 'Legal column: the human chimera patent initiative', *Lahey Clinic Medical Ethics*, **9**(1), 4, 7, http://www.lahey.org/NewsPubs/Publications/Ethics/JournalWinter 2002/Journal_Winter 2002_Legal.asp.
126. *Re Stuart Newman* (2005), US Patent Office decision, <http://patentlaw.typepad.com/patent/files/chimera_final_rejection.pdf>.
127. There has been much policy debate over this appropriations rider: Wilkie, D. (2004), 'Stealth stipulation shadows stem cell research: a provision in the US Appropriations Bill bans patents on human organisms', *The Scientist*, **18**(4), 42, 1 March.
128. *Re Stuart Newman* (2005), US Patent Office decision, <http://patentlaw.typepad.com/patent/files/chimera_final_rejection.pdf>.
129. *Re Stuart Newman* (2005), US Patent Office decision, <http://patentlaw.typepad.com/patent/files/chimera_final_rejection.pdf>.
130. *Re Stuart Newman* (2005), US Patent Office decision, <http://patentlaw.typepad.com/patent/files/chimera_final_rejection.pdf>.
131. Editorial (2005), 'Hybrid too human to patent: case highlights lack of criterion for genetically modified organisms', *Nature Review Drug Discovery*, http://www.nature.com/news/2005/050328/full/nrd1710.html, 31 March.
132. Editorial (2003), 'Patenting pieces of people', *Nature Biotechnology*, **21**, 341, 1 April.
133. Ibid.
134. Ibid.
135. Ibid.
136. Rabin, S. (2006), 'The human use of humanoid beings: chimeras and patent law', *Nature Biotechnology*, **24**, 517–19, 1 May.
137. Ibid.
138. Ibid.
139. Ibid.
140. Ibid.
141. Stanković, B. (2005), 'Patenting the Minotaur', *Richmond Journal of Law and Technology*, **12** (2), 1–39, http://law.richmond.edu/jolt/v12i2/article5.pdf.
142. Ibid., p. 36.
143. Ibid., p. 36.
144. Edwards, B. (2001), ' "And on his farm he had a geep": patenting transgenic animals', *Minnesota Intellectual Property Review*, **2**, 89–118 at 118.
145. Bagley, M. (2004–05), 'Stem cells, cloning and patents: what's morality got to do with it?', *New England Law Review*, **39**, 501–9 at 508–9.
146. Ibid.
147. Bagley, M. (2003), 'Patent first, ask questions later: morality and biotechnology in patent law', *William and Mary Law Review*, **45**, 469–547 at 546.
148. Gold, E.R. (2003–04), 'The reach of patent law and institutional competence', *The University of Ottawa Law and Technology Journal*, **1**, 263–84 at 282.

4. The storehouse of knowledge: patent law, scientific discoveries and products of nature

> What if each generation of scientists was forbidden to use – or even think about – the theorems, principles, and natural phenomena that had been discovered or proven by the previous generation of scientists? (Lori Andrews et al., 'When patents threaten science')[1]

Historically, patent examiners, courts and legislatures sought to draw a sharp distinction between inventions, which were patentable, and scientific discoveries, products of nature, and methods of human treatment, which belonged to the public domain, the 'storehouse of knowledge', which could be drawn upon by anyone.[2]

In 1873, George T. Curtis described the question as 'how far a discovery or invention which may first disclose and practically embody some truth in physics or some law in the operation of the forces of nature, for a useful purpose, is capable of being carried in the exclusive privileges secured by the grant of letters patent'.[3] In 1890, treatise-writer William Robinson observed of the basis for the prohibition of natural phenomena: 'A principle, in this sense, is a necessary factor in every means which produces physical effects, whether such means be natural or artificial, and it is generally this which makes the chief impression on the senses of the observer; but it is in itself no true invention, nor can it be protected by a patent.'[4]

In the landmark Telegraph Case, *O'Reilly v Morse*, the Supreme Court of the United States considered the validity of a patent held by Samuel Morse for an apparatus for accomplishing the transmission of signals from a distance to an electromagnetic telegraph.[5] In the Eighth Claim of the patent, Morse claimed rights to all uses of electromagnetism for sending signals over distance. The court ruled that this claim was invalid, because it sought an exclusionary property right in all uses of the underlying natural phenomenon of electromagnetism, not in his specific invention itself.

In the *Telephone Cases*, the Supreme Court of the United States considered the patentability of Alexander Graham Bell's invention, the telephone.⁶ The Fifth Claim related to '[t]he method of, and apparatus for, transmitting vocal or other sounds telegraphically, as herein described, by causing electrical undulation, similar in form to the vibrations of the air accompanying the said vocal or other sounds, substantially set forth [in the patent]'. The Supreme Court of the United States held that the patent claim was valid, because it did not claim the principle of converting electricity to sound waves, but rather a particular method and apparatus for utilizing that principle in a new and useful way.

In the 1862 case of *Morton v New York Eye Infirmary*, Shipman J rejected a patent application in respect of a method of performing surgery by applying ether to render the patient insensitive to pain.⁷ The judge acknowledged the testimony of 'distinguished surgeons' who ranked the idea of employing ether in surgery as 'among the great discoveries of modem times'.⁸ Nonetheless, His Honour insisted upon the fine distinction between scientific discoveries and inventions:

> A discovery may be brilliant and useful, and not patentable. No matter through what long, solitary vigils, or by what importunate efforts, the secret may have been wrung from the bosom of Nature, or to what useful purpose it may be applied. Something more is necessary. The new force or principle brought to light must be embodied and set to work, and can be patented only in connection or combination with the means by which, or the medium through which, it operates.⁹

The judge emphasized: 'A discovery of a new principle, force or law operating, or which can be made to operate, on matter, will not entitle the discoverer to a patent.'¹⁰ Shipman J stressed: 'It is only where the explorer has gone beyond the mere domain of discovery, and has laid hold of the new principle, force, or law, and connected it with some particular medium or mechanical contrivance by which, or through which, it acts on the material world, that he can secure the exclusive control of it under the patent laws.'¹¹

In *Mackay Radio & Tel. Co. v Radio Corporation of America*, the Court held, regarding a patent application for a certain type of antennae, that '[w]hile a scientific truth, or the mathematical expression of it, is not patentable invention, a novel and useful structure created with the aid of knowledge of scientific truth may be'.¹²

In *Funk Bros. Seed Co. v Kalo Inoculant Co.*, the Supreme Court of the United States held that an inventor claiming a bacterial species that exhibited the property of mutual non-inhibition could not patent the bacteria, because it was a claim for a natural phenomenon itself.¹³ Memorably, Douglas J observed that the qualities of bacteria were a natural phenomenon, which belonged to the 'storehouse of knowledge':

Bond does not create state of inhibition or of non-inhibition in the bacteria. Their qualities are the work of nature. Those qualities are of course not patentable. For patents cannot issue for the discovery of the phenomena of nature. The qualities of these bacteria, like the heat of the sun, electricity, or the qualities of metals, are part of the storehouse of knowledge of all men. They are manifestations of laws of nature, free to all men and reserved exclusively to none. He who discovers a hitherto unknown phenomenon of nature has no claim to a monopoly of it which the law recognizes. If there is to be invention from such a discovery, it must come from the application of the law of nature to a new and useful end.[14]

Douglas J observed: 'there is no invention here unless the discovery that certain strains of the several species of these bacteria are non-inhibitive and may thus be safely mixed is invention'.[15] He concluded: 'We cannot so hold without allowing a patent to issue on one of the ancient secrets of nature now disclosed.'[16]

In *Gottschalk v Benson*, the Court held that a mere procedure for solving a mathematical problem was not patentable, as it was a method for a general, non-specific, and non-inventive purpose.[17] In *Parker v Flook*, the Court disallowed a patent claiming methods for updating alarm limits and held that 'if a claim is directed essentially to a method of calculating, using a mathematical formula, even if the solution is for a specific purpose, the claimed method is nonstatutory'.[18] *In re Bergy*, Rich J commented: '[Patent] claims which directly or indirectly preempt natural laws or phenomena are proscribed, whereas claims which merely utilize natural phenomena via explicitly recited manufactures, compositions of matter or processes to accomplish new and useful end results define statutory inventions.'[19]

In the 1981 decision of *Diamond v Diehr*, the Supreme Court of the United States considered the meaning of section 101 of the *Patents Act 1952* (US) which provided a broad statement of patentable subject matter: 'Whoever invents or discovers any new and useful process, machine, manufacture, or composition of matter, or any new and useful improvement thereof, may obtain a patent therefor, subject to the conditions and requirements of this title.'[20] The context was a patent application in respect of claims for a process for curing rubber. In the majority judgment, Rehnquist J discussed the limits of patentable subject matter:

> This Court has undoubtedly recognized limits to §101 and every discovery is not embraced within the statutory terms. Excluded from such patent protection are laws of nature, natural phenomena, and abstract ideas.[21] 'An idea of itself is not patentable.'[22] 'A principle, in the abstract, is a fundamental truth; an original cause; a motive; these cannot be patented, as no one can claim in either of them an exclusive right.'[23] Only last Term, we explained: '[A] new mineral discovered in the earth or a new plant found in the wild is not patentable subject matter. Likewise, Einstein could not patent his celebrated law that $E = mc^2$; nor could Newton have patented the law of gravity. Such discoveries are manifestations

of . . . nature, free to all men and reserved exclusively to none.'[24] Our recent holdings in *Gottschalk v. Benson*, and *Parker v. Flook*, both of which are computer-related, stand for no more than these long-established principles.[25]

His Honour noted: 'We recognize, of course, that when a claim recites a mathematical formula (or scientific principle or phenomenon of nature), an inquiry must be made into whether the claim is seeking patent protection for that formula in the abstract.'[26] The judge held that the patent claims fell within the scope of patentable subject matter: 'We view respondents' claims as nothing more than a process for molding rubber products and not as an attempt to patent a mathematical formula.'[27]

In 1982, Congress established the Court of Appeals for the Federal Circuit (CAFC) to hear all patent appeal cases. Kenneth Dam noted that the creation of the Federal Circuit had displaced, as a practical matter, the Supreme Court of the United States jurisdiction in patent cases.[28] Scholars Peter Drahos and John Braithwaite have doubted, though, whether the Federal Circuit has increased the doctrinal stability and unity of patent law:

> Whether it has done this is open to question. Analysts have pointed to the large number of times the court has flatly contradicted itself, as well as its distortion of patent law in the context of biotech patenting in order to better serve the private sector. What it has done is to increase the chances of a patent holder succeeding in litigation.[29]

The Court of Appeals for the Federal Circuit has systematically eliminated time-honoured categorical exclusions from patent eligibility in relation to products of nature, plants, agriculture, surgical and medical treatments, mathematical algorithms, business methods. The Public Patent Foundation has complained: 'Through a series of decisions, the Federal Circuit has abandoned the substantive based standard established by [the Supreme] Court for determining patentable subject matter and replaced it with a more expansive formalistic approach that looks only to see whether a patent claim contains some structure or has some minimal practical utility.'[30] The public interest group comments: 'The Federal Circuit's form-over-substance approach has come to include virtually anything within patentable subject matter.'[31]

This chapter considers the significance of the Supreme Court of the United States decision in *Laboratory Corp. of America Holdings v Metabolite Laboratories, Inc.* which considers the limits of patentable subject matter.[32] There was much debate as to whether a patent for a method for detecting a form of vitamin B deficiency was invalid because it was beyond the limits of patentable subject matter. Brooks Gifford comments that there is conflict between the Supreme Court of the United States and the Federal Circuit on subject matter eligible for patent protection:

Diehr is the Supreme Court's most recent pronouncement of patent eligibility law and as such must be treated as authoritative. The CAFC has gradually eroded its eligibility requirements to the point they are almost non-existent. The CAFC's analysis and holding in Metabolite exemplifies the problems of liberal eligibility standards. The codification of the CAFC's holding in *State Street* in recent PTO examination guidelines only exacerbates the problem.[33]

Section one considers the history of the dispute between Metabolite Laboratories and LabCorp. Section two evaluates the various submissions of the amicus curiae in the case to the Supreme Court of the United States. The litigation attracted a wide diversity of opinions, from scientists and researchers, to doctors and the medical fraternity, and even financial service providers and information technology companies. Section three explores the majority decision that special leave to appeal had been improvidently granted. It also analyses the dissenting, minority decision of Breyer J, which expressed concern about the rapid expansion of the scope of patentable subject matter by the United States Patent and Trademark Office (USPTO) and the Court of Appeals for the Federal Circuit.

I LABCORP

In 1986, University Patents Incorporated filed for a patent at the USPTO.[34] The patent claimed a method for detecting a form of vitamin B deficiency, which focused upon a correlation in the human body between elevated levels of certain amino acids and deficient levels of vitamin B. Claim 13 provides for 'A method for detecting a deficiency of cobalamin or folate in warm-blooded animals comprising the steps of: assaying a body fluid for an elevated level of total homocysteine; and correlating an elevated level of total homocysteine in said body fluid with a deficiency of cobalamin or folate.'[35]

University Patents Incorporated licensed Metabolite, which sublicensed LabCorp to use the patent. From 1992 until 1998, LabCorp paid Metabolite a royalty every time it supplied a doctor with a homocysteine test. After 1998, LabCorp began using a test developed by another company. In response, Metabolite sued LabCorp for patent infringement.

A United States District Court for the District of Colorado

In the United States District Court for the District of Colorado, a jury found that LabCorp indirectly infringed Metabolite Laboratories, Inc.'s U.S. Patent No. 4,940,658.[36] The jury also found that LabCorp partially breached its contract with Metabolite. Based on this verdict, the district court assessed damages of $US 3 652 724.61 for breach of contract and $US 1 019 365.01 for indirect infringement. Subsequently, the District Court doubled the infringement award for wilful infringement and issued a permanent injunction.

B United States Court of Appeal for the Federal Circuit

In the United States Court of Appeal for the Federal Circuit, Rader J for the majority upheld the jury's verdicts and the District Court's rulings.[37] His Honour affirmed finding of indirect patent infringement and breach of contract, and affirmed the district court's award of damages to Metabolite. Rader J observed: 'The record shows that physicians order assays and correlate the results of those assays, thereby directly infringing.'[38] The judge held accordingly that 'a reasonable jury could find intent to induce infringement because LabCorp's articles state that elevated total homocysteine correlates to cobalamin/folate deficiency'.[39] His Honour affirmed the finding of indirect infringement based on the inducement analysis.

Rader J rejected the arguments of LabCorp that Claim 13 was invalid on grounds of indefiniteness, lack of written description and enablement, anticipation and obviousness. In particular, he emphasized that the patent was valid in terms of its novelty and inventiveness: 'Even if the secondary references disclosed total homocysteine, the record does not contain evidence showing that one of skill in the art would have been motivated to combine the various references.'[40] Rader J observed that the record contained evidence of other indicia, which supported the jury's verdict. His Honour noted that skilled artisans were initially sceptical about the invention, and that Metabolite had licensed the invention to eight companies. Rader J also refused to consider the contention of LabCorp that Claim 18, directed to the panel test, was also invalid on grounds of indefiniteness, and lack of written description and enablement.

In a separate judgment, Schall J agreed with the majority's conclusions with respect to validity, the absence of a case or controversy regarding infringement of Claim 18, breach of contract, enhanced damages and the district court's injunction. However the judge dissented from the majority's construction of Claim 13 of the '658 patent: 'Because I think Claim 13 covers only the correlation of elevated levels of homocysteine, I would remand the case for a recalculation of the damages resulting from indirect infringement.'[41]

C Writ of Certiorari

Appealing for a writ of certiorari to the Supreme Court of the United States, LabCorp argued that the patent was invalid because it 'claim[s] a monopoly over a basic scientific relationship used in medical treatment such that any doctor necessarily infringes the patent merely by thinking about the relationship after looking at a test result'.[42] The petitioner observed that the Supreme Court of the United States has 'long recognized that scientific facts and laws of nature are outside the scope of patentable inventions'.[43] The company commented:

> In this case, the Federal Circuit improperly broadened this limited scope of the patent laws in at least two ways. First, the court incorrectly construed the Patent as conferring a monopoly over the thought processes of doctors, by holding that direct infringement occurs whenever a doctor looks at a homocysteine test result and thinks about a possible connection to vitamin deficiencies – regardless of what testing method is used and without requiring any further confirmation of an actual deficiency. But the court took its flawed holding a step further, by holding that a third party such as LabCorp, which indisputably committed no direct infringement, can be held liable for indirect infringement merely by reciting a medical fact.[44]

LabCorp observed that the question was a matter of national importance: 'Although LabCorp is the party that bears the judgment in this case, if the Federal Circuit's decision is not reversed the ultimate losers will be thousands of doctors and millions of their patients.'[45] The company pleaded: 'The Court should therefore grant review to clarify that the patent laws do not permit a party to gain a monopoly over the thought processes of doctors or prevent anyone from simply disseminating truthful information about a basic scientific fact critical to patient care.'[46]

By contrast, Metabolite argued that the petitioner's contention that Claim 13 does not recite patentable subject matter was not properly presented.[47] The company maintained that, in any event, such arguments were without merit. Expanding upon such themes, Metabolite commented:

> The expansive language of Section 101 allows a patent on 'anything under the sun that is made by man.'[48] In addition to machines, manufactures, and compositions of matter, Section 101 makes 'processes' patentable. 'A process is a mode of treatment of certain materials to produce a given result.'[49] Claim 13 fits this description precisely: It claims a process for treating certain materials (i.e., assaying body fluids and correlating the assay results with vitamin status) to achieve a desired result (i.e., detecting cobalamin or folate deficiencies).[50]

Metabolite contended: 'The havoc caused by a decision questioning the validity of the '658 patent would extend far beyond the realm of medical diagnoses to every patented invention that incorporates a natural relationship (including most drugs, many medical devices, and a host of computer software and hardware applications).'[51]

II FRIENDS OF THE COURT

The Supreme Court of the United States granted a writ of certiorari to consider whether the patent was invalid on the grounds that it sought to 'claim a monopoly over a basic scientific relationship', namely, the relationship between homocysteine and vitamin deficiency.

The Supreme Court of the United States heard a large number of amicus curiae submissions on the question of the limits of patentable subject matter. There were stark divisions between those parties wishing to revisit patent eligibility, and those wishing to preserve the current status quo. Academic Rebecca Eisenberg commented upon such divisions amongst the friends of the court:

> The chorus of amici that have seized upon this opening to voice their competing views about the expanding reach of the patent system can only confirm the Court's suspicion that the issue is important and timely. But the failure of the Federal Circuit to address the issue of patent eligibility in this case leaves the Supreme Court with a poor record on which to consider the issue. This is particularly troubling because, if and when the Court takes up the topic of patent eligibility, it will be hard pressed to find any guidance on how to draw reasonable subject matter boundaries for the patent system in decisions of the past 25 years.[52]

The United States Government and a number of law bars sought to dissuade the Supreme Court of the United States from considering the limits of patentable subject matter. Such entities maintained that the writ of certiorari had been improvidently granted. A number of public interest groups called upon the Supreme Court of the United States to respond to the Federal Circuit's rapid expansion of eligible subject matter for patent protection. Representatives of medical professionals and patients questioned the wisdom of patenting medical treatments and diagnostics. Financial organizations debated the merits of patenting business methods.

A Scientific Discoveries

The United States Solicitor-General sought to dissuade the Supreme Court of the United States from intervening in the matter, protesting that the record was not sufficiently developed to permit an assessment of the patent claim's validity.[53] The Solicitor-General stressed the petitioner did not argue in the lower courts that the patent was non-patentable subject matter. He argued that the petition for the writ of certiorari should be denied: 'This case does not provide a suitable vehicle for considering the question posed by this court because the petitioner either failed to preserve that issue in the lower courts or to develop a complete record, and the correct resolution of this case might not turn on the choice of legal standard in any event.'[54]

The Intellectual Property Owners Association argued that the current standards for patentable subject matter, as set forth by the Court in *Diamond v Diehr*,[55] correctly delineated between those innovations that should be eligible for patent protection and those that should not.[56] The Association suggested that the current case should not serve as a vehicle for overturning or altering those standards: 'Rather, this case should reinforce the standards of *Diehr* and, thus, support the expectation that innovations in yet unknown areas of technology will be eligible for patent protection.'[57]

The Federal Circuit Bar Association maintained that the Supreme Court of the United States should dismiss the writ of certiorari as improvidently granted.[58] Not surprisingly, the Association was an enthusiastic defender of the Federal Circuit's progressive enlargement of the scope of patentable subject matter:

> Extensive Federal Circuit jurisprudence in this area confirms the Congressional judgment that broadly patentable subject matter fosters innovation in technology-based industries such as computer software, pharmaceuticals, and biotechnology, where subject matter is often unpredictable. Nor, during the twenty-five years that have passed since *Diehr*, has Congress seen fit to limit the scope of §101, even in the wake of decisions confirming the patentability of ever more diverse subject matter.[59]

The Association maintained: 'Restricting the scope of patentable subject matter is not an appropriate means for controlling what some perceive as improvidently granted patents involving methods of doing business, methods of practicing medicine, diagnostic methods, and the like.'[60] The Association was confident that the USPTO was entirely capable of responding adequately to increased filings in an area of newly recognized subject matter through the application of standard patent criteria. The Association observed that exemptions from patent infringement was a better alternative to limiting the categories of subject matter eligible for patent protection: 'Rather than limiting the potential scope of patents, Congress, when it has seen fit to act, has exempted certain actions or actors from charges of infringement.'[61]

The Boston Patent Law Association supported the respondents, arguing that the patent claim was indeed patentable subject matter.[62] The Association denied that the claim amounted to an unwarranted monopoly over scientific principles or medical treatments:

> Correlating test results, as prescribed by Claim 13, does not impose a monopoly over a basic scientific relationship used in medical treatment. Any doctor

looking at a test result of an assay suitable for complying with the assaying step of Claim 13 does not infringe the patent merely by thinking about a relationship between an elevated level of total homocysteine and a deficiency of cobalamin or folate. The monopoly conferred by a valid method patent, which is based on discovery of a basic scientific relationship, extends only to those who conduct a test for the purpose of correlating the test result with a condition associated with the relationship.[63]

The Association concluded that 'confirming the patentability of Claim 13 would not preempt the use of a scientific principle or remove the existence of knowledge from the public domain'.[64]

The Bar of the City of New York commented: 'The standards for determining whether a patent claim covers patentable subject matter as enunciated in *Diamond v Diehr* should be reaffirmed by the Court.'[65] The Bar noted: 'The examination of patentable subject matter under Section 101 should not be confused or conflated with an analysis of the other requirements for patentability under the Patent Act.'[66] The Bar commented: 'The circumstances in which a Section 101 analysis is needed should not be restricted because it is impossible to predict the advances in technology that may make a Section 101 analysis necessary.'[67]

B Products of Nature

There was much debate in the submissions to the Supreme Court of the United States about the relevance of the 'products of nature' doctrine in the light of twenty-first century developments in respect of medicine and biotechnology.[68]

Professor Craig Jepson of the Franklin Pierce Law Centre also supported the respondents.[69] He argued that the Supreme Court of the United States should expand upon the pronouncements on patentable subject matter in *Diamond v Chakrabarty* and *Diamond v Diehr*:

> This Court should discard the prohibition against patent claims directed to so-called laws of nature. The prohibition arose from dicta contained in old, English, common law decisions. It has no statutory support in our law. It is unnecessary and nearly impossible to apply because it is error-prone, redundant, and obsolete . . . The prohibition against claims directed to laws of nature is a relic, an arbitrary bar to patentability, such as Congress and this Court consistently have acted to eliminate.[70]

Jepson comments: 'The distinction between patentable subject matter, on the one hand, and a law of nature, on the other hand, has never shimmered with lucidity.'[71] He warns: 'Consequently, the prohibition is difficult to apply, and such application leads to arbitrary and artificial line drawing and patent claim drafting gamesmanship.'[72] Jepson concluded: 'This Court should use the present case to build on *Chakrabarty* and *Diehr* by discarding the arbitrary and unworkable prohibition.'[73]

The Public Patent Foundation pleaded for the Supreme Court of the United States to set firm limits as to the boundaries of patentable subject matter:

> Almost twenty-five years have passed since this Court last addressed the core issues of patentable subject matter. In that time, the Court of Appeals for the Federal Circuit has replaced this Court's substantive standard with a more formalistic approach that has expanded the definition of patentable subject matter to include virtually anything. This expansion by the Federal Circuit conflicts with this Court's precedent and, as such, merits remediation.[74]

The Public Patent Foundation made two further submissions. First, the Foundation argued that allowing claims that effectively cover all uses of a law of nature or abstract idea frustrated the patent system's goal of disclosure: 'There can be no "inventing around" a patent unless the patent discloses the basic scientific principles upon which the invention relies – those principles being broader than the invention claimed in the patent.'[75] Second, the Foundation maintained that patent claims that restrict communication regarding abstract ideas or laws of nature were contrary to the First Amendment: 'Were laws of nature and abstract ideas to be patentable subject matter, scientific expression could be seriously restricted in violation of the First Amendment.'[76]

Affymetrix is a supplier of commercial DNA microarrays and 'has an interest in ensuring that patents not issue on basic laws of nature so as to impede scientific progress in analyzing DNA and gene expression'.[77] Affymetrix argued that the fact that 'elevated levels of an amino acid in the blood correlated to a vitamin deficiency' is a natural phenomenon that leaves Claim 13 unpatentable under current precedent. Affymetrix was concerned about the broader ramifications of the patent application for the field of biotechnology:

> Allowing Claim 13 to stand would damage such future research and scientific progress. Claim 13 and others like it allow no room to design around, imitate, or improve upon the so-called 'invention' of a law of nature. DNA technology has opened up a vast array of tests based on naturally occurring biochemical mechanisms. But if claims like Claim 13 are sustained, such tests will be blocked by patents on the law of nature on which they are based. This is especially harmful given the nature of modern genomic research, which focuses not on one gene or gene function at a time, but rather on complex interconnections among genes and gene functions. Such interconnections cannot be studied if portions of the larger genomic map are blocked out.[78]

Affymetrix encouraged the Supreme Court of the United States to reaffirm the principle in *Diamond v Diehr* and invalidate Claim 13: 'Invalidating Claim 13 . . . will simply restore the balance between natural phenomena and human-made inventions that Congress originally sought to strike in the patent laws – a balance that reflects the Constitution and has served the patent system and the progress of science very well. The decision below should be reversed or vacated.'[79]

C Medical Patents

The USPTO and the courts have shown scant regard for the venerable prohibition on patenting methods of human treatment.[80]

In *Arrhythmia Research Technology, Inc. v Corazonix Corp.*, the Federal Circuit upheld the validity of an apparatus and a method for analysing electrocardiograph signals to detect heart attack risks.[81]

In *Pallin v Singer*, an eye surgeon was sued in respect of a surgical operation by a patent holder.[82] Although the defendant was able to defeat the legal action on the grounds of prior use, there were concerns about litigation against medical practitioners. Journalist Seth Shulman, commented:

> The implications of Singer's situation troubled many disparate parties. If doctors started suing each other for royalties over new medical procedures, patients would have to either absorb the additional costs of royalties, or worse, be denied the latest advances in treatment by health care providers who were not licensed to offer them. Almost everyone recognized a problem brewing, especially medical practitioners and their public advocates, who envisioned a horrendous tangle of litigation that would pit colleagues against one another and draw further ire from a public already critical of the excesses and spiralling costs of medical care . . . Not only would patients suffer from the privatization and licensing of medical procedures, the ethics of the profession faced a significant threat.[83]

In response, the United States Congress passed legislation which provided limited immunity to medical practitioners from claims of patent infringement.[84]

The long-standing debate over medical patents was re-opened in the Supreme Court of the United States case of *Laboratory Corp. of America Holdings v Metabolite Laboratories, Inc.*[85] Intervening as an amicus curiae, the American Medical Association was concerned about the ramifications of the decision of the Court of Appeal for the Federal Circuit for medical professionals:

> Since the time of Hippocrates, a basic tenet of medical ethics has been that discoveries and advances in medical care should be freely shared and openly

disseminated. This ethical principle has served to make such discoveries readily available, at minimal cost, for use in the diagnosis and treatment of patients. It also has helped physicians fulfill their fundamental obligation to act in their patients' best interests.[86]

The Association argued: 'The Federal Circuit's construction of Claim 13 contravenes limitations on the scope of patentable subject matter that have existed for more than two centuries.'[87] The doctors' lobby warned: 'Allowing a private party to enforce a patent on a scientific fact prevents physicians from exercising their best medical judgment in treating their patients and thereby inhibits the sound practice of medicine.'[88]

The People's Medical Society is the largest patients' advocacy group in the United States and is a recognized authority on health care issues.[89] The organization submitted an amicus brief because it was worried about the implications of the decision for health care. The brief was written by the well-known bioethics expert, Professor Lori Andrews of the Chicago-Kent School of Law. The People's Medical Society submitted:

> The Federal Circuit's holding that physicians are culpable of direct infringement of Claim 13 every time they think about or diagnose the relationship between homocysteine levels and a vitamin deficiency would have a chilling effect on protected speech and thought . . . Enforcement of Claim 13 will have a negative impact on medical care, the dissemination of medical information, and medical innovation. Upholding Claim 13 of the '658 patent would also deter people from participating in medical research.[90]

The Society thought it appropriate to overturn the decision of the Federal Circuit and hold invalid Claim 13 of the '658 patent. 'Metabolite's success in transmuting the Patent Act into an exclusive governmentally-issued license to preclude physicians, health care publishers, bioscience researchers, and citizens alike from merely thinking about a principle of human physiology underscores the importance of the issue presented.'[91] The Society urged the court to reaffirm its rule in *Diamond v Diehr*[92] that one cannot patent laws of nature, natural phenomena and abstract ideas.

The American Heart Association expressed concerns that patenting natural phenomenona could have an adverse impact upon medical professionals, patient care, and health care more generally: 'The Court's decision in this case will have profound effects on the way in which healthcare professionals are able to render healthcare advice to patients dealing with cardiovascular diseases and stroke (among other disorders)'.[93]

The American Clinical Laboratory Association (ACLA) expressed its concern that patents of the ilk of the one in the *Metabolite* case would cripple 'laboratories' ability to provide new lifesaving tests and patients' access to those tests would suffer'.[94] The Association commented: 'A decision by this Court upholding Claim 13, however, would permit those who make medical advances in the future to patent every potential application of the natural phenomena that underlie their inventions.'[95] The Association concluded: 'A ruling upholding [Claim 13 of the patent] would be a major shift in our patent regime, one that would do incalculable damage to the business of ACLA's members and – more important – to the ability and incentives of researchers to make advances in the state of the art of medical diagnosis using laboratory tests.'[96]

Perlegen, a company focused on personalized medicine, also supported the respondents, arguing: 'A holding that the diagnostic method identified in Claim 13 is unpatentable could significantly diminish Perlegen's incentive to engage in research and to develop diagnostic methods for determining the patient population for which particular drugs are safe and effective.'[97]

The American Association for Retired Persons (AARP) contended: 'Patents that claim the mental process of recognizing a medical phenomenon of nature legally prohibit diagnosis and treatment, and discourage communication of medical information.'[98] The Association concluded: 'In order to best promote progress, protect public health, and prevent unwarranted liability, this Court should invalidate Claim 13 and clarify indirect liability standards.'[99]

The amicus curiae, Patients not Patents, is a nonprofit organization dedicated to ensuring access to health care through litigation, advocacy and education.[100] The group is organized around the principle that medical treatment should not be denied or restricted because of the existence of patents or other intellectual property barriers. In support of the petitioner, the amicus curiae submitted: 'Upholding the claim would lead to a proliferation of patents for purely mental processes, which would in turn harm both individual patients as well as public health.'[101] The group asked the court to find that Claim 13 was invalid under 35 USC §101, because it claimed unpatentable subject matter: 'The threat posed by patent applicants to individual patients and the public health and research is real.'[102]

III 'MONOPOLIES BEYOND BELIEF': THE SUPREME COURT OF THE UNITED STATES

Granting leave to appeal, the Supreme Court of the United States listened to oral argument in March 2006.[103] The bench sought opinion as to whether the patent was invalid on the grounds that one cannot patent 'laws of nature, natural phenomena, and abstract ideas'.[104] The judges appeared to be divided on this central issue.

In oral argument, Scalia J suggested that the invention was a discovery of a natural principle. His honour asked the rhetorical question:

> What was made by man here? I mean, if you're talking about the type of assay that your client developed, which was involved in other claims, not in 13, they might say, yeah, that was made by man. But here, what 13 involves is simply discovery of the natural principle that when one, when there is the presence of one substance in a human being, there is a deficiency of two other ones. That's just a natural principle. What's made by man about that?[105]

Kennedy J was reluctant to make a ruling, noting: 'It seems imprudent of us to discuss it here if it hasn't been discussed in the Court of Appeals.' Unfortunately, the majority of the Supreme Court of the United States did not avail itself of the opportunity to make a modern ruling on the limits of patentable subject matter.

Breyer J expressed concerns that allowing doctors, scientists and computer experts to patent every 'useful idea' could establish 'monopolies beyond belief'.[106] His Honour asked if it would make sense to send the case back to the lower courts. The judge questioned the rubric that 'anything under the sun' is patentable:

> I mean, I can't resist pointing [out], as one of these briefs did, the phrase that anything under the sun that is made by man comes from a committee report that said something different. It said a person may have invented a machine or a manufacture, which may include anything under the sun that is made by man. So referring to that doesn't help solve the problem where we're not talking about a machine or a manufacture. Rather we are talking about what has to be done in order to make an abstract idea fall within the patent act. Now, sometimes you can make that happen by connecting it with some physical things in the world and sometimes you can't.[107]

The legal counsel responded by commenting that the remark that 'anything under the sun' came 'from a committee report that has already been incorporated in this Court's cases in *Chakrabarty* and in *Diehr* as exemplary of Congress's determination to have the mouth of the funnel be very wide'.[108]

Breyer J's concerns are long-standing. He has observed previously: 'The most difficult question is deciding [whether] products of research reflect only discovery of an existing aspect of nature, like Einstein's discovery of the principles of relativity, [or whether] they amount to a protectable invention or useful device.'[109]

A   Improvidently Granted Writ of Certiorari

In *Laboratory Corp. of America Holdings v Metabolite Laboratories Inc*, the majority of five judges of the Supreme Court of the United States ruled that the writ of certiorari had been improvidently granted, and dismissed the action.[110] This decision reflected the view of the judges that the written record had been insufficiently developed to consider the question of patentable subject matter. The majority judges of the Supreme Court of the United States may have also been disinclined to write a judgment, because the matter was late in the court's term.[111] The judges may have also tacitly accepted the broad, inclusive approach taken to patentable subject matter by the Court of Appeals for the Federal Circuit.

Nonetheless, Breyer J wrote a dissenting judgment, with the support of Stevens J and Souter J. His Honour emphasized that the case should not be dismissed by the Supreme Court of the United States, noting:

> The Court has dismissed the writ as improvidently granted. In my view, we should not dismiss the writ. The question presented is not unusually difficult. We have the authority to decide it. We said that we would do so. The parties and *amici* have fully briefed the question. And those who engage in medical research, who practice medicine, and who as patients depend upon proper health care, might well benefit from this Court's authoritative answer.[112]

Breyer J observed that the 'technical procedural objection is tenuous', especially given that LabCorp had argued the essence of its claim in the Federal Circuit. The judge observed that there was no good practical reason for refusing to decide the case: 'The record is comprehensive, allowing us to learn the precise nature of the patent claim, to consider the commercial and medical context (which the parties and *amici* have described in detail), and to become familiar with the arguments made in all courts.'[113] The judge noted that 'there is no indication that LabCorp's failure to cite §101 reflected unfair gamesmanship'.[114] Breyer J reflected that further consideration by the Federal Circuit might help reach a better decision. Nonetheless, he observed that 'the thoroughness of the briefing leads me to conclude that the extra time, cost, and uncertainty that further proceedings would engender are not worth the potential benefit'.[115] The judge noted that 'important considerations of the public interest – including that of clarifying the law in this area sooner rather than later – argue strongly for our deciding the question presented now'.[116]

On substantive issues, Breyer J expressed concern about the expansion of the scope of patentable subject matter by the USPTO, and the Court of Appeals for the Federal Circuit. His Honour questions the rubric that 'anything under the sun' is patentable. The judge affirmed that patent law protection does not extend to scientific discoveries. Breyer J reaffirmed that patent law excluded laws of nature, natural phenomena and abstract ideas. The judge expressed concerns about the impact of medical patents upon the medical profession and health care. Applying such principles, the judge ruled that Claim 13 of the Metabolite patent application was clearly invalid.

B Scientific Discoveries

Breyer J commented that scientific discoveries and ideas were excluded from the protection of patent law because they were part of the 'storehouse of knowledge':

> And scholars have noted that 'patent law['s] exclu[sion of] fundamental scientific (including mathematical) and technological principles,' (like copyright's exclusion of 'ideas') is a rule of the latter variety.[117] That rule reflects 'both . . . the enormous potential for rent seeking that would be created if property rights could be obtained in [those basic principles] and . . . the enormous transaction costs that would be imposed on would-be users.'[118] Thus, the Court has recognized that '[p]henomena of nature, though just discovered, mental processes, and abstract intellectual concepts are . . . the basic tools of scientific and technological work.'[119] It has treated fundamental scientific principles as 'part of the storehouse of knowledge' and manifestations of laws of nature as 'free to all men and reserved exclusively to none.'[120] And its doing so reflects a basic judgment that protection in such cases, despite its potentially positive incentive effects, would too often severely interfere with, or discourage, development and the further spread of useful knowledge itself.[121]

Breyer J recognized that such definitions were not always clear-cut: 'I concede that the category of non-patentable "phenomena of nature", like the categories of "mental processes" and "abstract intellectual concepts," is not easy to define.'[122] The judge commented: 'After all, many a patentable invention rests upon its inventor's knowledge of natural phenomena; many "process" patents seek to make abstract intellectual concepts workably concrete; and all conscious human action involves a mental process.'[123] He reflected: 'Nor can one easily use such abstract categories directly to distinguish instances of likely beneficial, from likely harmful, forms of protection.'[124] The judge, though, emphasized that the Metabolite case did not require a determination of the precise scope of the 'natural phenomenon' doctrine or any other difficult issue, because the application was clearly not patentable subject matter.

C Products of Nature

Breyer J emphasized the need to recognize the principle that laws of nature, natural phenomena and abstract ideas were excluded from patentable protection. The judge refers to the origins of this principle in English and American law.[125] Breyer J reaffirmed that precedent of *Diamond v Diehr*, which emphasized that patent law '[e]xclude[s] from . . . patent protection . . . laws of nature, natural phenomena and abstract ideas'.[126] His Honour also cites the comment in *Diamond v Chakrabarty* that 'Einstein could not have 'patent[ed] his celebrated law that $E = mc^2$; nor could Newton have patented the law of gravity.'[127] The judge also refers to rulings that one cannot patent 'a novel and useful mathematical formula,'[128] the motive power of electromagnetism or steam,[129] 'the heat of the sun, electricity, or the qualities of metals.'[130] Breyer J explains the guiding rationale behind this principle of patent law:

> The justification for the principle does not lie in any claim that 'laws of nature' are obvious, or that their discovery is easy, or that they are not useful. To the contrary, research into such matters may be costly and time-consuming; monetary incentives may matter; and the fruits of those incentives and that research may prove of great benefit to the human race. Rather, the reason for the exclusion is that sometimes *too much* patent protection can impede rather than 'promote the Progress of Science and useful Arts,' the constitutional objective of patent and copyright protection. U.S. Const., Art. I, §8, cl. 8.
>
> The problem arises from the fact that patents do not only encourage research by providing monetary incentives for invention. Sometimes their presence can discourage research by impeding the free exchange of information, for example by forcing researchers to avoid the use of potentially patented ideas, by leading them to conduct costly and time-consuming searches of existing or pending patents, by requiring complex licensing arrangements, and by raising the costs of using the patented information, sometimes prohibitively so.[131]

Pithily, Breyer J contends: 'Patent law seeks to avoid the dangers of overprotection just as surely as it seeks to avoid the diminished incentive to invent that underprotection can threaten.'[132] His Honour comments that the natural phenomena doctrine is a useful means of addressing such policy concerns: 'One way in which patent law seeks to sail between these opposing and risky shoals is through rules that bring certain types of invention and discovery within the scope of patentability while excluding others.'[133]

On the facts of the case, Breyer J observed that it was clear that the correlation between homocysteine and vitamin deficiency set forth in Claim 13 was a natural phenomenon:

> There can be little doubt that the correlation between homocysteine and vitamin deficiency set forth in Claim 13 is a 'natural phenomenon.' That is what the petitioners argue. It is what the Solicitor General has told us . . . .. Claim 13's process

instructs the user to (1) obtain test results and (2) think about them. Why should it matter if the test results themselves were obtained through an unpatented procedure that involved the transformation of blood? Claim 13 is indifferent to that fact, for it tells the user to use any test at all. Indeed, to use virtually any natural phenomenon for virtually any useful purpose could well involve the use of empirical information obtained through an unpatented means that might have involved transforming matter.[134]

Consequently, His Honour ruled that the claim was 'a product of nature', and, as such, not patentable subject matter.

D Medical Patents

The dissenting judges expressed concerns about granting patents in respect of medical products. Breyer J emphasized that the position of the majority 'threatens to leave the medical profession subject to the restrictions imposed by this individual patent and others of its kind'.[135] Indeed, he observed: 'Those restrictions may inhibit doctors from using their best medical judgment; they may force doctors to spend unnecessary time and energy to enter into license agreements; they may divert resources from the medical task of health care to the legal task of searching patent files for similar simple correlations; they may raise the cost of healthcare while inhibiting its effective delivery.'[136] Breyer J observed that it would be valuable to decide this case in order to diminish legal uncertainty in the area, affecting a substantial number of patent claims.

The judge noted that a ruling would 'permit those in the medical profession better to understand the nature of their legal obligations', and 'help Congress determine whether legislation is needed'.[137] In particular, Breyer J noted the limited liability of medical practitioners for the performance of certain medical and surgical procedures under section 287 (c) of the *Patent Act* 1952 (US). The judge concluded: 'In either event, a decision from this generalist Court could contribute to the important ongoing debate, among both specialists and generalists, as to whether the patent system, as currently administered and enforced, adequately reflects the "careful balance" that "the federal patent laws . . . embod[y]" '.[138] His Honour cited a range of case law, policy documents and academic literature which suggested a need to review the expansion of patentable subject matter.[139]

CONCLUSION

The decision of the Supreme Court of the United States in *Laboratory Corp. of America Holdings v Metabolite Laboratories, Inc.* was an anti-climax.[140]

After hearing oral argument and a range of amicus briefs about the limits of patentable subject matter, the majority of the court shied away from a consideration of such issues, declaring that the writ of certiorari was improvidently granted. Rebecca Eisenberg observed that such diffidence was not surprising:

> Since *Diamond v. Chakrabarty*, the Federal Circuit has gradually abdicated its authority to police these boundaries in favor of an approach that collapses the traditional restrictions on patent eligibility into a simple requirement that the invention be 'useful'. The PTO has ultimately acquiesced in this liberal approach to the threshold requirement of patent eligibility, diminishing the likelihood that the issue will come before the courts for review. To find authority for limitations on patentable subject matter, the Court would have to go back to its own decisions from the 1970s and earlier. These decisions are riddled with contradictions and were hardly up to the task of guiding examination of the patent claims that were arriving at the PTO 30 years ago.[141]

The ruling that the writ of certiorari was improvidently granted represents a lost opportunity to consider the dramatic expansion of patentable subject matter in the last two decades. The decision of the Supreme Court of the United States gives *carte blanche* to the Federal Circuit to continue in its expansive interpretation of patentable subject matter.

The passivity of the Supreme Court of the United States in defending its past precedents on patentable subject matter has been a source of consternation. Lori Andrews and her collaborators were frustrated and disappointed by the failure of the Supreme Court of the United States to intervene in the dispute: 'Patents can chill research if the patent holder forbids other researchers from using the scientific fact or natural phenomenon, or charges an excessive fee for access to that knowledge.'[142] The group called upon researchers and scientists to lobby for a reform of patent law: 'Scientists can be influential by helping policy-makers understand that open access to basic laws of nature, products of nature, and mathematical formulae is necessary for scientists to explore and innovate.'[143]

In response to case law, the USPTO published interim examination guidelines in respect of eligible patentable subject matter.[144] The Commissioner for Patents, John Doll, called for public comment on the guidelines, making particular reference to the litigation in the case of *Laboratory Corp. of America Holdings v Metabolite Laboratories, Inc.* The inquiry attracted a modest number of comments – by 21 entities in all.[145] A few of the submissions commented directly upon the issues arising in the *Laboratory Corp. of America Holdings v Metabolite Laboratories, Inc.* The National Institutes of Health (NIH) expressed dissatisfaction that the Supreme Court of the United States had not availed itself of the opportunity to make a ruling on patentable subject matter.[146] The Association of American Medical Colleges endorsed the dissenting judgment of Breyer J.[147] The Association commented: 'As medical science and practice together enter the still dawning "Age of Genomics", progress in both will increasingly require unfettered access to and understanding of information describing a vast number of correlations and associations between genomic variations and diverse pathological phenotypes.'[148] The Institute for Science, Law, and Technology submitted that the interim guidelines failed to take into account the concerns raised in *Laboratory Corp. of America Holdings v Metabolite Laboratories, Inc*.: 'The USPTO Guidelines are too broad and would allow patents on laws of nature or products of nature that would be prohibited by the Patent Clause of the Constitution, the Patent Act, and U.S. Supreme Court precedent.'[149] The Institute concluded: 'Given the USPTO's mission of encouraging technology and innovation, directives such as the Interim Guidelines can help examiners avoid the grant of broad monopolies which inevitably overcompensate patentees and stymie creation of the useful arts.'[150]

In addition to clarifying the boundaries of patentable subject matter, there is a need to reform exceptions to patent infringement. Brooks Gifford suggests that the litigation in *Laboratory Corp. of America Holdings v Metabolite Laboratories, Inc*. highlights the need to expand the United States statutory defence for limited liability for medical practitioners, which was forged in response to the case of *Pallin v Singer*: 'Another way to avoid the absurd results of the Metabolite case is to create statutory immunity from infringement liability for physicians and the laboratories they use.'[151] Gifford suggests that 'the exemption could be extended to cover laboratory testing companies such as LabCorp specifically'.[152] He commented that, 'if such an exemption existed, although the physicians and LabCorp would still be infringers of Metabolite's patent, they would not be subject to liability or injunction'.[153]

At an international level, member states of the World Trade Organization retain the flexibility under Article 27 (3)(a) of the *TRIPS Agreement* 1994 to exclude from patentability 'diagnostics, therapeutic and surgical methods for the treatment of humans or animals'.[154] There has been a diversity of approaches to the question of the patentability of methods of human treatment. In Australia, the Federal Court of Australia has overturned the old prohibition of the High Court of Australia that methods of human treatment are not a manner of manufacture protected by patent law.[155] In Canada, the Patent Office and the courts have refused patents which claim a method of medical treatment which would be invalid.[156] However, the Supreme Court of Canada has observed that the notion of methods of human treatment does not include matters related to the 'professional skill and judgment of the medical profession'.[157] In the European Union, patent attorneys have used 'Swiss-type' claims to circumvent the prohibition against patenting methods of human treatment.[158] In New Zealand, there has been fierce resistance against lifting the prohibition upon the patenting of methods of human treatment.[159] The Court of Appeals observed that 'the complexity of this area of the law and the policy choices required are matters which are best left to legislative reform.'[160] Indeed, it noted that the parliamentary process 'would allow proper consultation with medical professionals and other organisations as well as the commercial interests which favour patentability, and the formulation of considered reform proposals after that consultation process has taken place'.[161]

NOTES

1. Andrews, L., J. Paradise, T. Holbrook and D. Bochneak (2006), 'When patents threaten science', *Science*, **314**, 1395–6 (1 December).
2. Indeed, there was an unsuccessful attempt by the League of Nations to draft a convention to provide for sui generis protection of scientific discoveries, because it was feared that scientists and researchers would be unable to obtain patent protection in respect of their basic research. A 38 1923 XII (League of Nations Documents); Hamson, C.J. (1930), *Patent Rights for Scientific Discoveries*, Bobbs-Merrill: Indianapolis; Williamson, J. (1937), 'Scientific property', *Journal of Scientific Instruments*, **14**(3), 73–5, http://ej.iop.org/links/q08/RTvVxpHr8QVNaF2A7hhvVw/siv14i3p73.pdf; Merges, Robert (1996), 'Property rights theory and the commons: the case of scientific research', in Ellen Frankel Paul, Fred Miller and Jeffrey Paul (eds), *Scientific Innovation, Philosophy and Public Policy*, Cambridge: Cambridge University Press, http://www.law.berkeley.edu/institutes/bclt/pubs/merges/rpmart4.pdf.
3. Curtis, George (1873), *A Treatise on the Law of Patents For Useful Inventions* §124, 4th edn.
4. Robinson, William (1890) *Robinson on Patents* §§133–43.
5. *O'Reilly v Morse*, 56 U.S. (15 How.) 62 (1853).
6. *Telephone Cases*, 126 U.S. 1 (1888).
7. *Morton v New York Eye Infirmary* 2 Am. Law Reg. (N.S.) 672 C.C.N.Y. 1862.
8. *Morton v New York Eye Infirmary* 2 Am. Law Reg. (N.S.) 672 C.C.N.Y. 1862.
9. *Morton v New York Eye Infirmary* 2 Am. Law Reg. (N.S.) 672 C.C.N.Y. 1862.
10. *Morton v New York Eye Infirmary* 2 Am. Law Reg. (N.S.) 672 C.C.N.Y. 1862.
11. *Morton v New York Eye Infirmary* 2 Am. Law Reg. (N.S.) 672 C.C.N.Y. 1862.
12. *Mackay Radio & Tel. Co. v Radio Corp. of Am.*, 306 U.S. 86, 94 (1939).
13. *Funk Bros. Seed Co. v Kalo Inoculant Co.*, 333 U.S. 127 (1948).
14. *Funk Bros. Seed Co. v Kalo Inoculant Co.*, 333 U.S. 127 (1948), at 130.
15. *Funk Bros. Seed Co. v Kalo Inoculant Co.*, 333 U.S. 127 at 130 (1948).
16. *Funk Bros. Seed Co. v Kalo Inoculant Co.*, 333 U.S. 127 at 130 (1948).
17. *Gottschalk v Benson*, 409 U.S. 63 (1972).
18. *Parker v Flook*, 437 U.S. 584 (1978).
19. *In re Bergy*, 596 F.2d 952, 988 (C.C.P.A. 1979) 444 U.S. 1028 (1980).
20. *Diamond v Diehr* 450 U.S. 175 (1981). For a history of the decision, see Maureen O'Rourke (2006), 'The story of *Diamond v Diehr*: toward patenting software', in Jane Ginsburg and Rochelle Cooper Dreyfuss (eds), *Intellectual Property Stories*, New York: Foundation Press, pp. 194–219.
21. *Parker v Flook*, 437 U.S. 584, 98 S.Ct. 2522, 57 L.Ed.2d 451 (1978); *Gottschalk v Benson*, 409 U.S. 63 (1972); and *Funk Bros. Seed Co. v Kalo Inoculant Co.*, 333 U.S. 127, 130, 68 S.Ct. 440, 441, 92 L.Ed. 588 (1948).

22. *Rubber-Tip Pencil Co. v Howard*, 20 Wall. 498, 507, 22 L.Ed. 410 (1874).
23. *Le Roy v Tatham*, 14 How. 156, 175, 14 L.Ed. 367 (1853).
24. *Diamond v Chakrabarty*, 447 U.S., at 309, 100 S.Ct., at 2208, quoting *Funk Bros. Seed Co. v Kalo Inoculant Co.*, 333 U.S. 127, 130, 68 S.Ct. 440, 441, 92 L.Ed. 588 (1948).
25. *Diamond v Diehr* 450 U.S. 175 (1981).
26. *Diamond v Diehr* 450 U.S. 175 (1981).
27. *Diamond v Diehr* 450 U.S. 175 (1981).
28. Dam, K. (1994), 'The economic underpinnings of patent law', *Journal of Legal Studies*, **23**, 247, 270.
29. Drahos, Peter and John Braithwaite (2002), *Information Feudalism*, London: Earthscan Publications, p. 162.
30. Public Patent Foundation (2005), 'Brief of the Public Patent Foundation as amicus curiae in support of petitioner in *Laboratory Corp. of America Holdings v Metabolite Laboratories, Inc.*', 2005 WL 3597813 (Appellate Brief), 23 December.
31. Ibid.
32. *Laboratory Corp. of America Holdings v Metabolite Laboratories, Inc.* 126 S.Ct. 2921 (2006).
33. Gifford, B. (2006), 'Oh, Diehr: the CAFC's troubling patent eligibility jurisprudence as applied in Metabolite Laboratories *v* Labcorp', *Biotechnology Law Report*, **25**, 129–46.
34. Allen, R., S. Stabler and J. Lindenbaum (1986), 'Assay for sulfhydryl amino acids and methods for detecting and distinguishing cobalamin and folic acid deficiency', US Patent No: 4,940,658.
35. Allen, R., S. Stabler and J. Lindenbaum (1986), 'Assay for sulfhydryl amino acids and methods for detecting and distinguishing cobalamin and folic acid deficiency', US Patent No: 4,940,658.
36. *Metabolite Laboratories v Laboratory Corp. of America Holdings* Not Reported in F.Supp.2d, 2001 WL 34778749 D.Colo., 2001.
37. *Metabolite Laboratories, Inc. v Laboratory Corp. of America Holdings* 370 F.3d 1354 (C.A.Fed. (Colo.), 2004).
38. *Metabolite Laboratories, Inc. v Laboratory Corp. of America Holdings* 370 F.3d 1354 (C.A.Fed. (Colo.), 2004).
39. *Metabolite Laboratories, Inc. v Laboratory Corp. of America Holdings* 370 F.3d 1354 (C.A.Fed. (Colo.), 2004).
40. *Metabolite Laboratories, Inc. v Laboratory Corp. of America Holdings* 370 F.3d 1354 (C.A.Fed. (Colo.), 2004).
41. *Metabolite Laboratories, Inc. v Laboratory Corp. of America Holdings* 370 F.3d 1354 (C.A.Fed. (Colo.), 2004).
42. Laboratory Corp. of America Holdings (2004), 'Petition for a writ of certiorari in *Laboratory Corp. of America Holdings v Metabolite Laboratories, Inc.*', 2004 WL 2505526, 3 November.
43. Ibid., 19.
44. Ibid.
45. Ibid.
46. Ibid.
47. Metabolite Laboratories Inc. (2004), 'Brief for respondents in opposition in *Laboratory Corp. of America Holdings v Metabolite Laboratories, Inc.*', 2004 WL 2803464, 3 December.
48. *Diamond v Chakrabarty*, 447 U.S. at 308–9.
49. *Cochrane v Deener*, 94 U.S. 780, 788 (1877); accord *Diamond v Diehr*, 450 U.S. at 183.
50. Metabolite Laboratories Inc. (2006), 'Brief for respondents in *Laboratory Corp. of America Holdings v Metabolite Laboratories, Inc.*', 2006 WL 303905, 6 February, 28–30.
51. Ibid., 46.
52. Eisenberg, R. (2006), 'Biotech patents: looking backward while moving forward', *Nature Biotechnology*, **24**(3), 317–19.

53. United States Solicitor-General (2005), 'Brief for the United States as amicus curiae in *Laboratory Corp. of America Holdings v Metabolite Laboratories, Inc.*', 2005 WL 3533248, 23 December.
54. Ibid.
55. *Diamond v Diehr* 450 U.S. 175 (1981).
56. Intellectual Property Owners Association (2005), 'Brief of amicus curiae intellectual property owners association in support of neither party in *Laboratory Corp. of America Holdings v Metabolite Laboratories, Inc.*', 2005 WL 3476621, 15 December.
57. Ibid.
58. Federal Circuit Bar Association (2006), 'Amicus curiae brief of the Federal Circuit Bar Association in support of respondents in *Laboratory Corp. of America Holdings v Metabolite Laboratories, Inc.*', 2006 WL 303906, 6 February.
59. Ibid., 3
60. Ibid., 3-4.
61. Ibid.
62. Boston Patent Law Association (2006), 'Brief of amicus curiae Boston Patent Law Association in support of respondents in *Laboratory Corp. of America Holdings v Metabolite Laboratories, Inc.*', 2006 WL 303909, 6 February.
63. Ibid., 13.
64. Ibid., 13.
65. Bar of the City of New York (2005), 'Amicus curiae brief of the Association of the Bar of the City of New York in support of neither party in *Laboratory Corp. of America Holdings v Metabolite Laboratories, Inc.*', 2005 WL 3597808, 23 December.
66. Ibid.
67. Ibid.
68. *Ex parte Latimer*, 1889 Comm'r Dec 123 (1889).
69. Franklin Pierce Law Center (2006), 'Brief amicus curiae Franklin Pierce Law Center in support of the respondents in *Laboratory Corp. of America Holdings v Metabolite Laboratories, Inc.*', 2006 WL 304571, 6 February.
70. Ibid., 3-4.
71. Ibid.
72. Ibid.
73. Ibid.
74. Public Patent Foundation (2005), 'Brief of the Public Patent Foundation as amicus curiae in support of petitioner in *Laboratory Corp. of America Holdings v Metabolite Laboratories, Inc.*', 2005 WL 3597813, 23 December, 1.
75. Ibid., 11.
76. Ibid., 19.
77. Affymetrix (2005), 'Brief for amici curiae Affymetrix, Inc. and Professor John H. Barton in Support of Petitioner in *Laboratory Corp. of America Holdings v Metabolite Laboratories, Inc.*', 2005 WL 3597814, 23 December.
78. Ibid., 3-4.
79. Ibid., 4.
80. *Morton v New York Eye Infirmary* 17 F. Cas 879 (S.D.N.Y. 1862) (No. 9865), and *Ex parte Brinkerhoff* 24 Dec Comm'n Patent 349 (1883).
81. *Arrhythmia Research Technology, Inc. v Corazonix Corp.*, 958 F.2d 1053, 1064 (Fed. Cir. 1992).
82. *Pallin v Singer* (1995) 36 USPQ (2d) 1050.
83. Shulman, Seth (1999), 'The new medical licenses', *Owning The Future*. Boston: Houghton Mifflin Company, p. 35.
84. S. 287 (c) of the *Patent Act* 1952 (US).
85. *Laboratory Corp. of America Holdings v Metabolite Laboratories, Inc.*, 126 S.Ct. 2921 (2006).
86. American Medical Association et al. (2005), 'Brief for the American Medical Association, the American College of Medical Genetics, the American College of Obstetricians and Gynecologists, the Association for Molecular Pathology, the

Association of American Medical Colleges, and the College of American Pathologists as amici curiae in support of petitioner in *Laboratory Corp. of America Holdings v Metabolite Laboratories, Inc.*', 2005 WL 3597812 (Appellate Brief), 23 December.
87. Ibid.
88. Ibid.
89. People's Medical Society (2005), 'Brief of amicus curiae People's Medical Society in support of petitioner in *Laboratory Corp. of America Holdings v Metabolite Laboratories, Inc.*', 2005 WL 3597702, 22 December, 4–5.
90. Ibid.
91. Ibid.
92. *Diamond v Diehr*, 450 U.S. 175, 185, 101 S.Ct. 1048, 67 L.Ed.2d 155 (1981).
93. American Heart Association (2005), 'Brief of the American Heart Association as amicus curiae in support of petitioner in *Laboratory Corp. of America Holdings v Metabolite Laboratories, Inc.*', 2005 WL 3561169, 23 December.
94. American Clinicial Laboratory Association (2005), 'Brief of the American Clinical Laboratory Association as amicus curiae in support of petitioner in *Laboratory Corp. of America Holdings v Metabolite Laboratories, Inc.*', 2005 WL 3543098 23 December, 2.
95. Ibid.
96. Ibid.
97. Perlegen Sciences (2006), 'Brief for amici curiae Perlegen Sciences, Inc. and Mohr, Davidow Ventures in support of respondents in *Laboratory Corp. of America Holdings v Metabolite Laboratories, Inc.*', 2006 WL 303908, 6 February.
98. American Association for Retired Persons (2005), 'Brief Amicus Curiae of AARP in support of petitioner in *Laboratory Corp. of America Holdings v Metabolite Laboratories, Inc.*', 2005 WL 3597809, 23 December.
99. Ibid.
100. Patients not Patents, Inc. (2005), 'Brief of Patients not Patents, Inc., as amicus curiae in support of petitioner in *Laboratory Corp. of America Holdings v Metabolite Laboratories, Inc.*', 2005 WL 3597811, 23 December.
101. Ibid., 1–2.
102. Ibid.
103. *Laboratory Corp. of America Holdings v Metabolite Laboratories, Inc.*, Slip Copy United States Supreme Court Official Transcript. Oral Argument 2006 WL 711253 (U.S.), 74 USLW 3558.
104. *Laboratory Corp. of America Holdings v Metabolite Laboratories, Inc.*, Slip Copy United States Supreme Court Official Transcript. Oral Argument 2006 WL 711253 (U.S.), 74 USLW 3558.
105. *Laboratory Corp. of America Holdings v Metabolite Laboratories, Inc.*, Slip Copy United States Supreme Court Official Transcript. Oral Argument 2006 WL 711253 (U.S.), 74 USLW 3558.
106. *Laboratory Corp. of America Holdings v Metabolite Laboratories, Inc.*, Slip Copy United States Supreme Court Official Transcript. Oral Argument 2006 WL 711253 (U.S.), 74 USLW 3558.
107. *Laboratory Corp. of America Holdings v Metabolite Laboratories, Inc.*, Slip Copy United States Supreme Court Official Transcript. Oral Argument 2006 WL 711253 (U.S.), 74 USLW 3558 at 43.
108. *Laboratory Corp. of America Holdings v Metabolite Laboratories, Inc.*, Slip Copy United States Supreme Court Official Transcript. Oral Argument 2006 WL 711253 (U.S.), 74 USLW 3558 at 44.
109. Breyer, S. (2000), 'Genetic advances and legal institutions', *Journal of Law, Medicine & Ethics*, **28**, 23–8 at 27.
110. *Laboratory Corp. of America Holdings v Metabolite Laboratories, Inc.*, 126 S.Ct. 2921 (2006).
111. *Laboratory Corp. of America Holdings v Metabolite Laboratories, Inc.*, 126 S.Ct. 2921 (2006).

112. *Laboratory Corp. of America Holdings v Metabolite Laboratories, Inc.*, 126 S.Ct. 2921 at 2922 (2006).
113. *Laboratory Corp. of America Holdings v Metabolite Laboratories, Inc.*, 126 S.Ct. 2921 at 2925 (2006).
114. *Laboratory Corp. of America Holdings v Metabolite Laboratories, Inc.*, 126 S.Ct. 2921 at 2926 (2006).
115. *Laboratory Corp. of America Holdings v Metabolite Laboratories, Inc.*, 126 S.Ct. 2921 at 2926 (2006).
116. *Laboratory Corp. of America Holdings v Metabolite Laboratories, Inc.*, 126 S.Ct. 2921 at 2926 (2006).
117. Landes, William and Richard Posner (2003), *The Economic Structure of Intellectual Property Law*, Cambridge (Mass.) and London (UK): Harvard University Press, p. 305.
118. Cf. *Nichols v Universal Pictures Corp.*, 45 F.2d 119, 122 (C.A.2 1930) (L.Hand, J).
119. *Gottschalk v Benson*, 409 U.S. 63, 67, 93 S.Ct. 253, 34 L.Ed.2d 273 (1972).
120. *Funk Bros. Seed Co. v Kalo Inoculant Co.*, 333 U.S. 127 (1948).
121. *Laboratory Corp. of America Holdings v Metabolite Laboratories, Inc.*, 126 S.Ct. 2921 at 2922–2923 (2006).
122. *Laboratory Corp. of America Holdings v Metabolite Laboratories, Inc.*, 126 S.Ct. 2921 at 2923 (2006). In the foundation case of *NRDC v The Commissioner of Patents* (1959) 102 CLR 252, the High Court of Australia makes a similar observation about the difficulty of distinguishing between an invention and scientific discoveries: 'The truth is that the distinction between discovery and invention is not precise enough to be other than misleading in this area of discussion. There may indeed be a discovery without invention – either because the discovery is of some piece of abstract information without any suggestion of a practical application of it to a useful end, or because its application lies outside the realm of "manufacture". But where a person finds out that a useful result may be produced by doing something which has not been done by that procedure before, his claim for a patent is not validly answered by telling him that although there was ingenuity in his discovery that the materials used in the process would produce the useful result no ingenuity was involved in showing how the discovery, once it had been made, might be applied.'
123. *Laboratory Corp. of America Holdings v Metabolite Laboratories, Inc.*, 126 S.Ct. 2921 at 2923 (2006).
124. *Laboratory Corp. of America Holdings v Metabolite Laboratories, Inc.*, 126 S.Ct. 2921 at 2923 (2006).
125. *Neilson v Harford*, Webster's Patent Cases 295, 371 (1841); *Le Roy v Tatham*, 14 How. 156, 175, 14 L.Ed. 367 (1853); *O'Reilly v Morse*, 15 How. 62, 14 L.Ed. 601 (1854); *The Telephone Cases*, 126 U.S. 1, 8 S.Ct. 778, 31 L.Ed. 863 (1888).
126. *Diamond v Diehr*, 450 U.S. 175, 185, 101 S.Ct. 1048, 67 L.Ed.2d 155 (1981).
127. *Diamond v Chakrabarty*, 447 U.S. 303, 309, 100 S.Ct. 2204, 65 L.Ed.2d 144 (1980).
128. *Parker v Flook*, 437 U.S. 584, 585, 98 S.Ct. 2522, 57 L.Ed.2d 451 (1978).
129. *O'Reilly v Morse*, 56 U.S. (15 How.) 62 (1853).
130. *Funk Brothers Seed Co. v Kalo Inoculant Co.*, 333 U.S. 127, 130, 68 S.Ct. 440, 92 L.Ed. 588 (1948).
131. *Laboratory Corp. of America Holdings v Metabolite Laboratories, Inc.*, 126 S.Ct. 2921 at 2922 (2006).
132. *Laboratory Corp. of America Holdings v Metabolite Laboratories, Inc.*, 126 S.Ct. 2921 at 2922 (2006).
133. *Laboratory Corp. of America Holdings v Metabolite Laboratories, Inc.*, 126 S.Ct. 2921 at 2922 (2006).
134. *Laboratory Corp. of America Holdings v Metabolite Laboratories, Inc.*, 126 S.Ct. 2921 at 2927 (2006).
135. *Laboratory Corp. of America Holdings v Metabolite Laboratories, Inc.*, 126 S.Ct. 2921 at 2929 (2006).
136. *Laboratory Corp. of America Holdings v Metabolite Laboratories, Inc.*, 126 S.Ct. 2921 at 292 (2006).

137. *Laboratory Corp. of America Holdings v Metabolite Laboratories, Inc.*, 126 S.Ct. 2921 at 2929 (2006).
138. *Laboratory Corp. of America Holdings v Metabolite Laboratories, Inc.*, 126 S.Ct. 2921 at 2929 (2006).
139. *Bonito Boats, Inc. v Thunder Craft Boats, Inc.*, 489 U.S. 141 (1989); *eBay Inc. v MercExchange, L.L.C.*126 S.Ct. 1837 (2005), Federal Trade Commission (2003), *To Promote Innovation: The Proper Balance of Competition and Patent Law and Policy*, Federal Trade Commission, http://www.ftc.gov/os/2003/10/innovationrpt.pdf, October; Pollack, M. (2002), 'The multiple unconstitutionality of business method patents: common sense, Congressional consideration, and constitutional history', *Rutgers Computer and Technology Law Journal*, **28**, 61–120; and Pitofsky, R. (2001), 'Antitrust and intellectual property: unresolved issues at the heart of the new economy', Berkeley Center for Law and Technology, http://www.ftc.gov/speeches/pitofsky/ipf301.htm, 1 March.
140. *Laboratory Corp. of America Holdings v Metabolite Laboratories, Inc.*, 126 S.Ct. 2921 (2006).
141. Eisenberg, R. (2006), 'Biotech patents: looking backward while moving forward', *Nature Biotechnology*, **24**(3), 317–19.
142. Andrews, L., J. Paradise, T. Holbrook and D. Bochneak (2006), 'When patents threaten science', *Science*, **314**, 1395–6 (1 December).
143. Ibid.
144. United States Patent and Trademark Office (2005), 'Interim guidelines for examination of patent applications for patent subject matter eligibility', Official Gazette of the Patent Office, **1300**, 142, http://www.uspto.gov/web/offices/com/sol/notices/70fr 75451.pdf, 22 November.
145. United States Patent and Trademark Office (2006), 'Regarding interim guidelines for examination of patent applications for patent subject matter eligibility', http://www.uspto.gov/web/offices/pac/dapp/opla/comments/ab98/ab98.html, 11 August.
146. National Institutes of Health (2006) 'Comments on the USPTO's interim guidelines for examination of patent applications for patent subject matter eligibility', http://www.uspto.gov/web/offices/pac/dapp/opla/comments/ab98/nih.pdf, 31 July.
147. *Laboratory Corp. of America Holdings v Metabolite Laboratories, Inc.*, 126 S.Ct. 2921 (2006).
148. The Association of American Medical Colleges (2006), 'Re: request for comments on interim guidelines for examination of patent applications for patent subject matter eligibility', http://www.uspto.gov/web/offices/pac/dapp/opla/comments/ab98/aamc.pdf, 31 July.
149. Institute for Science, Law, and Technology (2006), 'Comments on the interim guidelines for examination of patent applications for patent subject matter eligibility', http://www.uspto.gov/web/offices/pac/dapp/opla/comments/ab98/islat.pdf, 31 July.
150. Ibid.
151. Gifford, B. (2006), 'Oh, Diehr: the CAFC's troubling patent eligibility jurisprudence as applied in Metabolite Laboratories *v* Labcorp', *Biotechnology Law Report*, **25**, 129–46.
152. Ibid., 129–46.
153. Ibid., 129–46.
154. Faunce, T. and P. Drahos (1998), 'Trade related aspects of intellectual property rights (TRIPS) and the threat to Patients: a plea for doctors to respond internationally', *Medicine And Law*, **17**, 299–310; and Martin, T. (2000), 'Patentability of methods of medical treatment: a comparative study', *Journal of the Patent and Trademark Office Society*, **82**(6), 381–423.
155. *NRDC v Commissioner Of Patents* (1959) 102 CLR 252; *Joos v Commissioner Of Patents* (1972) 46 ALJR 438; *Anaesthetic Supplies Pty Ltd v Rescare Ltd* (1994) 28 IPR 383; and *Bristol-Myers Squibb v FH Faulding* (2000) 46 IPR 553. For secondary commentary, see Pila, J. (2001), 'Methods of medical treatment within Australian and United Kingdom patents law', *University of New South Wales Law Journal*, **24**, 420–61; and Urquijo, A. (2004), 'The restriction of access to healthcare by patent law: fact or fiction?', *University of New South Wales Law Journal*, **27**(1), 170–97.

156. *Tennessee Eastman Co. v Commissioner of Patents*, [1974] S.C.R. 111.
157. *Apotex Inc. v Wellcome Foundation Ltd.*, [2002] 4 S.C.R. 153, 2002 SCC 77 (CanLII).
158. Armstrong, D. (2001), 'The arguments of law, policy and practice against Swiss-type patent claims', *Victoria University of Wellington Law Review*, **32**(1), 201–54.
159. *Pfizer v Commissioner of Patents (NZ)* (2004) 60 IPR 624.
160. *Pfizer v Commissioner of Patents (NZ)* (2004) 60 IPR 624.
161. *Pfizer v Commissioner of Patents (NZ)* (2004) 60 IPR 624.

5. The book of life: patent law and the human genome project

> You are on a mission to discover
> Why the human heart still slows
> When divers break the surface
> Why mermaids still swim in our dreams.
> (Michael Symmons Roberts, 'Mapping the Genome')[1]

The Human Genome Project represented the first foray into 'Big Science' by the medical and the biological science communities.[2] The initiative garnered a great of deal of both public and private support. The Human Genome Project was a grand scientific enterprise which attracted both hyperbole and ridicule alike. The project was lauded as 'the moon shot of the life sciences', the 'holy grail of man', 'the code of codes' and 'the book of life'.[3] Francis Collins, a leader of the public consortium behind the Human Genome Project, observed:

> As you will hear today, this Book of Life is actually at least three books. It's a history book: a narrative of the journey of our species through time. It's a shop manual: an incredibly detailed blueprint for building every human cell. And it's a transformative textbook of medicine: with insights that will give health care providers immense new powers to treat, prevent and cure disease. We are delighted by what we've already seen in these books. But we are also profoundly humbled by the privilege of turning the pages that describe the miracle of human life, written in the mysterious language of all the ages, the language of God.[4]

Such lofty rhetoric and sanctimony has also received scorn and scepticism. Richard Lewontin has sought to debunk the pretensions of the Human Genome Project.[5] He observed: 'The big irony of the sequencing of the human genome is that the result turns out not to provide the answer to the chief question that motivated the project.'[6] Lewontin suggested that the publication of the Human Genome Project was a terrific anti-climax: 'Now that we have the complete sequence of the human genome we do not, alas, know anything more than we did before about what it is to be human.'[7] He wondered whether to trust calls for new 'Big Science' initiatives: 'The reaction to the discovery that human beings do not have much more genomic information than plants and worms has been to call for a new and even more grandiose project.'[8]

Nobel Laureate James Watson and the director of the National Institutes of Health, Bernadine Healy, had acrimonious altercations over the Human Genome Project. The National Institutes of Health sought to patent expressed sequence tags (ESTs)[9] that were archived by J. Craig Venter in 1991.[10] This action caused great ructions in the scientific community and led to James Watson resigning from the leadership of the public consortium. On 20 August 1992, the United States Patent and Trademark Office (USPTO) rejected the patent applications over the ESTs. It ruled that the claims failed to meet the patent criteria (novelty, non-obviousness and utility) because they were 'vague, indefinite, misdescriptive, incomplete, inaccurate, and incomprehensible'.[11] The National Institutes of Health case was apparently damaged by the identification of 15-letter segments in some of the ESTs in other genes in the database. In the wake of the scandal, Venter resigned from the National Institutes of Health in the wake of personal criticism of his work: for instance, the Nobel scientist, James Watson, denounced his patents as 'monkey work'.[12]

In 1998, Venter established Celera Genomics with the help of the Parker Elmer Corporation.[13] He announced that his company would use high-powered sequencing machines to map the human genome. The organization's slogan was 'Speed matters. Discovery can't wait.'[14] The approach of Celera Genomics on intellectual property and biotechnology is illuminated by James Shreeve's *The Genome War*.[15] The company's patent attorney, Robert Millman, provided the following advice:

> What we do first is go for the low-hanging fruit. Celera's business strategy is based on the fact that we have the largest gene pipeline on earth. The first product of the company is the database itself, which we sell to subscribers. But our second core asset is internal gene discovery, beginning with the stuff within easy reach. The low-hanging fruit.[16]

The company was conscious that its competitors, Human Genome Sciences and Incyte, were also seeking patents in respect of genes. Millman warned, though, that Celera Genomics had to take care not to outrage the public: 'Seriously, we need to get across that we're not building an evil empire.'[17] He advised that the company should stress the greater public benefit: 'Sure, we'll make a little money along the way, but this is not to gouge the public.'[18]

In response to such criticism from the public consortium behind the Human Genome Project, Venter emphasized that Celera Genomics would only seek a small portfolio of patents in respect of medically significant genes:

> Since its founding we have said that Celera will seek to develop on its own 100–300 medically important genes for use by pharmaceutical and biotechnology

companies from among the 100,000 human genes. We will give preference in licensing these potential therapeutic targets to our subscribers and we will license them on a non-exclusive basis. As I said at the earlier hearing, we are not attempting to patent the human genome, any of its chromosomes, or any random sequence.[19]

Venter stressed that the company supported the administrative approach of the USPTO: 'Fundamental patent requirements of utility, novelty, and non-obviousness are complete and effective protections to the fear propagated that the human genome will be patented or that the revolution will be slowed.'[20] He insisted: 'Consistent with long-established principles of patent law, we do expect that patents and other protections for subsequent inventions using the genome alphabet and showing utility, novelty, and non-obviousness are not only appropriate, but required to assure that incentives continue to fuel the genomic revolution.'[21]

John Sulston, the director of the Sanger Centre from 1993 to 2000 and the leader of the British component of the Human Genome Project, was outraged by the patent applications filed by Venter and Celera Genomics. He recounts in his book, *The Common Thread*, that he was of the firm view that genomic information should not be patented because it was part of the common heritage of humankind: 'The genome sequence is a discovery, not an invention.'[22] Sulston was concerned that J. Craig Venter wanted 'to establish a monopoly position on the human sequence'.[23]

Francis Collins, the leader of the United States component of the Human Genome Project, was somewhat more circumspect in his criticism of J. Craig Venter and Celera Genomics. Rather than calling for a prohibition on human genetic patents, he contended that the USPTO should have a more stringent application of the patent criteria of novelty, inventive step and utility:

> I think everybody agrees that the raw fundamental genome sequence of us folks ought not to be the subject of constraints on its accessibility and, so, it ought to be in the public domain and it ought to not be patented. When it comes to a gene that has a function then, in fact, patenting makes a lot more sense. But the raw fundamental information, the stuff that we're putting on the Internet every 24 hours really just ought to be out there because the public is only going to benefit if scientists put their best energies to figuring out what it all means. And that will most likely happen if there are no constraints on their ability to do.[24]

Collins found it reassuring, though, that the USPTO was engaging in public discussion as to where the dividing line ought to be drawn between patentable and non-patentable subject matter in the field of the life sciences.[25]

On 14 March 2000, President Bill Clinton and British Prime Minister Tony Blair made a joint announcement, stating that the human genome should remain in the public domain:

To realize full promise of the research, raw fundamental data on the human genome – including the human DNA sequence and its variations – should be made freely available to scientists everywhere. Unencumbered access to this information will promote discoveries that will reduce the burden of disease, improve health around the world and enhance the quality of life for all human kind. Intellectual property protection for gene-based inventions will also play an important role in stimulating the development of important new health care products. We applaud the decision by scientists working on the Human Genome Project to release raw fundamental information about the human DNA sequence and its variants rapidly into the public domain, and we commend other scientists around the world to adopt this policy.[26]

That announcement sparked a crash in biotechnology stocks, bleeding the market valuations of companies such as Celera Genomics and Human Genome Sciences.[27] Clinton and Blair clarified their positions soon after the announcement, maintaining that the patenting of specific human genes would still be legal and appropriate.[28] The USPTO issued a press release, stating that 'United States patent policy remains unaffected by Tuesday's historic joint statement [by Clinton and Blair].'[29] Todd Dickinson of the USPTO emphasized that 'genes and other genomic inventions remain patentable so long as they meet the statutory criteria of utility, novelty and non-obviousness'.[30] He added: 'Genes and genomic inventions that were patentable last week continue to be patentable this week, under the same set of rules.'[31]

In February 2001, *Nature* and *Science* published papers reporting the sequence of the 3.2 billion base pair human genome. The *Nature* paper was by the publicly funded International Human Genome Sequencing Consortium.[32] The *Science* paper was by the private company Celera Genomics, led by J. Craig Venter.[33] The public consortium deposited its sequence material in GenBank as soon as was possible. However, Celera Genomics posted its data on its own website on publication, and limited free access to the genetic information.

This chapter considers the administrative and legal responses to the flood of gene patents filed in the wake of the human genome project. It explores the recurring legal debates over the patentability of ESTs. It is argued that heightened utility guidelines alone will fail to regulate gene patent applications properly; there is a need to raise the thresholds of novelty and inventive step. Section one explores the development of administrative guidelines by the USPTO in respect of utility and biotechnological inventions. Section two analyses the divisions amongst the judges, Michel CJ and Rader J, in the Court of Appeals for the Federal Circuit *In re Fisher*.[34] In particular, it focuses upon the debate as to whether ESTs could be best conceived of as research tools, and the proper threshold for determinations of utility. This section examines the interpretation and application of the decision of *In re Fisher* by the Board of Patent Appeals and Interferences in a number of recent rulings.[35] The conclusion considers the implications of the decision of the Court of Appeals for the Federal Circuit of *In re Fisher* for the patent applications in respect of genes and gene sequences, in the wake of the human genome project. It considers legislative proposals put forward by renegade members of the United States Congress in respect of gene patents, including *The Genomic Research And Diagnostic Accessibility Act* 2002 (US), *The Genomic Science And Technology Innovation Act* 2002 (US) and the *Genomic Research and Accessibility Act* 2007 (US).

I RAISING THE BAR: THE UNITED STATES PATENT AND TRADEMARK OFFICE UTILITY GUIDELINES

In the United States, the courts have sought to define the requirement of utility under patent law. In *Brenner v Manson*, the Supreme Court of the United States took a restrictive view of utility.[36] It held that a chemical product with no known use, or useful for merely further research, was not a patentable invention. Fortas J emphasized the importance of the requirement of utility in patent law:

> The basic quid pro quo contemplated by the Constitution and the Congress for granting a patent monopoly is the benefit derived by the public from an invention with substantial utility. Unless and until a process is refined and developed to this point – where specific benefit exists in currently available form – there is insufficient justification for permitting an applicant to engross what may prove to be a broad field.[37]

The judge observed: 'Whatever weight is attached to the value of encouraging disclosure and of inhibiting secrecy, we believe a more compelling consideration is that a process patent in the chemical field, which has not been developed and pointed to the degree of specific utility, creates a monopoly of knowledge which should be granted only if clearly commanded by the statute.'[38] Fortas J warned: 'Such a patent may confer power to block off whole areas of scientific development, without compensating benefit to the public.'[39] Emphasizing that 'a patent is not a hunting license', His Honour observed: 'It is not a reward for the search, but compensation for its successful conclusion.'[40] The judge concluded: '[A] patent system must be related to the world of commerce rather than to the realm of philosophy.'[41]

A Revised Utility Examination Guidelines

In 2001, the USPTO issued revised examination guidelines explaining how the utility requirement should be applied by patent examiners.[42] The guidelines required patent applicants to explicitly identify, unless already well established, a specific, substantial and credible utility for all inventions. In effect, it raised the bar to ensure that patent applicants demonstrate a 'real world' utility. The Director of the USPTO, Todd Dickinson, explained the administrative reforms to Congress:

> The issue of the utility of an invention is one that the USPTO takes very seriously. That is why we continue to take steps to ensure that genomic patent applications are meticulously scrutinized for an adequate written description, sufficiency of the disclosure, and enabled utilities, in accordance with the standards set forth by our reviewing courts. In order to meet the utility requirement of 35 U.S.C. §101, our new utility guidelines require patent applicants to explicitly identify, unless already well-established, a specific, substantial and credible utility for all inventions. In effect, we have raised the bar to ensure that patent applicants demonstrate a 'real world' utility. One simply cannot patent a gene itself without also clearly disclosing a use to which that gene can be put. As a result, we believe that hundreds of genomic patent applications may be rejected by the USPTO, particularly those that only disclose theoretical utilities.[43]

He observed: 'An asserted utility is credible unless the logic underlying the assertion is seriously flawed, or the facts upon which the assertion is based are inconsistent with the logic underlying the assertion.'[44] Dickinson noted: 'A utility is specific when it is particular to the subject matter claimed.'[45] Finally, he observed: 'A substantial utility is one that defines a "real world" use.'[46] Dickinson noted: 'Utilities that require or constitute carrying out further research to identify or reasonably confirm a "real world" context of use are not substantial utilities.'[47]

B Public Comment

The utility guidelines attracted 51 public comments from a range of private companies, research institutions and individuals.[48] The USPTO vigorously responded to various criticisms of the guidelines in its public consultation process.[49] It dismissed a host of objections to the patenting of genes and gene sequences.

First, a number of critics argued that genes should not be considered to be patentable subject matter because they were scientific discoveries, not inventions. The USPTO responded that 'an inventor's discovery of a gene can be the basis for a patent on the genetic composition isolated from its natural state and processed through purifying steps that separate the gene from other molecules naturally associated with it'.[50] Second, several comments stated that a gene was not a new composition of matter because it was a 'product of nature'. In response, the USPTO asserted: 'Patenting compositions or compounds isolated from nature follows well-established principles, and is not a new practice.'[51]

Several comments suggested that the USPTO should seek guidance from Congress as to whether naturally occurring genetic sequences were patentable subject matter. The Patent Office declined such an invitation: '[T]he intent of Congress with regard to patent eligibility for chemical compounds has already been determined: DNA compounds having naturally occurring sequences are eligible for patenting when isolated from their natural state and purified, and when the application meets the statutory criteria for patentability.'[52]

Third, several comments stated that patents should not issue for genes because the human genome was part of the common heritage of humankind, and should not be open to private ownership. Other comments stated that patents should be for marketable inventions and not for discoveries in nature. The USPTO rejected this argument that life forms should not be able to be commercialized: 'The patent system promotes progress by securing a complete disclosure of an invention to the public, in exchange for the inventor's legal right to exclude other people from making, using, offering for sale, selling, or importing the composition for a limited time.'[53]

C Stakeholder Perspectives

There has been much policy discussion about the new USPTO examination guidelines for the requirement of utility.[54] The leader of the United States component of the Human Genome Project, Francis Collins, provided qualified praise for the utility guidelines:

> The Patent Office is seeing fewer of what they call 'generation one' patents, where there's just a sequence and no clue as to what it does. PTO intends to reject those. They are seeing a reasonable number of 'generation two' applications, where there's a sequence, and homology suggests a function. NIH views such applications as problematic, since homology often provides only a sketchy view of function. Increasingly, PTO is seeing more in the 'generation three' category, which I think most people would agree is more appropriate for patent protection. These are gene sequences for which you have biochemical, or cell biological, or genetic data describing function.[55]

Collins concluded: 'I think the Patent Office deserves credit for moving toward a stronger requirement for utility.'[56]

Harold Varmus, the President of Memorial Sloan-Kettering Cancer Center in New York City, and the Director of the NIH from 1993 to 1999, testified about the utility guidelines to the House of Representatives Subcommittee on Courts and Intellectual Property:

> Recently, to the relief of many of us, the PTO has considered raising the bar to gene patenting, especially for the utility standard. Although the new proposal is an improvement and the final position of the PTO has not yet been announced, I believe that the bar may still not be raised high enough. Under the new proposal, a patent could be issued for a gene or a portion of a gene based on still quite superficial and potentially misleading information about the properties of the gene or about how it might be used to diagnose, prevent, or treat disease. Such information may be dependent only on the similarity between the new gene and others previously described. Establishing the legitimacy of such claims, even if the predictions were confirmed experimentally, would doubtless require legal proceedings, such as those that follow accusations of infringement.[57]

He commented that 'overly enthusiastic protection of intellectual property, too early in the process of product development, can impede the delivery of public health benefits from discoveries in many important fields, including genomics'.[58]

The Nuffield Council on Bioethics has argued that the USPTO has set the requirement of utility too low:

> While we welcome the new USPTO guidelines, we take the view that where 'credibility' means no more than 'theoretical possibility' (ie where something is credible simply where it is not incredible) the threshold for utility is still set too low. The current state of genetics and biochemistry does not make it difficult to suggest functions for DNA sequences that are 'theoretically possible', in the sense that they are not ruled out by what is already known; but this should not suffice for the award of a patent. Instead, what is required is some evidence that the DNA sequence actually has the claimed 'specific' utility and that the claimed utility is truly 'substantial'.[59]

The Council recommended that the USPTO should monitor the impact of the Guidelines on the examination of patents to ensure that the criterion for utility was rigorously applied so that the grant of a patent more properly reflects the inventor's contribution. If this proves not to be the case, the Guidelines should be reviewed and strengthened to achieve this purpose as soon as is practicable.[60]

Andrea Ryan, president-elect of the American Intellectual Property Law Association, observed: 'The patent issues surrounding biotechnology and specifically genes and gene-related technology are less than 20 years old and it will take time to sort out the application of the patent laws to this technology.'[61] She submitted: 'We believe the Revised Written Description

Guidelines and the Utility Guidelines as published by the Office have taken great steps forward in the complex area of the written description requirements for a biotechnology patent.'[62]

Charles Ludlam, vice president for government relations at the Biotechnology Industry Organization (BIO), was broadly supportive of the utility guidelines. He observed that there was a lack of consensus amongst the members of BIO as to the threshold required for utility:

> There is a difference of opinion among BIO members as to whether different types of inventions will or will not satisfy the utility requirement. For example, some BIO members believe that utility of most proteins cannot be conclusively demonstrated until the protein has been expressed and biologically characterized. Other BIO members believe that utility can be based on a prediction of biological activity made on the basis of homology to existing classes of polypeptides and proteins.[63]

Ludlam commented: 'Rather than attempting to dictate one standard or the other, BIO encourages the PTO to evaluate carefully the rationale presented in support of an asserted utility, particularly with respect to the specificity of the recited utility, and the scientific credibility of the basis for that specifically recited utility.'[64]

Randal Scott of the biotechnology firm, Incyte Genomics, argued that patent law should be applied to genetic inventions, much the same as it had been applied to other technologies: 'Incyte believes that the application of existing patent law principles to genomic inventions will support the continued acceleration of genomic research, resulting in an increase in the pipeline of new drugs that are safer and less expensive than has previously been possible.'[65] The company generally supported the efforts of the USPTPO in clarify the existing law in the utility guidelines, particularly as they applied to ESTs: 'We favor this clarification of the proper application of current law to a new category of genomic inventions.'[66] Nonetheless, the company was somewhat concerned that examiners from the USPTO could apply such guidelines in an over-zealous fashion: 'Incyte has concerns, however, about unattributed quotes that purport to announce a Patent Office "decision" to limit the issuance of gene patents . . . [and] suggest that the Patent Office will issue patents on genes only if the specific biological activity of the genes is disclosed in the patent application.'[67]

The patent litigator, Gerald Dodson, observed that his university clients wanted a wide perimeter in which to protect inventions whose potential for use was uncertain.[68] He believed that the utility guidelines of the USPTO were too restrictive and could hit universities with a devastating economic blow. Dodson would rather the court system make these decisions instead of patent examiners:

Bring a lawsuit and let the court decide if something has utility. The court could give a small damage award if they thought the utility was small. To the extent that people have received patents on inventions or devices perceived as having insignificant or de minimus utility, the system will remedy that.[69]

Dodson maintains that the characterization of the utility directives as guidelines, rather than rules, is ultimately a meaningless distinction. He observes, 'They will spill over into court challenges.'[70] Dodson adds: 'The patent system should be allowed to work, and it makes more sense to give people patents for inventions that have utility, even if the utility may not seem significant at the time of application.'[71]

II THE FISHER KING

In the case of *In re Fisher*, Dane K. Fisher and Raghunath V. Lalgudi of the agricultural biotechnology firm, Monsanto, filed a patent application in 2001 claiming compounds and compositions related to molecules derived from maize, corn and plant tissue.[72] The application included a 'Sequence Listing' disclosing partial sequences for 32 236 nucleic acid molecules extracted from corn plants. Claim 1 of the application recited: 'A substantially purified nucleic acid molecule that encodes a maize protein or fragment thereof comprising a nucleic acid sequence selected from the group consisting of SEQ ID NO: 1 through SEQ ID NO: 5.'[73] Claim 2 was directed to proteins, and Claims 3 to 7 related to transformed plants.

In January 2001, the Patent Examiner issued a restriction requirement ordering the applicants to elect certain claims and to limit their invention to 'no more than five of the individual sequences for examination'.[74] In response, Monsanto withdrew claims 2 to 7 and limited claim 1 to five nucleic acid sequences.

Monsanto claimed that the patent application disclosed that the five claimed ESTs may be used in a variety of ways, including:

(1) serving as a molecular marker for mapping the entire maize genome, which consists of ten chromosomes that collectively encompass roughly 50 000 genes;
(2) measuring the level of mRNA in a tissue sample via microarray technology to provide information about gene expression;
(3) providing a source for primers for use in the polymerase chain reaction ('PCR') process to enable rapid and inexpensive duplication of specific genes;
(4) identifying the presence or absence of a polymorphism;
(5) isolating promoters via chromosome walking;
(6) controlling protein expression; and
(7) locating genetic molecules of other plants and organisms.[75]

The biotechnology company maintained that these were specific, credible and substantial uses, to use the language of the utility guidelines promulgated by the USPTO.

In September 2001, the Patent Examiner issued a final rejection of claim 1 of the '643 application, finding that the claim lacked utility under 35 U.S.C. §101; failed to satisfy the enablement and written description requirements of 35 U.S.C. §112; and was anticipated by two prior art references under 35 U.S.C. §102. In the examiner's opinion, the alleged uses of the ESTs are 'non-specific uses that are applicable to nucleic acids in general and not particular or specific to the nucleic acids being claimed'.[76] Monsanto appealed against the ruling to the Board of Patent Appeals and Interferences.

In its March 2004 decision, the Board of Patent Appeals and Interferences affirmed the Examiner's final rejection of claim 1 for failure to satisfy the utility requirement of Section 101 and the enablement requirement of Section 112.[77] However, it reversed the Examiner's written description rejection. The Board was unconvinced by the analogies drawn between ESTs and microscopes:

> This argument has been reviewed but is not convincing because the microscope provides information to the scientist which is automatically useful. For example, the microscope may be used for identification and differentiation between gram-positive and gram-negative bacteria. The differentiation of bacteria facilitates in the administration of proper antibiotics. For example, if the microscope is used to determine whether Staph is present or whether Strep is present provides valuable information to the scientist and/or doctor for treating patients. The instant invention, however, provides no information to this extent. If the scientist determines that SEQ ID NO: 1 is present, the scientist does not know how to use this information. Thus, the identification of SEQ ID NO: 1 is not a substantial utility.[78]

Accordingly, Monsanto appealed to the Court of Appeals for the Federal Circuit. First, it asked whether 'the Board erred by concluding that an EST is subject to a heightened standard of utility under 35 US 101 that hinges upon some undefined "spectrum" of knowledge about the function of the gene that corresponds to the EST.'[79] Second, it questioned 'whether the Board erred by concluding that ESTs corresponding to genes of unknown function are incapable of satisfying the utility requirement of 35 USC 101, even though all ESTs, including each of the claimed ESTs, can be used as research tools to provide one or more specific, substantial, and commercially valuable benefits to the scientific community'.[80] It is worth noting that the company had six other appeals pending on the same legal issue.[81]

David Korn of the Association of American Medical Colleges speculated: 'I wouldn't be amazed if somebody from Monsanto said that they were doing this to test the guidelines.'[82] The USPTO defended the decision of the Board of Patent Appeals and Interferences, with the support of a number of amicus curiae, including Genentech Inc., Affymetrix, Eli Lilly and various research organizations.[83]

In the course of oral argument, the Court of Appeals for the Federal Circuit relied upon various metaphors to make sense of ESTs.[84] Equating genetics with literature, one of the judges sought to compare ESTs to a page of a book within a vast library:

> Isn't this the equivalent of claiming a single page of a book in the middle of a library? The library as a whole will be very valuable once it's complete, but one page out of the library would not seem to be enough for a patentable invention.[85]

Monsanto's eminent counsel, Seth Waxman, replied that ESTs could be used in ways having nothing to do with a library: 'The mapping is desired to establish a statistical correlation between identified sequences and plant traits identified by cross breeding, he explained.'[86] There was also much debate in argument as to whether ESTs could be likened to microscopes – a well-accepted research tool.

The Court of Appeals for the Federal Circuit was deeply divided over the patent application by Monsanto in respect of ESTs in maize.[87] For the majority, Michel CJ held that the invention lacked specific and substantial utility and, in any case, the application failed because there was a lack of enablement. Bryson J supported this opinion. Dissenting, Rader J argued that the patent application satisfied the requirements of utility under United States patent law. In addition, he held that the ruling on enablement should be reversed because it was consequential upon the other findings in respect of utility. In addition to divisions as to the application of patent doctrine, the judges of the Court of Appeals for the Federal Circuit expressed larger philosophical differences of opinion as to the nature of research tools, the level of scientific progress and the role of patent policy.

A Michel CJ

In the lead judgment, Michel J agreed, first of all, with the submission of the United States Government and the amici that none of Fisher's seven asserted uses meets the utility requirement of §101. The judge applied the ruling of the Supreme Court of the United States in *Brenner v Manson*,[88] and held that Fisher's application lacked utility:

> We agree with the Board that the facts here are similar to those in Brenner. There, as noted above, the applicant claimed a process for preparing compounds of unknown use. Similarly, Fisher filed an application claiming five particular ESTs

which are capable of hybridizing with underlying genes of unknown function found in the maize genome. The Brenner court held that the claimed process lacked a utility because it could be used only to produce a compound of unknown use. The Brenner court stated: 'We find absolutely no warrant for the proposition that although Congress intended that no patent be granted on a chemical compound whose sole "utility" consists of its potential role as an object of use-testing, a different set of rules was meant to apply to the process which yielded the unpatentable product.' Applying that same logic here, we conclude that the claimed ESTs, which do not correlate to an underlying gene of known function, fail to meet the standard for utility intended by Congress.[89]

The judge held that Fisher had failed to provide any evidence to prove that his claimed ESTs could be successfully used in the seven ways disclosed in the patent application: 'All of Fisher's asserted uses represent merely hypothetical possibilities, objectives which the claimed ESTs, or any EST for that matter, *could* possibly achieve, but none for which they have been used in the real world.'[90] Michel CJ concluded that Fisher had only disclosed general uses for its claimed ESTs, not specific ones that satisfy §101: 'Any EST transcribed from any gene in the maize genome has the potential to perform any one of the alleged uses.'[91]

Second, Michel CJ denied that there were strong analogies between ESTs and other patentable research tools, such as microscopes:

> Fisher compares the claimed ESTs to certain other patentable research tools, such as a microscope. Although this comparison may, on first blush, be appealing in that both a microscope and one of the claimed ESTs can be used to generate scientific data about a sample having unknown properties, Fisher's analogy is flawed. As the government points out, a microscope has the specific benefit of optically magnifying an object to immediately reveal its structure. One of the claimed ESTs, by contrast, can only be used to detect the presence of genetic material having the same structure as the EST itself. It is unable to provide any information about the overall structure let alone the function of the underlying gene. Accordingly, while a microscope can offer an immediate, real world benefit in a variety of applications, the same cannot be said for the claimed ESTs. Fisher's proposed analogy is thus inapt.[92]

His Honour concluded that Fisher's asserted uses were insufficient to meet the standard for a 'substantial' utility under §101.

Third, Michel CJ noted that proof of a utility may be supported when a claimed invention meets with commercial success. However, His Honour rejected the arguments of Monsanto that the database of ESTs had a significant commercial value:

> Fisher's reliance on the commercial success of general EST databases is also misplaced because such general reliance does not relate to the ESTs at issue in this case. Fisher did not present any evidence showing that agricultural companies

have purchased or even expressed any interest in the claimed ESTs. And it is entirely unclear from the record whether such business entities ever will.[93]

Again, the judge was concerned that Monsanto had failed to provide factual evidence to support its arguments that there was a commercial market for the ESTs at issue in the case.

Finally, Michel CJ held that the policy concerns raised by the amicus curiae were beyond the purview of the Court of the Appeals for the Federal Circuit:

> The concerns of the government and amici, which may or may not be valid, are not ones that should be considered in deciding whether the application for the claimed ESTs meets the utility requirement of §101. The same may be said for the resource and managerial problems that the PTO potentially would face if applicants present the PTO with an onslaught of patent applications directed to particular ESTs. Congress did not intend for these practical implications to affect the determination of whether an invention satisfies the requirements set forth in 35 U.S.C. §§101, 102, 103 and 112. They are public policy considerations which are more appropriately directed to Congress as the legislative branch of government, rather than this court as a judicial body responsible simply for interpreting and applying statutory law.[94]

This judicial disavowal of policy considerations is disappointing. The concerns of the government and the amici about the development of unwarranted monopolies in respect of genes and gene fragments are pertinent to the issues at hand in the litigation. The court is unwise and short-sighted to discount the resource and managerial problems of the USPTO. An administrative failure to deal properly with the flood of gene patent applications will have flow-on impact for the judiciary. A better approach would be to adjust the settings of patent criteria, in light of the policy directions of the Congress.

It is doubtful that the USPTO utility guidelines will be an effective means of addressing some of the problems with gene patents, if the courts remain blind to the public policy considerations that are being targeted.

B   Rader J

In his forceful dissent, Rader J maintained that the ESTs satisfied the requirements of the utility guidelines, because they constituted research tools: 'While I agree that an invention must demonstrate utility to satisfy §101, these claimed ESTs have such a utility, at least as research tools in isolating and studying other molecules.'[95]

First, Rader J submitted that the ESTs satisfied the demands of the Supreme Court of the United States in *Brenner v Manson*:

Several, if not all, of Fisher's asserted utilities claim that ESTs function to study other molecules. In simple terms, ESTs are research tools. Admittedly ESTs have use only in a research setting. However, the value and utility of research tools generally is beyond question, even though limited to a laboratory setting. (Many research tools such as gas chromatographs, screening assays, and nucleotide sequencing techniques have a clear, specific and unquestionable utility (e.g., they are useful in analyzing compounds.)) Thus, if the claimed ESTs qualify as research tools, then they have a 'specific' and 'substantial' utility sufficient for §101. If these ESTs do not enhance research, then *Brenner* v. *Manson* controls and erects a §101 bar for lack of utility. For the following reasons, these claimed ESTs are more akin to patentable research tools than to the unpatentable methods in *Brenner*.[96]

Rader J observed that the cases of *Brenner v Manson* and *In re Kirk*[97] 'share a common underpinning – a method of producing a compound with no known use has no more benefit to society than the useless compound itself'.[98] His Honour contended that the factual matrix contained *In re Fisher* was very different: 'Unlike the methods and compounds in *Brenner* and *Kirk*, Fisher's claimed EST's *are* beneficial to society.'[99] Approvingly, the judge observed that the ESTs would help scientists obtain a better understanding of the maize genome.

Second, Rader J argued that the analogies between microscopes and ESTs were persuasive, because both were research tools, which led to incremental improvements in scientific knowledge:

> These research tools are similar to a microscope; both take a researcher one step closer to identifying and understanding a previously unknown and invisible structure. Both supply information about a molecular structure. Both advance research and bring scientists closer to unlocking the secrets of the corn genome to provide better food production for the hungry world. If a microscope has §101 utility, so too do these ESTs . . .
>
> Even with a microscope, significant additional research is often required to ascertain the particular function of a 'revealed' structure. To illustrate, a cancerous growth, magnified with a patented microscope, can be identified and distinguished from other healthy cells by a properly trained doctor or researcher. But even today, the scientific community still does not fully grasp the reasons that cancerous growths increase in mass and spread throughout the body, or the nature of compounds that interact with them, or the interactions of environmental or genetic conditions that contribute to developing cancer.
>
> Significant additional research is required to answer these questions. Even with answers to these questions, the cure for cancer will remain in the distance. Yet the microscope still has 'utility' under §101. Why? Because it takes the researcher one step closer to answering these questions. Each step, even if small in isolation, is nonetheless a benefit to society sufficient to give a viable research tool 'utility' under §101. In fact, experiments that fail still serve to eliminate some possibilities and provide information to the research process.[100]

Such comparisons are deft and cunning (if not wholly convincing). By drawing affinities with microscopes, Rader J seeks to legitimize ESTs, and make them seem worthy of protection under patent law.

Scathingly, Rader J remarks: 'Nonetheless, this court, oblivious to the challenges of complex research, discounts these ESTs because it concludes (without scientific evidence) that they do not supply enough information.'[101] His Honour doubts the conclusions of his fellow judges: 'This court reasons that a research tool has a "specific" and "substantial" utility *only* if the studied object is readily understandable using the claimed tool – that no further research is required.'[102] Rader J reasons: 'Otherwise, only the final step of a lengthy incremental research inquiry gets protection.'[103]

Third, Rader J is critical that the USPTO does not, in his view, recognize the gradual and incremental nature of scientific development:

> Science always advances in small incremental steps. While acknowledging the patentability of research tools generally (and microscopes as one example thereof), this court concludes with little scientific foundation that these ESTs do not qualify as research tools because they do not 'offer an immediate, real world benefit' because further research is required to understand the underlying gene. This court further faults the EST research for lacking any 'assurance that anything useful will be discovered in the end'. These criticisms would foreclose much scientific research and many vital research tools. Often scientists embark on research with no assurance of success and knowing that even success will demand 'significant additional research'.[104]

Rader J contended: 'The United States Patent Office, above all, should recognize the incremental nature of scientific endeavour.'[105] He questioned the distinction drawn between research tools, which provided 'substantial' and 'insubstantial' advances: 'How does the Patent Office know which "insubstantial" research step will contribute to a substantial breakthrough in genomic study?'[106] Answering this rhetorical question for himself, Rader J concludes: 'Quite simply, it does not.'[107]

Finally, Rader J is willing to address some of the concerns of the USPTO about the administrative difficulties posed by the flood of applications in the field of biotechnology. He comments insightfully that the patent doctrine of utility is a poor instrument to discriminate between such applications:

> In truth, I have some sympathy with the Patent Office's dilemma. The Office needs some tool to reject inventions that may advance the 'useful arts' but not sufficiently to warrant the valuable exclusive right of a patent. The Patent Office has seized upon this utility requirement to reject these research tools as contributing 'insubstantially' to the advance of the useful arts. The utility requirement is ill suited to that task, however, because it lacks any standard for assessing

the state of the prior art and the contributions of the claimed advance. The proper tool for assessing sufficient contribution to the useful arts is the obviousness requirement of 35 U.S.C. §103.[108]

Although his advocacy for patent protection for ESTs is questionable, Rader J's concerns about the utility standard should be taken seriously. A more rigorous application of the requirements of novelty and inventive step (through according greater creative problem-solving capacities to the person skilled in the art) would ultimately be a better means of regulating patent law in the field of biotechnology.[109] In the May 2007 case of *KSR International Co. v Teleflex, Inc*, the Supreme Court of the United States emphasized the need for the USPTO and lower courts to set a high threshold for the standard of non-obviousness.[110]

C USPTO Board of Appeals and Interferences

After the long-standing controversy over ESTs, the decision of the Court of Appeals for the Federal Circuit in *In re Fisher* has received an enthusiastic reaction. Montreal academic Yann Joly commented that the decision represents an important shift in United States jurisprudence on patent law and biotechnology. 'It seems that with the *Fisher* case, the American judiciary has made another important step toward ending the abuse of some biotechnology companies and relieving the concerns of a majority of actors in this dynamic research field.'[111] Dianne Nicol from the University of Tasmania was similarly enthused: 'This decision lends support to the view that, as a general rule, it will be extremely difficult to overcome the utility hurdle for EST claims in the US.'[112]

Paula Davis, James Kelley and Steven Caltrider from Eli Lilly, and Stephen Heinig from the Association of American Medical Colleges, were delighted by the majority decision:

> The majority in the *Fisher* case would require patent applicants seeking to protect their ESTs to first identify the function of the underlying protein-encoding sequences ... What is very clear from *Fisher* is that filing as soon as the EST is sequenced, as was the norm previously, is not sufficient. It is thus evident that many of the EST applications currently filed with the PTO will not meet the threshold for utility and will likely be abandoned. The ESTs disclosed within these applications (many of which have been published) will be available freely for use in research.[113]

Jim Brogan from Cooley Goddard LLP cautioned that the decision of *In re Fisher* was not necessarily a definitive one on ESTs, as there remained scope for patent attorneys to develop a stronger factual record to support the utility of their patent claims.[114] Nonetheless, Brogan was of the view that the ruling by the Court of Appeals for the Federal Circuit would allay the concerns of the biotechnology industry about patents being granted on ESTs, where there was no knowledge of function.

The decision in *In re Fisher* is already proving to be influential in the administrative practice of the USPTO in reviewing patent applications in respect of biological inventions.[115] This small sample of decisions from the Board of Appeals and Interferences suggests that the USPTO will apply the decision *In re Fisher* with vigour and purpose. However, the limitations of the utility doctrine should be recognized. As the academic, David Resnik, has observed: 'While it may be a good idea to "raise the bar" on gene patents, this new PTO policy is little more than a temporary and limited solution to some of the difficult economic, legal, scientific, and medical issues relating to gene patents.'[116] The USPTO utility guidelines are a makeshift and stop-gap measure to address the glut of biotechnology patent applications. This administrative response is no substitute for full-bodied legislative reform in respect of intellectual property and biotechnology.

CONCLUSION

In the United States Congress, there have been a number of efforts by maverick representatives and senators to implement legislative reforms in respect of gene patents. Such action has been fuelled by a number of critical reports about the impact of gene patents upon scientific research, innovation, health care and competition.[117]

In 2002, Michigan Democrat Congresswoman Lynn Rivers introduced legislation into the House of Representatives of the United States Congress aimed at preserving research innovation, and quality patient care in the field of genetic testing.[118] The Congresswoman declared: 'Evidence is mounting that the patenting of human genes is both inhibiting important biomedical research and interfering with patient care.'[119]

The Genomic Research and Diagnostic Accessibility Act 2002 (US) had three major provisions. Section 2 exempts from patent infringement those individuals who use patented genetic sequence information for non-commercial research purposes. Section 3 would exempt medical practitioners utilizing genetic diagnostic tests from patent infringement remedies. This section builds on an existing legislative reform in the United States which exempts health care providers from patent infringement suits when they use a patented medical or surgical procedure.[120] Such a measure was put in place after an uproar over the case of *Pallin v Singer*,[121] in which an eye surgeon was sued for patent infringement in respect of a surgical procedure. Section 4 of the bill would require public disclosure of genomic sequence information contained within a patent application when public funds were used in the development of the invention.

The Genomic Science and Technology Innovation Act 2002 (US) called for an in-depth study by the White House Office of Science and Technology Policy on the impact of Federal patent policies on the rate of innovation, the cost, and the availability of genomic technologies. The two bills, *The Genomic Research and Diagnostic Accessibility Act* 2002 (US) and *The Genomic Science and Technology Innovation Act* 2002 (US), lapsed after Congresswoman Lynn Rivers failed to be re-elected to Congress.

In 2007, United States Congressmen Xavier Becerra, a Democrat of Southern California, and Dave Weldon, a Republican of Florida and a medical doctor, introduced the *Genomic Research and Accessibility Act* 2007 (US). The bill was intended to put an immediate end to the practice of patenting any and all portions of the human genome. The legislation would amend the *Patent Act* 1952 (US) and introduce a new section 106, providing a prohibition on the patenting of human genetic material: 'Notwithstanding any other provision of law, no patent may be obtained for a nucleotide sequence, or its functions or correlations, or the naturally occurring products it specifies.' The legislation, though, would not be retrospective, and would only apply to patents issued after the Act was amended. Becerra noted: 'Thus, if we enact this bill into law quickly, we will reach balance in less than two decades – a patent-free genome that does not hinder scientific research, business enterprise, or human morality.'[122]

Representative Becerra observed that genes were a product of nature, and should not be patented:

> One-fifth of the blueprint that makes up you . . . me . . . my children . . . your children . . . all of us . . . is owned by someone else . . . And we have absolutely no say in what those entities do with our genes. This cannot be what Watson and Crick intended.
>
> We seek simply to fix a regulatory mistake. Genes are a product of nature; they were not created by man, but instead are the very blueprint that creates man, and thus, are not patentable. Gene patenting would be the analogous equivalent to patenting water, air, birds or diamonds.
>
> Enacting the *Genomic Research and Accessibility Act* does not hamper invention, indeed, it encourages it. The proliferation of scientific prowess, medical innovation, and economic advancement will all occur if the study of genes is allowed to happen unabated. Incredible manifestations of intellectual property will result: medicines, machines, processes – most deserving of recognition, some potentially life-saving, and all worthy of a patent.[123]

His colleague Representative Weldon added: 'The practice of gene patenting is preventing critical research from advancing because scientists are wary of trespassing on patent laws.'[124] He observed: 'This not only violates the spirit of the Human Genome Project, it hinders the discovery of medical breakthroughs that could save lives.'[125] Weldon concluded: 'Our bill is a common sense measure to ensure that genes yet unpatented remain the province of science.'[126]

The author of *Jurassic Park*, Michael Crichton, lent his support to this legislative initiative, writing a stinging opinion piece in *The New York Times*. The science-fiction writer observed, with typical melodrama: 'You, or someone you love, may die because of a gene patent that should never have been granted in the first place.'[127] He contended: 'Gene patents are now used to halt research, prevent medical testing and keep vital information from you and your doctor.'[128] Crichton contended that patents should not be granted in respect of human genes, because they were natural features of the world and part of the common heritage of humankind:

> Humans share mostly the same genes. The same genes are found in other animals as well.
> Our genetic makeup represents the common heritage of all life on earth.
> You can't patent snow, eagles or gravity, and you shouldn't be able to patent genes, either.
> Yet by now one-fifth of the genes in your body are privately owned.[129]

The author observed: 'The United States Patent Office misinterpreted previous Supreme Court rulings and some years ago began – to the surprise of everyone, including scientists decoding the genome – to issue patents on genes.'[130] He commended the efforts of Congressmen Becerra and Dave Weldon, to ban the practice of patenting genes found in nature, through the *Genomic Research and Accessibility Act* 2006 (US): 'This bill will fuel innovation, and return our common genetic heritage to us.'[131]

The legislative proposal has renewed hostilities over the legitimacy of gene patents. The bioethicist, Lori Andrews, a law professor at Chicago-Kent College of Law, argued that patenting genes runs counter to the purpose of the patent system: to reward innovation and further research. A supporter of the Weldon bill, she said,

> What's driving this legislation now is a sense that allowing the patenting of our genes is a mistake, and it's an easy fix. The original geneticists didn't need financial incentives and genes differ from drugs. There is much more public funding involved, and we, the public, should be shareholders in genetic discoveries. Scientists say: 'We're not patenting a product of nature', which is prohibited under patent law, but they take out the parts of the gene that have nothing to do with function, and then say: 'This is a new invention'.[132]

By contrast, Robert Cook-Deegan commented that gene patents were not necessarily evil: 'I think some gene patents are useful and socially productive.'[133] He observed, for instance, that erythropoietin, for one, 'proved useful, not only in kidney failure, but also in treating cancer after high-dose chemotherapy'.[134] He did not believe that patenting genes hampered other people's research, as long as patent-holders act 'responsibly'. Cook-Deegan stressed: 'I think this is a better way to go than congressional legislation.'[135] In any case, he noted that 'the juicy parts of the genome already have been patented', and would be unaffected by the proposed legislation.[136]

NOTES

1. Roberts, Michael Symmons (2004), 'Mapping the genome', *Corpus*, London: Cape Poetry.
2. The term was popularized by A. Weinberg (1961), 'Impact of large-scale science on the United States', *Science*, **134**(3473), 161–4; and Peter Galison and Bruce Hevly (eds) (1992), *Big Science: The Growth of Large-Scale Research*, Stanford, Calif.: Stanford University Press.
3. Cook-Deegan, R. (2001), 'Hype and hope', *American Scientist*, **89**, 62.
4. Collins, F. (2001), 'Remarks at the press conference announcing sequencing and analysis of the human genome', http://www.genome.gov/10001379, 12 February.
5. Lewontin, Richard (2000), *It Ain't Necessarily So: The Dream of the Human Genome and Other Illusions*, New York: New York Review of Books.
6. Lewontin, R. (2001), 'They got the wrong key of life', *The Sunday Times*, 8 July.
7. Ibid.
8. Ibid.
9. 'An EST is a short nucleotide sequence that represents a fragment of a cDNA clone. It is typically generated by isolating a cDNA clone and sequencing a small number of nucleotides located at the end of one of the two cDNA strands. When an EST is introduced into a sample containing a mixture of DNA, the EST may hybridize with a portion of DNA. Such binding shows that the gene corresponding to the EST was being expressed at the time of mRNA extraction': Michel CJ, *In re Fisher* 421 F.3d 1365 at 1367 (C.A.Fed.,2005).
10. Davies, Kevin (2001), *The Sequence: Inside the Race for the Human Genome*, London: Weidenfeld & Nicolson.
11. Davies, Kevin (2001), *The Sequence: Inside the Race for the Human Genome*, London: Weidenfeld and Nicolson, p. 63.
12. Ibid., p. 62.
13. Celera Genomics, http://www.celera.com/.
14. Davies, Kevin (2001), *The Sequence: Inside the Race for the Human Genome*, London: Weidenfeld & Nicolson, p. 62.
15. Shreeve, James (2004), *The Genome War: How Craig Venter Tried to Capture the Code of Life and Save the World*, New York: Ballantine Books.
16. Ibid., 230.
17. Ibid., 235.
18. Ibid., 235.
19. Venter, J.C. (2000), 'Prepared statement', the Subcommittee on Energy and Environment, United States House of Representatives Committee on Science, http://www.ostp.gov/html/00626_4.html, 6 April.
20. Ibid.
21. Ibid.

22. Sulston, John and Georgina Ferry (2002), *The Common Thread: A Story of Science, Politics, Ethics and the Human Genome*, London: Bantam Press, pp. 266–7.
23. Ibid.
24. Collins, F. (2000), 'Mapping the genome', Online Newshour with Jim Lehrer, http://www.pbs.org/newshour/bb/health/jan-june00/extended_collins.html.
25. Ibid.
26. Clinton, W. and T. Blair (2000), 'Joint statement by President William Clinton and Prime Minister Tony Blair of the United Kingdom', http://ipmall.info/hosted_resources/ippresdocs/ippd_44.htm, 14 March.
27. Davies, Kevin (2001), *The Sequence: Inside the Race for the Human Genome*, London: Weidenfeld & Nicolson, pp. 205–7; and Shreeve, James (2004), *The Genome War: How Craig Venter Tried to Capture the Code of Life and Save the World*, New York: Ballantine Books, pp. 322–6.
28. Gaglioti, F. (2000), 'Wall Street and the commercial exploitation of the human genome', http://www.wsws.org/articles/2000/apr 2000/gene-a10.shtml, 10 April; see also Moody, Glyn (2004), *Digital Code of Life: How Bioinformatics is Revolutionizing Science, Medicine, and Business*, Hoboken, NJ: John Wiley & Sons.
29. Ibid.
30. Ibid.
31. Ibid.
32. The International Human Genome Sequencing Consortium (2001), 'Initial sequencing and analysis of the human genome', *Nature*, **409**, 860–921.
33. Venter, J.C. et al. (2001), 'The sequence of the human genome', *Science*, **291**, 16 February, 1301–4.
34. *In re Fisher* 421 F.3d 1365 (C.A.Fed.,2005).
35. *Ex Parte Raymond H. Boutin* 2006 WL 2822238, *4+ (Bd.Pat.App & Interf. Sep 28, 2006) (NO. APL 2006-1879, APP 10/010,114); *Ex Parte Preeti Lal, Neil Corley et al.* 2006 WL 2710996, *3+ (Bd.Pat.App & Interf. Sep 18, 2006) (NO. APL 2006-1035, APP 09/925,140); *Ex Parte d. Wade Walke* 2006 WL 2711006, *2+ (Bd.Pat.App & Interf. Sep 18, 2006) (NO. APL 2006-2131, APP 10/309,422); *Ex Parte Preeti Lal, Jennifer Hillman*, et al., 2006 WL 1665364, *3+ (Bd.Pat.App & Interf. Jan 01, 2006) (NO. APL 2005-0102, APP 09/840,787); *Ex Parte Gary C. Starling*, 2006 WL 1665405, *2+ (Bd.Pat.App & Interf. Jan 01, 2006) (NO. APL 2005-2121, APP 09/745,605).
36. *Brenner v Manson*, 383 US 519 (1966).
37. *Brenner v Manson*, 383 US 519 at 534 (1966).
38. *Brenner v Manson*, 383 US 519 at 534 (1966).
39. *Brenner v Manson*, 383 US 519 at 534 (1966).
40. *Brenner v Manson*, 383 US 519 at 536 (1966).
41. *Brenner v Manson*, 383 US 519 at 536 (1966).
42. United States Patent and Trademark Office (2001), 'Utility examination guidelines', *Federal Register*, **66**, 1092, http://www.uspto.gov/web/offices/com/sol/notices/utilexmguide.pdf.
43. Dickinson, T. (2000), 'Statement of Todd Dickinson before the Subcommittee on Courts and Intellectual Property of the House Committee on the Judiciary', http://www.house.gov/judiciary/dick0713.htm, 13 July.
44. Ibid.
45. Ibid.
46. Ibid.
47. Ibid.
48. United States Patent and Trademark Office (2000), 'Public comments on the United States Patent and Trademark Office revised interim utility examination guidelines', *Federal Register*, **65**, FR 3425, http://www.uspto.gov/web/offices/com/sol/comments/utilguide/index.html, 21 January.
49. United States Patent and Trademark Office (1999), 'Revised interim utility examination guidelines', *Federal Register*, **64**(244), http://www.uspto.gov/web/offices/com/sol/notices/utilexmguide.pdf, 21 December.

50. Ibid.
51. Ibid.
52. Ibid.
53. Ibid.
54. Grisham, J. (2000), 'New rules for gene patents', *Nature Biotechnology*, **18**, 921; Benson, J. (2002) 'Resuscitating the patent utility requirement, again: a return to *Brenner v Manson*', *University of California Davis Law Review*, **36**, 267–95; Breen Smith, M. (2002), 'An end to gene patents? The human genome project versus the United States Patent and Trademark Office's 1999 utility guidelines', *University of Colorado Law Review*, **73**, 747–85; Wei, T. (2003), 'Patenting genomic technology – 2001 utility examination guidelines: an incomplete remedy in need of prompt reform', *Santa Clara Law Review*, **44**, 307–33; Lopez-Beverage, C. (2005), 'Should congress do something about upstream clogging caused by the deficient utility of expressed sequence tag patents?', *Journal of Technology Law and Policy*, **10**, 35–92; and Pippen, S. (2006), 'Dollars and lives: finding balance in the patent "gene utility" doctrine', *Boston Journal of Science and Technology Law*, **12**, 193–226.
55. Marshal, E., E. Pennisi and L. Roberts (2000), 'In the crossfire: Collins on genomes, patents, and "rivalry"', *Science*, **287**(5462), 2396–8, 31 March.
56. Ibid.
57. Varmus, H. (2000), 'Testimony for hearing on gene patents and other genomic inventions', House of Representatives Subcommittee on Courts and Intellectual Property, http://judiciary.house.gov/Legacy/varm0713.htm, 13 July.
58. Ibid.
59. Nuffield Council on Bioethics (2002), *The Ethics of Patenting DNA, A Discussion Paper*, London: Nuffield Council on Bioethics, http://www.nuffieldbioethics.org/go/ourwork/patentingdna/publication_310.html, 45.
60. Ibid., 60.
61. Ryan, A. (2000), 'Testimony for hearing on gene patents and other genomic inventions', House of Representatives Subcommittee on Courts and Intellectual Property, http://www.aipla.org/Content/ContentGroups/Legislative_Action/106th_Congress/Testimony4/Statement_at_oversight_hearing_on_gene_patents_and_other_genomic_inventions,_by_M_Andrea_Ryan,_July_.htm, 7 July.
62. Ibid.
63. Ludlam, C. (2000) 'Biotechnology Industry Organization comment on interim utility and written description guidelines', http://www.uspto.gov/web/offices/com/sol/comments/utilitywd/bio.pdf, 22 March.
64. Ibid.
65. Scott, R. (2000), 'Testimony for hearing on gene patents and other genomic inventions', House of Representatives Subcommittee on Courts and Intellectual Property, http://judiciary.house.gov/Legacy/scot0713.htm, 7 July.
66. Ibid.
67. Ibid.
68. Slind For, V. (2001), 'Both sides now: MoFo IP Partners Kate Murashige and Gerald Dodson diverge on the PTO's new guidelines for biotech patents', IP Worldwide, 14 June.
69. Ibid.
70. Ibid.
71. Ibid.
72. Fisher, D. and R. Lalgudi (2001), 'Nucleic acid molecules and other molecules associated with plants', US Patent Application Serial No: 09/619,643.
73. Ibid.
74. As cited in *In re Fisher* 421 F.3d 1365 at 1374 (C.A.Fed.,2005).
75. Monsanto elaborates upon such uses in its briefs: Monsanto Inc. (2004), 'Corrected brief for appellants Dane K. Fisher and Raghunath V. Lalgudi', 2004 WL 4996614, 27 September, 12–20.

76. As reported in *Ex parte Fisher* 72 U.S.P.Q.2d 1020, 2004 WL 2185929 (United States Patent and Trademark Office Board of Patent Appeals and Interferences), http://patentlaw.typepad.com/patent/Ex_20Parte_20Fisher.pdf.
77. *Ex parte Fisher* 72 U.S.P.Q.2d 1020, 2004 WL 2185929 (United States Patent and Trademark Office Board of Patent Appeals and Interferences), http://patentlaw.typepad.com/patent/Ex_20Parte_20Fisher.pdf.
78. *Ex parte Fisher* 72 U.S.P.Q.2d 1020, 2004 WL 2185929 (United States Patent and Trademark Office Board of Patent Appeals and Interferences), http://patentlaw.typepad.com/patent/Ex_20Parte_20Fisher.pdf.
79. Monsanto Inc. (2004), 'Corrected Brief for Appellants Dane K. Fisher and Raghunath V. Lalgudi *In re Fisher*', 2004 WL 4996614, 27 September.
80. Ibid.
81. *In re Kovalic*, No. 05-1007; *In re Lalgudi*, No. 05-1010; *In re Byrum*, No. 1011; *In re Anderson*, No. 1012; *In re Adab*, No. 05-1013; and *In re Boukharov*, No. 05-1014.
82. Lawrence, S. (2005), 'US Court Case to Define EST Patentability', *Nature Biotechnology*. http://www.nature.com/news/2005/050502/full/nbt0505-513.html, 3 May.
83. United States Patent and Trademark Office (2004), 'Brief and addendum for appellee. Director of the United States Patent and Trademark Office *In re Fisher*', 7 December; Genentech Inc. (2004), 'Brief of Genentech Inc. as amicus curiae supporting affirmance and supporting the United States Patent and Trademark Office *In re Fisher*', 15 December; Affymetrix Inc. (2004), 'Brief for amicus curiae Affymetrix, Inc. in support of Appellee *In re Fisher*', 2004 WL 4996615, 14 December; and Eli Lilly et al. (2004), 'Brief for amici curiae Eli Lilly and company, Association of American Medical Colleges, Baxter Healthcare Corporation, National Academy of Sciences Dow AgroSciences LLC, and American College of Medical Genetics in Support of the United States Patent and Trademark Office in Support of Affirmance *In re Fisher*', 2004 WL 4996616, 14 December.
84. For a theoretical discussion of the use of metaphor in the biological sciences, see Fox Keller, Evelyn (1995), *Refiguring Life: Metaphors of Twentieth-Century Biology*, New York: Columbia University Press; Fox Keller, Evelyn (2000), *The Century of the Gene*, Cambridge, Mass.: Harvard University Press; Fox Keller, Evelyn (2002), *Making Sense of Life: Explaining Biological Development with Models, Metaphors, and Machines*, Cambridge, Mass.: Harvard University Press; and Jardine, Lisa (1999), *Ingenious Pursuits: Building the Scientific Revolution*, London: Abacus History.
85. *In re Fisher*, Fed. Cir., No. 04-1465, 5/3/05 oral argument.
86. *In re Fisher*, Fed. Cir., No. 04-1465, 5/3/05 oral argument.
87. *In re Fisher* 421 F.3d 1365 at 1374 (C.A.Fed.,2005).
88. *Brenner v Manson*, 383 US 519 (1966).
89. *In re Fisher* 421 F.3d 1365 at 1374 (C.A.Fed.,2005).
90. *In re Fisher* 421 F.3d 1365 at 1373 (C.A.Fed.,2005).
91. *In re Fisher* 421 F.3d 1365 at 1374 (C.A.Fed.,2005).
92. *In re Fisher* 421 F.3d 1365 at 1373 (C.A.Fed.,2005).
93. *In re Fisher* 421 F.3d 1365 at 1377–1378 (C.A.Fed.,2005).
94. *In re Fisher* 421 F.3d 1365 at 1378 (C.A.Fed.,2005).
95. *In re Fisher* 421 F.3d 1365 at 1379 (C.A.Fed.,2005).
96. *In re Fisher* 421 F.3d 1365 at 1379 (C.A.Fed.,2005).
97. *In re Kirk* 54 C.C.P.A. 1119, 376 F.2d 936 (1967).
98. *In re Fisher* 421 F.3d 1365 at 1380 (C.A.Fed.,2005).
99. *In re Fisher* 421 F.3d 1365 at 1380 (C.A.Fed.,2005).
100. *In re Fisher* 421 F.3d 1365 at 1380-1381 (C.A.Fed.,2005).
101. *In re Fisher* 421 F.3d 1365 at 1380 (C.A.Fed.,2005).
102. *In re Fisher* 421 F.3d 1365 at 1380 (C.A.Fed.,2005).
103. *In re Fisher* 421 F.3d 1365 at 1380 (C.A.Fed.,2005).
104. *In re Fisher* 421 F.3d 1365 at 1380 (C.A.Fed.,2005).
105. *In re Fisher* 421 F.3d 1365 at 1381 (C.A.Fed.,2005).
106. *In re Fisher* 421 F.3d 1365 at 1381 (C.A.Fed.,2005).

107. *In re Fisher* 421 F.3d 1365 at 1381 (C.A.Fed.,2005).
108. *In re Fisher* 421 F.3d 1365 (C.A.Fed.,2005).
109. Gold, E.R. and K. Durell (2005), 'Innovating the skilled reader: tailoring patents to new technologies', *Intellectual Property Journal*, **19**(1), 189–226.
110. The matter concerns the validity of S. Engelgau (2000), 'Adjustable pedal assembly with electronic throttle control', US Patent No: 6,237,565 B1. At issue is claim 4, which comprises (i) a pre-existing 'adjustable pedal assembly', combined with (ii) a pre-existing 'electronic control'. In his decision, Kennedy J for the Supreme Court of the United States cautioned:

> We build and create by bringing to the tangible and palpable reality around us new works based on instinct, simple logic, ordinary inferences, extraordinary ideas, and sometimes even genius. These advances, once part of our shared knowledge, define a new threshold from which innovation starts once more. And as progress beginning from higher levels of achievement is expected in the normal course, the results of ordinary innovation are not the subject of exclusive rights under the patent laws. Were it otherwise patents might stifle, rather than promote, the progress of useful arts.

KSR International Co. v Teleflex, Inc. 2007 WL 1237837, 82 U.S.P.Q.2d 1385 (2007).
111. Joly, Y. (2006), 'Wind of change: *In re Fisher* and the evolution of American biotechnology patent law', in M. Rimmer (ed.), *Patent Law and Biological Inventions, Law in Context*, **24**(1), 67–84 at 80.
112. Nicol, D. (2005), 'On the legality of gene patents', *The University of Melbourne Law Review*, **29**(3), 809–41 at 835.
113. Davis, P., J. Kelley, S. Caltrider and S. Heinig (2005), 'ESTs stumble at the utility threshold', *Nature Biotechnology*, **23**, 1227–9.
114. Brogan, J. (2005), 'Federal Court rules gene fragments not patentable', Cooley Alert, http://www.cooley.com/files/tbl_s24News%5CPDFUpload152%5C1646%5CALERT_GeneFragPatents.pdf.
115. *Ex Parte Raymond H. Boutin* 2006 WL 2822238, *4+ (Bd.Pat.App & Interf. Sep 28, 2006) (NO. APL 2006-1879, APP 10/010,114); *Ex Parte Preeti Lal, Neil Corley et al.* 2006 WL 2710996, *3+ (Bd.Pat.App & Interf. Sep 18, 2006) (NO. APL 2006-1035, APP 09/925,140); *Ex Parte d. Wade Walke* 2006 WL 2711006, *2+ (Bd.Pat.App & Interf. Sep 18, 2006) (NO. APL 2006-2131, APP 10/309,422); *Ex Parte Preeti Lal, Jennifer Hillman*, et al., 2006 WL 1665364, *3+ (Bd.Pat.App & Interf. Jan 01, 2006) (NO. APL 2005-0102, APP 09/840,787); *Ex Parte Gary C. Starling*, 2006 WL 1665405, *2+ (Bd.Pat.App & Interf. Jan 01, 2006) (NO. APL 2005-2121, APP 09/745,605).
116. Resnik, D. (2000), 'Comments on the United States Patent and Trademark Office utility guidelines', http://www.uspto.gov/web/offices/com/sol/comments/utilguide/dresnick.pdf, 16 March.
117. Federal Trade Commission (2003), *To Promote Innovation: The Proper Balance of Competition and Patent Law and Policy*, Washington, DC: Federal Trade Commission, http://www.ftc.gov/os/2003/10/innovationrpt.pdfl; National Academy of Sciences (2004), *A Patent System for the 21st Century*, Washington, DC: National Academy of Sciences, http://books.nap.edu/catalog.php?record_id=10976; and National Research Council Committee on Intellectual Property Rights in Genomic and Protein Research and Innovation (2005), *Reaping the Benefits of Genomic and Proteomic Research: Intellectual Property Rights, Innovation and Public Health*, Washington, DC: National Academies Press.
118. Rivers, L. (2002), 'Introduction of *The Genomic Research And Diagnostic Accessibility Act* of 2002 H.R. 3967 and *The Genomic Science And Technology Innovation Act* of 2002 H.R. 3966', Congressional Record, 14 March, E353.
119. Ibid.
120. *A Bill to Limit The Issuance Of Patents On Medical Procedures*, House of Representatives 112, 104th Congress, 1st Session, 3 March 1995.
121. *Pallin v Singer* 36 USPQ (2d) 1050 (1995); and Shulman, Seth (1999), 'The new medical licenses', *Owning The Future*, Boston: Houghton Mifflin Company, pp. 33–59.

122. Becerra, X. and D. Weldon (2007), 'Representatives Becerra and Weldon introduce bill to ban the practice of gene patenting', United States Congress, http://weldon.house.gov/News/DocumentSingle.aspx?DocumentID=57930, 9 February.
123. Ibid.
124. Ibid.
125. Ibid.
126. Ibid.
127. Crichton, M. (2007), 'Patenting life', *The New York Times*, 13 February.
128. Ibid.
129. Ibid.
130. Ibid.
131. Ibid.
132. Jenks, S. (2007), 'Debate grows over patenting of genes', *Florida Today*, 10 March.
133. Ibid.
134. Ibid.
135. Ibid.
136. Ibid.

6. The dilettante's defence: patent law, research tools and experimental use

In 1854, the American transcendentalist poet and writer, Henry David Thoreau, wrote his classic text, *Walden*, about the two years that he spent living in Walden Pond, near Concord, Massachusetts.[1] He rhapsodized about the life of the natural philosopher:

> To be a philosopher is not merely to have subtle thoughts, nor even to found a school, but so to love wisdom as to live according to its dictates, a life of simplicity, independence, magnanimity, and trust. It is to solve some of the problems of life, not only theoretically, but practically. The success of great scholars and thinkers is commonly a courtier-like success, not kingly, not manly.[2]

This romantic vision of the scientist as a natural philosopher and amateur thinker has been remarkably powerful in patent jurisprudence. Memorably, the defence of experimental use in the United States has been described as a 'dilettante affair'.[3]

In the early part of the nineteenth century, Story J, a legal polymath, an associate justice of the Supreme Court of the United States and the Dane Professor of Harvard University, devised the defence of experimental use in respect of patent law. In the 1813 appellate decision of *Whittemore v Cutter*, Story J considered whether a party had infringed the patent assigned on a machine used to produce playing cards. His Honour observed: 'It could never have been the intention of the legislature to punish a man, who constructed such a machine merely for philosophical experiments, or for the purpose of ascertaining the sufficiency of the machine to produce its desired effects.'[4] In the subsequent 1813 decision in *Sawin v Guild*, Story J considered the issue again in the context of an action for patent infringement against a deputy sheriff who had seized a patented machine for cutting brad nails to settle a debt.[5] Clarifying his opinion in *Whittemore v Cutter*, His Honour observed that 'the making of patented machine to be an offence within the purview of it, must be the making with intent to use for profit, and not for the mere purpose of philosophical experiment, or to ascertain the verity and exactness of the specification'.[6] The judge elaborated: 'In other words, that the making must be with intent to infringe the patent right, and deprive the owner of the lawful rewards of his discovery.'[7] In 1861, the decision in *Peppenhausen v Falke* elaborated upon the nature of the common law defence of experimental use: 'It has been held, and no doubt is now well settled, that an experiment with a patented article for the sole purpose of gratifying a philosophical taste, or curiosity, or for mere amusement, is not an infringement of the rights of the patentee.'[8] The research exemption has been compared to the defence of fair use in copyright law, which was also a creation of Story J, in *Folsom v Marsh*.[9] However, the defence of experimental use in patent law has not been codified in legislation as its counterpart, the defence of fair use in copyright law.

With its origins in mechanical inventions, the defence of experimental use has had to be adapted to deal with a range of technologies and scientific fields in the twentieth and twenty-first centuries.[10] The research exemption has had a particular application in a number of important industries, such as agriculture, biotechnology and pharmaceutical drugs. The defence of experimental use was first adapted to deal with the sui generis regime of plant breeders' rights.[11] The research exemption has since become a live issue in the context of patents in respect of agricultural chemicals.[12] There has been much concern about the impact of gene patents upon the conduct of research and scientific communication in the life sciences.[13] The status of research tools in biotechnology has been particularly troublesome. The defence of experimental use has posed particular issues in the field of pharmaceutical drugs.[14] There has been much debate as to whether generic manufacturers can engage in clinical testing with a pharmaceutical drug prior to the expiry of a patent. There has been much discussion about the application of defence of experimental use to particular fields of scientific endeavour, such as information technology,[15] stem cell research,[16] vaccines for infectious diseases[17] and nanotechnology.[18]

This chapter considers how a number of modern United States precedents have explored the scope and the limits of the common law defence of experimental use, and the statutory safe harbour for research on pharmaceutical drugs. It is contended that the research exemption should be codified, modernized and streamlined, so that it can comfortably accommodate a broad range of new frontier technologies in the fields of agriculture, biotechnology and pharmacology. It is argued that, in the absence of legislative reform, the defence of experimental use is in danger of becoming a historical anachronism – of entirely no use to modern scientific researchers who do not conform to Story J's pastoral ideal of the natural philosopher. Section one considers the development of the common law defence of experimental use in the United States. In *Madey v Duke University*, the Court of Appeals for the Federal Circuit considered the defence of experimental use in the context of a dispute between an academic inventor and a university.[19] It stressed that 'use in keeping with the legitimate business of the alleged infringer does not qualify for the experimental use defence'.[20] It is recommended that there is a need for the United States Congress to codify the defence of experimental use, and broaden the scope of its operation. Section two examines the development of the Bolar exception in respect of experimentation on pharmaceutical drugs. It explores the decision of the Supreme Court of the United States in *Merck KGaA v Integra Lifesciences I, Ltd.* on the scope of the safe harbour for pharmaceutical drugs.[21] The conclusion highlights the competing norms, and lack of harmonization, between a number of jurisdictions, including the United States, the European Union and Australia.

I 'A NOBILITY OF SCIENCE, AND A MAGIC OF ACADEMIA': *MADEY v DUKE UNIVERSITY*

There has been much controversy over the decision of the Federal Circuit in *Madey v Duke University* over patent law and experimental use.[22] The facts of the case are instructive. John Madey, an academic physicist, invented the free electronic laser as a tenured professor at Stanford University in the 1980s. The invention was a component of the so called 'Star Wars' missile defence programme, proposed by President Ronald Reagan. Madey obtained patents for two related inventions, one concerning a microwave electron gun[23] and the other a free electronic laser oscillator.[24] Madey was recruited as a tenured Professor in the Faculty of Physics at Duke University. He was the director of a laboratory which contained two free electronic lasers, which incorporated the inventions covered by the patents.

Duke University was unhappy with the performance of Madey as the director of the laboratory, accusing the scientist of administrative incompetence:

> Madey was responsible for managing the laboratory. His laboratory management skills, however, turned out to be poor. Madey's managerial incompetence forced Duke to take remedial measures . . . The review committee found the lab in disarray, and unanimously recommended that Duke replace Madey as Director of the DFELL. The Duke provost concurred in this appraisal and removed Madey as Director.[25]

Duke University offered Madey the position of Chief Scientist; however, he declined to accept this demotion.

After an academic dispute, Madey resigned from his position at Duke University and continued his academic and research activities at the University of Hawaii. Despite his departure, the laboratory at Duke

University continued to use the free electronic lasers. In response, Madey claimed that Duke University had infringed the patents by using the inventions without his permission. He portrayed himself as an innocent victim: 'Unless you have the good fortune to be at a place that looks after your every need, and not many people do, you need to be versed and skilled in the business and contractual aspects of what you do.'[26] The university claimed that it was protected by the defence of experimental use because the equipment was being used for academic purposes, including instruction and research, and not commercial purposes. Duke University alleged that 'this case is really an effort by a disgruntled ex-faculty member to punish Duke for removing him from two purely administrative posts in the Duke Free Electron Laser Laboratory'.[27]

A United States District Court for North Carolina

At first instance, the United States District Court for North Carolina entered summary judgment for Duke University.[28] In the course of the judgment, Beaty J discussed the history of the common law defence of experimental use:

> Although the 'basic law of patents establishes that unauthorized use of a patented product or process constitutes infringement,' for well over a century, United States 'patent jurisprudence has paid homage to . . . an exception from infringement liability for . . . unauthorized uses of patented inventions [,]' where the uses were solely for research, academic, or experimental purposes. As expressed by Justice Story in *Whittemore v. Cutter*, the case in which the experimental use doctrine originated, the underlying rationale for exempting such uses from liability is that 'it could never have been the intention of the legislature to punish a man, who constructed a . . . [patented device] merely for philosophical experiments . . .'. Although the scope of the defense has recently been the issue of much debate, to date, the experimental use defense remains viable and may be asserted in those cases in which the allegedly infringing use of the patent is made for experimental, non-profit purposes only.[29]

On the facts of the case, the judge concluded that the common law experimental use defence was a complete answer to Madey's claim of patent infringement. The judge commented that 'that Defendant's primary purpose is to teach, research, and expand knowledge, and to not engage in patent development for the purpose of commercial benefit'.[30] The judge noted in a footnote that 'all users to date, even by Plaintiff's own acknowledgment, have been faculty members of Duke, other academics, or non-industrial users seeking to pursue experimental interests'.[31]

In the appeal, the counsel for Madey argued that Duke University should not be able to avail itself of the defence of experimental use:

> A modern research laboratory like the Duke FEL Lab, with funding of tens of millions of dollars and employing hundreds of people, bears no resemblance to Justice Story's 19th century amateur philosopher pondering the nature of life in his home laboratory. During his tenure at Duke, Dr. Madey brought over $40 million in federal grants and contracts to the Duke FEL Lab to support its activities. Despite Dr Madey's departure from Duke, the FEL Lab continues to receive millions of dollars in federal funding for Duke, and the two major lasers are essential to virtually all of the Lab's activities.[32]

The counsel argued that there were three primary legal errors related to experimental use. First, Madey claimed that the district court improperly shifted the burden to Madey to prove that Duke's use was not experimental. Second, Madey argued that the district court applied an overly broad version of the very narrow experimental use defence inconsistent with past precedents. Third, Madey attacked the supporting evidence relied on by the district court as overly general and not indicative of the specific propositions and findings required by the experimental use defence.

In response, Duke University argued that the patent infringement action must fail because all of the claimed infringing uses were experimental in nature and therefore not actionable:

> The undisputed facts show that the DFELL engaged in performing research that has *no* commercial implication whatsoever. Duke's use of the devices at issue, therefore, falls under the common law 'experimental use' exception to patent infringement liability. Apparently realizing this, Madey takes the untenable position that because Duke is in the 'business' of performing research, the research that Duke performs is not experimental. Madey's theory, if adopted, would effectively chill all research performed at universities. The experimental use exception guards against this harmful and unwanted effect. Accordingly, the district court's grant of summary judgment that Duke did not infringe the '103 and '994 patents because its research constituted experimental use should be affirmed.[33]

Duke University submitted: 'Justice Story's reference to "philosophical experiments" encompasses exactly what takes place in the DFELL, namely scholarly inquiry to advance knowledge.'[34] Duke University observed that the defence of experimental use should include laboratory experiments in the basic sciences as well as other scholarly inquiries: 'A "philosophical experiment" does not mean, as Madey claims, amateur philosophizing.'[35]

B   Court of Appeals for the Federal Circuit

The United States Court of Appeals for the Federal Circuit upheld the appeal by Madey against Duke University.[36] A one-time patent attorney and intellectual property specialist, Gajarsa J, observed on behalf of the court:

> Our precedent clearly does not immunize use that is in any way commercial in nature. Similarly, our precedent does not immunize any conduct that is in keeping with the alleged infringer's legitimate business, regardless of commercial implications. For example, major research universities, such as Duke, often sanction and fund research projects with arguably no commercial application whatsoever. However, these projects unmistakably further the institution's legitimate business objectives, including educating and enlightening students and faculty participating in these projects. These projects also serve, for example, to increase the status of the institution and lure lucrative research grants, students and faculty.
>
> In short, regardless of whether a particular institution or entity is engaged in an endeavor for commercial gain, so long as the act is in furtherance of the alleged infringer's legitimate business and is not solely for amusement, to satisfy idle curiosity, or for strictly philosophical inquiry, the act does not qualify for the very narrow and strictly limited experimental use defense. Moreover, the profit or non-profit status of the user is not determinative.[37]

The judge determined that the district court attached too great a weight to the non-profit, educational status of Duke, 'effectively suppressing the fact that Duke's acts appeared to be in accordance with any reasonable interpretation of Duke's legitimate business objectives'.[38] He stressed that 'Duke ... like other major research institutions of higher learning is not shy in pursuing an aggressive patent licensing program from which it derives a not insubstantial revenue stream.'[39] The judge directed that on remand the district court would have to revise and limit its conception of the experimental use defence: 'The correct focus should not be on the non-profit status of Duke but on the legitimate business Duke is involved in and whether or not the use was solely for amusement, to satisfy idle curiosity, or for strictly philosophical inquiry.'[40]

Professor Janice Mueller of the University of Pittsburgh has sought to place the decision in *Madey v Duke University* in the context of broader themes of patent jurisprudence at work in the Court of Appeals for the Federal Circuit.[41] She commented that a number of factors were behind the virtual nullification of the defence of experimental use: 'It is doubtful that any single theory can satisfactorily explain the seeming hostility to a meaningful experimental use defense by a majority of the judges of the Federal Circuit.'[42] First, Mueller commented that the decision reflected a deep-seated legal formalism: 'Patent law scholars have noted a rise of formalism in recent Federal Circuit decisions, evidenced by a preference for bright-line rules over more nuanced, multi-factored, "totality of the circumstances" standards.'[43] Second, she observed: 'The Federal Circuit's effective nullification of the common law experimental use defense to infringement in *Madey* and *Integra* can be seen as but a piece of a broader hostility to any judicial derogation of a patent owner's right to exclude others.'[44] Third, Mueller noted that the

Court of Appeals was resistant to limitations and exceptions to intellectual property rights: 'The Federal Circuit's general reluctance to derogate the exclusive rights of patent owners stands in sharp contrast to the growing number of statutory provisions that achieve that effect.'[45] Finally, Mueller commented that the Court of Appeals for the Federal Circuit was disinclined to take notice of international developments: 'The Federal Circuit's decisions are rarely influenced by international patent law norms.'[46]

C The Supreme Court of the United States

Duke University sought special leave from the Supreme Court of the United States to appeal against the decision. A range of universities and research institutions led by the Association of American Medical Colleges put forward an amicus brief in support of the petitioner.[47] The brief stressed that the decision would have a significant impact upon academic scientific research. Counsel Joseph Keyes and Keith Jones lamented:

> The Federal Circuit's decision limiting the scope of the common law experimental use exemption from liability for patent infringement is of immense importance to all universities whose faculties engage in scientific research. By effectively eliminating the exemption for even noncommercial academic scientific research, the decision erects a significant roadblock by the advancement of science. The amici curiae are deeply disturbed by this ruling. In the past, university-based research has been crucial to scientific progress on almost every front. The decision below threatens to stifle that research and thereby endanger this nation's continued leadership in science and technology. The question presented by this case is vital to the nation's scientific well-being.[48]

There were four themes to this case. First, the amicae brief submitted that the experimental use exemption had historically protected non-commercial research from claims of patent infringement. Second, it argued that the decision of the federal circuit represented a radical departure from prior case law. Third, it submitted that the decision of the Federal Circuit would have a significant chilling effect on academic scientific research, especially in biotechnology and biomedicine. In its view, 'universities will be forced to bear substantial administrative and financial costs to cover patent searches, infringement opinions, licensing agreements, and the inevitable litigation that will be engendered by the Federal Circuit's new rule of patent law'.[49] Finally, it called upon the Supreme Court of the United States to restore the federal common law experimental use exemption to its traditional role as a safe haven for non-commercial scientific inquiry.

Ralph Nader's Consumer Project on Technology and the advocacy group Public Knowledge submitted that the Supreme Court of the United

States should grant certiorari in this case.[50] Counsel Joshua Sarnoff, from an intellectual property law clinic in Washington University, stressed the importance of basic research:

> Basic research is foundational and leads to further discoveries and important inventions. Knowledge is transferred from basic research to additional basic research, applied research, and technology development. Access to patented technologies and other research inputs is essential for basic research. Patent licensing practices increasingly prevent or discourage access to these inputs, precluding or delaying basic research that leads to life-saving technologies and other benefits.[51]

The lawyer proffered three main arguments. First, Sarnoff argued that the decision of the Federal Circuit frustrates the constitutional goal of patent law to 'promote the Progress of Science and useful Arts'. Second, he maintained that the narrow interpretation of experimental use threatened basic research, scientific progress and public health. Finally, he submitted that the experimental use exception should progress, like the defence of fair use in copyright law. The Consumer Project on Technology and Public Knowledge pleaded: 'The court now has an opportunity to ensure that the experimental use exception functions to promote progress.'[52]

Significantly, though, a number of universities and research institutions declined to join this amicus brief. Most notably, the Wisconsin Alumni Research Foundation (WARF) refused to support the case of Duke University, because of concerns that the challenge would adversely affect industry investment in university research.[53] Managing director Carl Gulbrandsen observed: 'We believe it's a mistake to say [to industry] you need to pay us for intellectual property but we aren't going to pay you, because we're a university.'[54] General counsel Elizabeth Donley warned her scientists: 'Researchers should not rely on the research exemption when creating materials that are later disclosed and licensed or sold for commercial purposes.'[55] The technology transfer unit had significant patent holdings, particularly in the field of stem cell research. It manages 800 issued patents and 600 pending applications, as well as being involved in 30 spin-off companies.

The Solicitor General of the United States submitted that the petition for a writ of certiorari should be denied.[56] He presented a few arguments to support this case. First, the Solicitor General argued that the decision of the court of appeals was not directly contrary to prior case law applying the experimental use defence. Second, he contended that the petitioner's broad and speculative policy concerns were a matter for legislative, rather than judicial, consideration. Finally, he said that the interlocutory posture and unusual genesis of the case also counselled against granting the review. The Solicitor General summarized his case accordingly:

> To date, the common law experimental use defense has been applied infrequently by lower courts and only as a narrow exception to the general statutory prohibition on patent infringement. The Federal Circuit's treatment of that defense in this case is generally in line with the lower court case law that has developed in this area. While petitioner asserts that a more robust exception for experimental use is needed to accommodate university research in particular, the existing case law does not establish such an exception and any substantial altering of the balance between the goals of the patent laws and the demands of academic research calls for judgments that are legislative, not judicial, in nature.[57]

The Solicitor General doubted the capacity of the Supreme Court of the United States to fashion a comprehensive solution. It asserted that 'it seems improbable that a 190-year-old, judge-made defense with little rooting in any statutory text could anticipate the challenges of the modern academic and research environment and adequately accommodate the competing policy concerns raised by the parties in this case.'[58]

The Supreme Court of the United States refused to hear an appeal against the decision of the Court of Appeals. It is unfortunate that the judges declined to intervene in the matter. The Supreme Court missed an opportunity to modernize the defence of experimental use in light of developments in new technologies. The judges could have articulated the philosophical underpinnings of the research exemption. The Supreme Court could have also defined the nature and scope of the defence of experimental use. The judges could have offered further guidance and direction to the lower courts as to the distinction between experimental use and infringing activity. The Supreme Court could have reined in the Federal Circuit over its excessive emphasis on commercial intent. As the Dean of the Medical School at Duke University, R. Sanders Williams, observed: 'There's a nobility of science and a magic of academia, which could be threatened by putting all university activity in the same cold glare as the corporate world.'[59]

The Supreme Court of the United States refused to hear an appeal against the decision of the Court of Appeals. Duke University was disappointed by this decision. The Associate Director of News and Communications, David Jarmul, observed:

> The Supreme Court's denial of certiorari means that Duke and other universities must now confront the issue of what the Federal Circuit decision will mean for scientific research. Unless the Congress provides a legislative remedy, universities may need to alter their research practices to such an extent that basic scientific research cannot continue on a consistent course. This challenge, which some universities and national organizations have been exploring in a general way, now becomes more urgent. Although Duke regrets that the Supreme Court declined this opportunity to clarify a legal question of such importance to the research community and society generally, it remains optimistic that its position in the case will eventually prevail.[60]

The case returned to the federal district court for further litigation. Duke University sought to renew its defence through the experimental use issue and explore other possible legal arguments.

D Remand

On remand, in September 2004, Beaty J of the District Court of North Carolina applied the approach of the Court of Appeals for the Federal Circuit, and held that Duke University had failed to demonstrate that it was entitled to the defence of experimental use.[61] His Honour commented that the case of the university was weak:

> The Court finds that further motions or supplementation of the record with respect to the experimental use defense at this time would be unnecessary. Duke has offered absolutely no evidence showing that it is entitled to this defense. Rather, as discussed above, it has actually conceded that this research was done in furtherance of its legitimate business purposes, that is, educating its students, and it has offered no indication of any evidence it could offer to support this defense ... Given the Federal Circuit's extremely narrow conception of the experimental use defense and the total lack of evidence currently in the record, the Court has doubts about whether Duke will be able to provide any evidence in support of its experimental use defense.[62]

In addition to its reliance on the defence of experience use, Duke University also claimed that it had a licence to work the patents for government research purposes pursuant to the *Bayh-Dole Act* 1980 (US). However, Beaty J observed: 'Given the conflicting evidence presented by the parties, Duke has failed to establish its entitlement to the government license defense at this time.'[63] The trial judge therefore refused to grant the motions of Duke University for summary judgment. The court did allow the parties additional time to supplement their discovery responses and pre-trial disclosures.

In 2006, Duke University reached a settlement with John Madey. The patented laser was sent to Madey's laboratory at the University of Hawaii. Triumphant, Madey commented: 'I always knew it would happen. It was just a question of when.'[64]

II THE BOLAR EXCEPTION: *MERCK KGAA v INTEGRA LIFESCIENCES I, LTD.*

In *Roche Products, Inc. v Bolar Pharmaceuticals Co., Inc*, Bolar planned to market a generic version of a pharmaceutical drug, which was the subject of a patent about to expire.[65] It engaged in testing with the patented pharmaceutical drug in order to prepare a regulatory application to the

United States Food and Drug Administration. Roche sued Bolar for the infringement of its patents on the active ingredients of a prescription sleeping aid called Dalmane.[66] Amongst other things, Bolar argued that its activities were protected by the defence of experimental use.

The Court of Appeals for the Federal Circuit held that Bolar was liable for patent infringement, and could not avail itself of the defence of experimental use. Taking a literal reading of historical precedent, Nichols J adopted a narrow construction of the defence of experimental use:

> Bolar's intended 'experimental' use is solely for business reasons and not for amusement, to satisfy idle curiosity, or for strictly philosophical inquiry . . . It is obvious here that it is a misnomer to call the intended use *de minimis*. It is no trifle in its economic effect on the parties even if the quantity used is small. It is no dilettante affair such as Justice Story envisioned. We cannot construe the experimental use rule so broadly as to allow a violation of the patent laws in the guise of 'scientific inquiry', when that inquiry has definite, cognizable, and not insubstantial commercial purposes.[67]

The judge observed: 'Despite Bolar's argument that its tests are "true scientific inquiries" to which a literal interpretation of the experimental use exception logically should extend, we hold the experimental use exception to be truly narrow, and we will not expand it under the present circumstances.'[68] The Supreme Court of the United States refused to grant leave for an appeal against the decision of the United States Court of Appeals for the Federal Circuit.[69]

In the wake of this decision, the United States Congress passed the *Drug Price Competition and Patent Term Restoration Act (the Hatch–Waxman Act)* 1984 (US). The legislation created an exception in section 271 (e)(1) of the *Patent Act* 1952 (US), namely that 'it shall not be an act of infringement to make, use, offer to sell, or sell within the United States or import into the United States a patented invention . . . solely for uses reasonably related to the development and submission of information under a Federal law which regulates the manufacture, use or sale of drugs or veterinary biological products'.

The Supreme Court of the United States has since interpreted this exception to encompass not only regulatory data gathering on pharmaceuticals, but also the comparable testing of medical devices.[70] The United States Government has been assiduous in its efforts to export this scheme to other jurisdictions. The Canadian Government and the Australian Government have also linked their patent systems to the marketing regimes for pharmaceutical drugs. The European Union has also passed a directive, which creates a Bolar exception in respect of research on pharmaceutical drugs for regulatory approval.[71]

The Supreme Court of the United States ruled upon the scope of this safe harbour exemption in the case of *Merck v Integra Lifesciences I Ltd.*

A District Court and Court of Appeals for the Federal Circuit

In the case of *Integra Lifesciences I Ltd. v Merck KgaA*,[72] Integra Lifesciences I Ltd, the owner of patents for a pharmacologically useful peptide, sued competitors for patent infringement.[73] In response, Merck argued that the safe harbour in section 271(e)(1) of the *Patent Act* 1952 (US) was a defence to the allegations of patent infringement.

In the United States District Court for the Southern District of California, Fitzgerald J held that the patents were infringed, though one was invalid, and entered judgment on jury's damage award.[74]

In the Court of Appeals, Rader J for the majority held that the competitors' use of patented peptide in experiments to identify best drug candidate to subject to future clinical testing under Food and Drug Administration (FDA) processes did not come within statutory safe harbour. In addition, the district court correctly construed the term 'peptide' to have its full ordinary meaning in the art. However, the district court erred in denying Merck's motion for reconsideration of an appropriate reasonable royalty. Therefore, the court remanded for further consideration of the damages issue.

In the marginalia, Rader J was adamant that it was inappropriate to view the case in light of the defence of experimental use under patent law. His Honour observed:

> In her dissent, Judge Newman takes this opportunity to restate her dissatisfaction with this court's decision in *Madey* v. *Duke University*. However, the common law experimental use exception is not before the court in the instant case. The issue before the jury was whether the infringing pre-clinical experiments are immunized from liability via the 'FDA exemption', i.e., 35 U.S.C. §271(e)(1) ... Judge Newman's dissent, however, does not mention that the Patent Act does not include the word 'experimental', let alone an experimental use exemption from infringement. Nor does Judge Newman's dissent note that the judge-made doctrine is rooted in the notions of de minimis infringement better addressed by limited damages.[75]

Furthermore, Rader J challenged the argument that Integra's patented peptides did not constitute research tools. In footnote four, he provided this rebuttal: 'The dissent does not explain why one of those "certain uses" cannot embrace use of an RGD peptide as a laboratory tool to facilitate the identification of a new therapeutic.'[76]

In dissent, Newman J maintained that the case provided a good illustration of the issues at stake in respect of patent law and experimental use:

'The majority's prohibition of all research into patented subject matter is as impractical as it is incorrect.'[77] She protested that the virtual elimination of the common law defence of experimental use was 'ill-suited to today's research-founded, technology-based economy'.[78] Newman J observed:

> This case raises a question of the nature and application of the common law research exemption, an exemption from infringement that arose in judge-made law almost two centuries ago, and that recently has come into sharper focus. Its correct treatment can affect research institutions, research-dependent industry, and scientific progress. The question is whether, and to what extent, the patentee's permission is required in order to study that which is patented.[79]

Articulating the information function of patent law, Newman J observed: 'Study of patented information is essential to the creation of new knowledge, thereby achieving further scientific and technologic progress.'[80] Her Honour emphasized that there should be a broad defence of experimental use in light of such public purposes of patent law: 'The purpose of a patent system is not only to provide a financial incentive to create new knowledge and bring it to public benefit through new products; it also serves to add to the body of published scientific/technologic knowledge.'[81] Newman J contended that a narrow interpretation of the defence of experimental use would be detrimental to the dissemination of information, and the promotion of scientific understanding: 'Prohibition of research into such knowledge cannot be squared with the framework of the patent law.'[82]

In her dissent, Newman J noted that 'the common law exception is not unlimited' because 'it must preserve the patentee's incentive to innovate, an incentive secured only by the right to exclude'.[83] Newman J recoils from the position that a research exemption could not encompass certain commercial uses. She observes that, 'while that threshold invention may (as here) exact tribute from or enjoin commercial and pre-commercial activity, the patent does not bar all research that precedes such activity'.[84] Newman J acknowledged that determining the scope of the common law defence of experimental use was a difficult task, which is dependent upon a number of variables: 'Setting the boundaries of a common law exemption requires careful understanding of the mechanisms of the creation, development, and use of technical knowledge, and of today's complexity of interactions among invention and the innovating fruits of invention.'[85] She declines to define the boundaries of the research exemption for all purposes and all activities. Newman J does observe, though, 'that there is a generally recognized distinction between "research" and "development", as a matter of scale, creativity, resource allocation, and often the level of scientific/ engineering skill needed for the project; this distinction may serve as a useful divider, applicable in most situations'.[86] She concludes: 'Like "fair use" in copyright law, the great variety of possible facts may occasionally raise dispute as to particular cases. However, also like fair use, in most cases it will be clear whether the exemption applies.'[87]

B The Supreme Court of the United States

The company Merck obtained special leave from the Supreme Court of the United States to appeal against the decision.

The Supreme Court of the United States hearing on the scope of the safe harbour exception attracted a wide range of amicus curiae briefs from research organizations, pharmaceutical drug manufacturers, biotechnology companies, legal associations and public interest groups.

A number of public interest groups – including the Consumer Project on Technology, the Electronic Frontier Foundation and the Public Knowledge – put in an amicus curiae brief in support of the petitioner.[88] The attorney, Joshua Sarnoff, supported a broad reading of the safe harbour for research on pharmaceutical drugs.[89] He contended that the Supreme Court of the United States should confirm that both the particular safe harbour and the general defence of experimental use applied in this case. Sarnoff observed that it was important to consider the interplay between the particular safe harbour for research on pharmaceutical drugs with the general defence of experimental use:

> This Court should effectuate the legislative policy present since the inception of the Patent Act that the Federal Circuit's interpretations of the experimental use exception continue to subvert. By repudiating the Federal Circuit's narrow interpretations and by confirming that Congress intended a broad experimental use exception, the Court will better assure that patents do not chill scientific research and competitive evaluation through the threat of litigation and the tax of licensing. The Court will thereby effectuate the legislative balance designed to 'Promote the Progress of Science and useful Arts.'[90]

The amicus curiae commented: 'Scientific discovery, competitive improvement, and public health are being adversely affected by the Federal Circuit's narrow interpretations of the experimental use exception.'[91] The friends of the court noted that other jurisdictions had adopted a broader reading of the defence of experimental use: 'The narrow interpretations of the experimental use exception by the Federal Circuit also place the United States in conflict with the international community, and are likely to result in scientific research, patent rights, and wealth leaving the United States.'[92]

Several academic luminaries (Rochelle Cooper Dreyfuss, John Duffy, Arti Rai and Katherine Strandburg) put forward an amicus curiae submission.[93] The professoriate suggested that it was unnecessary for the Supreme

Court of the United States to rule upon the scope of the general defence of experimental use.[94] The academics expressed reservations that an overly narrow interpretation of the traditional experimental use exemption threatened to upset long-standing practices of the research community: 'The Federal Circuit's recent narrowing trend has thus been met with consternation by researchers, the intellectual property bar, and intellectual property scholars.'[95]

In *Merck KGaA v Integra Lifesciences I, Ltd.*, Scalia, J delivered a pithy judgment on behalf of a unanimous Supreme Court of the United States.[96] His Honour summed up the issue thus: 'This case presents the question whether uses of patented inventions in preclinical research, the results of which are not ultimately included in a submission to the Food and Drug Administration (FDA), are exempted from infringement by 35 U.S.C. §271(e)(1).'[97]

First, Scalia J emphasized that the safe harbour provision should be read broadly to include the use of patented inventions in a manner reasonably related to the federal regulatory process:

> Though the contours of this provision are not exact in every respect, the statutory text makes clear that it provides a wide berth for the use of patented drugs in activities related to the federal regulatory process. As an initial matter, we think it apparent from the statutory text that §271(e)(1)'s exemption from infringement extends to all uses of patented inventions that are reasonably related to the development and submission of *any* information under the FDCA. This necessarily includes preclinical studies of patented compounds that are appropriate for submission to the FDA in the regulatory process. There is simply no room in the statute for excluding certain information from the exemption on the basis of the phase of research in which it is developed or the particular submission in which it could be included.[98]

The judge endorsed the ruling in *Eli Lilly*,[99] which declined to limit §271(e)(1)'s exemption from infringement to submissions under particular statutory provisions that regulate drugs.

Second, Scalia J emphasized that the safe harbour exemption was not limited to only preclinical data pertaining to safety of drug in humans, for two reasons:

> First, the FDA's requirement that preclinical studies be conducted under 'good laboratory practices' applies only to experiments on drugs 'to determine their safety.' The good laboratory practice regulations do not apply to preclinical studies of a drug's efficacy, mechanism of action, pharmacology, or pharmacokinetics. Second, FDA regulations do not provide that even safety-related experiments not conducted in compliance with good laboratory practices regulations are not suitable for submission in an IND. Rather, such studies must include 'a brief statement of the reason for the noncompliance'.[100]

The judge rejected the argument of the respondents that the experiments in question here are necessarily disqualified because they were not conducted in conformity with the FDA's good laboratory practices regulations.

Third, Scalia J commented that the safe harbour exemption can extend to experimentation on drugs, which are not ultimately the subject of an FDA submission:

> Congress did not limit §271(e)(1)'s safe harbor to the development of information for inclusion in a submission to the FDA; nor did it create an exemption applicable only to the research relevant to filing an ANDA for approval of a generic drug. Rather, it exempted from infringement *all* uses of patented compounds 'reasonably related' to the process of developing information for submission under *any* federal law regulating the manufacture, use, or distribution of drugs. We decline to read the 'reasonable relation' requirement so narrowly as to render §271(e)(1)'s stated protection of activities leading to FDA approval for all drugs illusory. Properly construed, §271(e)(1) leaves adequate space for experimentation and failure on the road to regulatory approval.[101]

The judge recognized 'the reality that, even at late stages in the development of a new drug, scientific testing is a process of trial and error'.[102] His Honour observed that drug manufacturers and researchers needed the freedom to conduct experiments to test the viability of drug candidates: 'In the vast majority of cases, neither the drugmaker nor its scientists have any way of knowing whether an initially promising candidate will prove successful over a battery of experiments.'[103]

Fourth, Scalia J held that the safe harbour exemption can extend to the use of patented compounds in experiments that are not ultimately submitted to the FDA:

> The use of a patented compound in experiments that are not themselves included in a 'submission of information' to the FDA does not, standing alone, render the use infringing. The relationship of the use of a patented compound in a particular experiment to the 'development and submission of information' to the FDA does not become more attenuated (or less reasonable) simply because the data from that experiment are left out of the submission that is ultimately passed along to the FDA. Moreover, many of the uncertainties that exist with respect to the selection of a specific drug exist as well with respect to the decision of what research to include in an IND or NDA. As a District Court has observed, '[I]t will not always be clear to parties setting out to seek FDA approval for their new product exactly which kinds of information, and in what quantities, it will take to win that agency's approval.' This is especially true at the preclinical stage of drug approval.[104]

The judge supported the submission of the United States Government that the use of patented compounds in preclinical studies is protected under §271(e)(1) as long as there is a reasonable basis for believing that the experiments will produce 'the types of information that are relevant to an IND or NDA'.[105]

Finally, the judge addressed the controversial topic of research tools. In footnote 7, Scalia J emphasized that the facts of the case did not pertain to research tools:

> The Court of Appeals also suggested that a limited construction of §271(e)(1) is necessary to avoid depriving so-called 'research tools' of the complete value of their patents. Respondents have never argued the RGD peptides were used at Scripps as research tools, and it is apparent from the record that they were not.[106]

The judge noted the earlier comments of Newman J that 'use of an existing tool in one's research is quite different from the study of the tool itself'.[107] His Honour concluded: 'We therefore need not – and do not – express a view about whether, or to what extent, §271(e)(1) exempts from infringement the use of "research tools" in the development of information for the regulatory process.'[108] Such a decision would in part allay and cool the fears of various research tool developers and biotechnology firms.

C Reinstatement of Appeal

Taking heed of the decision of the Supreme Court of the United States, the Court of Appeals for the Federal Circuit reinstated the appeal, and sought submissions from the parties and amicus curiae.[109]

The decision of the Supreme Court of the United States in *Merck KGaA v Integra Lifesciences I, Ltd.*, has been hailed as a boon to consumers and patients by maintaining a vibrant environment in the United States for drug testing.[110] Some commentators, though, wondered whether the judges had left the exact boundaries of the safe harbour vague and ill-defined.[111] Harold Wegner comments: 'The jagged and amorphous upstream boundary of the safe harbor represents a business person's worst dream, an invitation to test such boundaries as opposed to the opportunity to operate under a clear set of defined rights.'[112] The intellectual property lawyer doubts whether there would be sufficient consensus for a legislative compromise on the issue: 'There are far too many diverse interests in the mix to create a consensus to support any one position.'[113]

The ruling of the Supreme Court of the United States in *Merck KGaA v Integra Lifesciences I, Ltd.*, has earned its detractors in sections of the biotechnology industry, the pharmaceutical drugs industry and the research tool industry.[114] The ruling has also caused some consternation amongst judges in lower courts. In an academic comment, Gajarsa J of United States Court of Appeals for the Federal Circuit, wondered whether the decision heralded a more interventionist approach by the Supreme Court of the United States: 'Although only time will tell, perhaps we are witnessing the beginning of what will become a comprehensive Supreme Court "reform" of this country's patent law jurisprudence.'[115] As the author of the decision in *Madey v Duke University*, Gajarsa J was no doubt wary that his judicial interpretation of the scope of the general defence of experience use could be subject to future scrutiny by the Supreme Court of the United States.

CONCLUSION

In the United States, there has been much controversy over both the common law defence of experimental use and the statutory safe harbour for research into pharmaceutical drugs, in a series of high profile decisions. In *Merck KGaA v Integra Lifesciences I, Ltd.*, the Supreme Court of the United States resurrected the statutory safe harbour for research in respect of pharmaceutical drugs. Curiously, Scalia J did not adequately address the larger question of the scope of the general defence of experimental use, which had been read down in a procrustean fashion by the Court of Appeals for the Federal Circuit in *Madey v Duke University*. The judge does not even provide guidance as to the proper relationship between the safe harbour for research on pharmaceutical drugs and the general defence of experimental use. This omission is regrettable. Academic commentator Rebecca Lynn laments:

> Unfortunately, *Merck* does not provide adequate guidance on the boundaries of the experimental use defense for pharmaceutical companies or research tool companies, nor does it provide protection for experimental use outside of the areas of biotechnology and medical devices. Given the inability of the courts to clearly define an experimental use exemption, and their attempt to stretch the current statutory exemption for generic drugs far beyond its intended purpose, it is time for Congress to enact a broad experimental use defense.[116]

The defence of experimental use has been put under strain by the increasing commercialization of universities and research institutions. The blurring between the private and public sectors should not be cause to abolish the defence altogether. Rather, such collaborations should provide an opportunity to reconceptualize the defence of experimental use. The proposals of the Federal Trade Commission provide an opportunity to clarify the nature and scope of the research exemption.[117] In the analysis of experimental use, there should be a shift away from the overweening emphasis upon the commercial use of a patented invention. There should be instead a new focus upon the relationship between a patented invention and the experimental use. As Tim Sampson has observed: 'It would . . . have been better for the United States to adopt the European approach to experimental use, which seeks to strike a balance between the non-commercial and commercial phases of research.'[118]

The *TRIPS Agreement* 1994 provides a fair degree of flexibility in respect of exceptions to patent rights. Article 30 noted that 'members may provide limited exceptions to the exclusive rights conferred by a patent, provided that such exceptions do not unreasonably conflict with a normal exploitation of the patent and do not unreasonably prejudice the legitimate interests of the patent owner, taking account of the legitimate interests of third parties'. The World Trade Organization Panel decision in the *Canada–Patent Protection* case provides some guidance on the allowable extent of research exemptions under the *TRIPS Agreement* 1994.[119] The Panel observed that 'practically all Members of the WTO had such an exception albeit drafted in a great variety of ways'.[120] The Panel observed that the exception was limited in character and narrowly defined: 'It only applied to typically one out of five patent rights referred to in Article 28.1 of the *TRIPS Agreement*, since only use was permissible, while offering for sale, selling and importing were not permissible.'[121] The Panel maintained that the patent holder's legitimate interests do not include a monopoly on research: 'Given that the "basic patent deal" required the patentee to disclose his invention to the public and to accept that it served as the basis for further research, it could be reasonably argued that a "research monopoly" was not included in his legitimate interests and, therefore, the interests of third parties and their balancing with the patentee's interests appeared to be redundant for the research exception.'[122] The emerging norms in the European Union would be the best international model for a broad-based research exemption. Ideally, such reforms should take place at a multilateral level at forums such as the World Trade Organization and the World Intellectual Property Organization.

NOTES

1. Thoreau, Henry David (1854), *Walden, or Life in the Woods*, New York: Signet Classic.
2. Ibid., p. 15.
3. Mueller, J. (2001), 'No "dilettante affair": rethinking the experimental use exception to patent infringement for biomedical research tools', *Washington Law Review*, **76**, 1–66; see also Mueller, J. (2004), 'The evanescent experimental use exemption from U.S. patent infringement liability: implications for university/nonprofit research and development', *Baylor Law Review*, **56**, 918–79.
4. *Whittemore v Cutter* 29 F. Cas. 1120 (1813).
5. *Sawin v Guild* 21 Fed. Cas. 554 (1813).

6. *Sawin v Guild* 21 Fed. Cas. 554 (1813).
7. *Sawin v Guild* 21 Fed. Cas. 554 (1813).
8. *Peppenhausen v Falke* 19 F. Cas. 1048, 1049 (C.C.S.D.N.Y. 1861).
9. *Folsom v Marsh* (1841) 9 Fed. Cas. 342.
10. Burk, D. and M. Lemley (2002), 'Is patent law technology-specific?', *Berkeley Technology Law Journal*, **17**, 1155–202.
11. Janis, M. (2003), 'Experimental use and the shape of patent rights for plant innovation'. Economics of Innovation and Science Policy, Department of Economics, Iowa State University, http://www.econ.iastate.edu/department/seminar/ispw/Janis-seminar-Fall-03.pdf.
12. *Monsanto v Stauffer Chemical Co* (UK) [1985] RPC 515; *Monsanto Co v Stauffer Chemical Co (NZ)* [1984] FSR 559; *Monsanto v Stauffer* [1987] FSR 57 and *Monsanto v Stauffer Japan* (1989) 20 IIC 91.
13. Cornish, William and David Llewelyn (2003), *Intellectual Property: Patents, Copyright, Trade Marks and Allied Rights*, London: Sweet and Maxwell, pp. 246–7.
14. Lentz, E. (2004), 'Pharmaceutical and biotechnology research after Integra and Madey', *Biotechnology Law Report*, **23**, 265–76.
15. Migliorini, R. (2006), 'The narrowed experimental use exception to patent infringement and its application to patented computer software', *Journal of the Patent and Trademark Office Society*, **88**, 523–46.
16. Freschi, G. (2005), 'Navigating the research exemption's safe harbor: Supreme Court to clarify scope implications for stem cell research in California', *Santa Clara Computer and High Technology Law Journal*, **21**, 855–99.
17. Brignati, M. (2006), 'Access to the safe harbor: bioterrorism, influenza, and the Supreme Court's Interpretation of the Research Exemption from Patent Infringement', *Journal of Intellectual Property Law*, **13**, 375–404.
18. Zovko, N. (2006), 'Nanotechnology and the experimental use defense to patent infringement', *McGeorge Law Review*, **37**, 129–56.
19. *Madey v Duke University* 307 F.3d 1351 (2002).
20. *Madey v Duke University* 307 F.3d 1351 (2002).
21. *Merck KGaA v Integra Lifesciences I, Ltd.* 545 U.S. 193 (2005).
22. Federal Trade Commission (2003), *To Promote Innovation: The Proper Balance of Competition and Patent Law and Policy*, Washington DC: Federal Trade Commission. http://www.ftc.gov/os/2003/10/innovationrpt.pdf; Eisenberg, R. (2003), 'Patent swords and shields', *Science*, **299**, 1018–19; Sampson, T. (2004), 'Madey, Integra and the Wealth Of Nations', *European Intellectual Property Review*, **26**(1), 1–6; Mueller, J. (2004), 'The evanescent experimental use exemption from U.S. patent infringement liability: implications for university/nonprofit research and development', *Baylor Law Review*, **56**, 918–80; Gitter, D. (2001), 'International conflicts over patenting human DNA sequences in the United States and the European Union: an argument for compulsory licensing and a fair use exemption', *New York University Law Review*, **76**, 1623–91; Janis, M. (2003), 'Experimental use and the shape of patent rights for plant innovation', Economics of Innovation and Science Policy, Department of Economics, Iowa State University, http://www.econ.iastate.edu/department/seminar/ispw/Janis-seminar-Fall-03.pdf; Resnik, D. (2003), 'Patents and the research exemption', *Science*, **299**, 821; Epstein, Richard (2003), 'Steady the course: property rights in genetic material', in F. Scott Kieff (ed.), *Perspectives on Properties of the Human Genome Project*, Amsterdam: Academic Press, Elsevier, p. 159; and Cooper Dreyfuss, Rochelle (2003), 'Varying the course in patenting genetic material: a counter-proposal to Richard Epstein's steady course', in F. Scott Kieff (ed.), *Perspectives on Properties of the Human Genome Project*, Amsterdam: Academic Press, Elsevier, p. 195.
23. Madey, J. and G. Westenskow (1984), 'A microwave electron gun', U.S. Patent No. 4,641,103.
24. Madey, J. and E. Szarmes (1991), 'Free-electron laser oscillator for simultaneous narrow spectral resolution and fast time resolution spectroscopy', U.S. Patent No. 5,130,994.

25. Duke University (2001), 'Appellate brief in *Madey v Duke University*', 2001 WL 34633573, 6.
26. Brickley, P. (2003), 'Patent rights wrangle puts law in question', *The Scientist*, **17**(42), http://www.the-scientist.com/yr 2003/mar/prof2_030310.html, 10 March.
27. Duke University (2001), 'Appellate brief in *Madey v Duke University*', 2001 WL 34633573, 11.
28. *Madey v Duke University* 266 F. Supp. 2d 420 (2001).
29. *Madey v Duke University* 266 F. Supp. 2d 420 at 425 (2001).
30. *Madey v Duke University* 266 F. Supp. 2d 420 at 423 (2001).
31. *Madey v Duke University* 266 F. Supp. 2d 420 at 423 (2001).
32. Madey (2001), 'Appellant's opening brief in *Madey v Duke University*', 22 October.
33. Duke University (2001), 'Appellate brief in *Madey v Duke University*', 2001 WL 34633573, 11–12.
34. Ibid., 18.
35. Ibid., 19.
36. *Madey v Duke University* 307 F.3d 1351 (2002).
37. *Madey v Duke University* 307 F.3d 1351 at 1362 (2002).
38. *Madey v Duke University* 307 F.3d 1351 at 1362 (2002).
39. *Madey v Duke University* 307 F.3d 1351 at 1362–1363 (2002).
40. *Madey v Duke University* 307 F.3d 1351 at 1363 (2002).
41. Mueller, J. (2004), 'The evanescent experimental use exemption from U.S. patent infringement liability: implications for university/nonprofit research and development', *Baylor Law Rev*iew, **56**, 918–80.
42. Ibid., 962.
43. Ibid., 963.
44. Ibid., 966.
45. Ibid., 968.
46. Ibid., 969.
47. Association of American Medical Colleges (2003), 'Brief for the Association of American Medical Colleges et al. petition for a writ of certiorari, *Duke University v Madey*', Supreme Court of the United States, http://www.aamc.org/newsroom/pressrel/patentbrief.pdf.
48. Ibid., p. 3.
49. Ibid., p. 3.
50. Consumer Project on Technology and Public Knowledge et al. (2003), 'Petition for a writ of certiorari by the Consumer Project on Technology and Public Knowledge in *Duke University v Madey*, Supreme Court of the United States', http://jurist.law.pitt.edu/amicus/duke_v_madey_cert_petition_au.pdf, January.
51. Ibid., p. 2.
52. Ibid., p. 15.
53. Wysocki, B. (2004), 'Cutting edge: a laser case sears universities' right to ignore patents', *The Wall Street Journal*, 11 October, A1.
54. Ibid.
55. Donley, B. (2003), 'Using patented materials in your research: what you should know', Wisconsin Alumni Research Foundation, 21 May.
56. United States Solicitor-General (2003), 'Brief for the United States as amicus curiae in Petition for a writ of certiorari in *Duke University v Madey*', Supreme Court of the United States, http://www.usdoj.gov/osg/briefs/2002/2pet/6invit/2002-1007.pet.ami.inv.html, May.
57. Ibid., 5.
58. Ibid., 16.
59. Wysocki, B. (2004), 'Cutting edge: a laser case sears universities' right to ignore patents', *The Wall Street Journal*, 11 October, A1.
60. Duke University (2003), 'Duke University statement on Supreme Court action in case involving academic research', 27 May.
61. *Madey v Duke University* WL 2148935 (2004).

62. *Madey v Duke University* WL 2148935 at 8 (2004).
63. *Madey v Duke University* WL 2148935 at 10 (2004).
64. Bhattacharjee, Y. (2007), 'In the courts', *Science*, **315**, 581b.
65. *Roche Products, Inc. v Bolar Pharmaceuticals Co., Inc* 733 F. 2d 858 (1984).
66. Archer, G., E. Fells, R. Fryer, W. Orange, E. Reeder and L.H. Strenbach (1964), 'Novel 1-and/or 4-Substituted Alkyl 5-Aromatic – 3H – 1,4 – Benzodiazepines and Benzodiazepine-2-Ones', US Patent No: 3,299,053.
67. *Roche Products, Inc. v Bolar Pharmaceuticals Co., Inc* 733 F. 2d 858 at 863 (1984).
68. *Roche Products, Inc. v Bolar Pharmaceuticals Co., Inc* 733 F. 2d 858 at 863 (1984).
69. *Bolar Pharmaceutical Co. Inc. v Roche Products Inc.* 469 U.S. 856 (1984).
70. *Eli Lilly & Co v Medtronic* 496 US 661 (1990).
71. *European Union Directive on the Community code relating to medicinal products for human use 2004/27/EC of 31 March 2004 amending Directive 2001/83/EC.* Article 10 (6) provides: 'Conducting the necessary studies and trials with a view to the application of paragraphs 1, 2, 3 and 4 and the consequential practical requirements shall not be regarded as contrary to patent rights or to supplementary protection certificates for medicinal products.'
72. *Integra Lifesciences Ltd v Merck KgaA* 331 F. 3d 860 (2003).
73. Ruoslahti, E. and M. Pierschbacher (1985), 'Tetrapeptide', U.S. Patent No: 4,792,525; Ruoslahti, E. and M. Pierschbacher (1985), 'Tetrapeptide', US Patent No: 5,695,997; Ruoslahti, E., E. Hayman and M. Pierschbacher (1985), 'Use of peptides in control of cell attachment and detachment', U.S. Patent No: 4,879,237; Ruoslahti, E., E. Hayman, and M. Pierschbacher (1987), 'Peptides in cell detachment and aggregation', US Patent No. 4,789,734; and Pierschbacher, M. (1985), 'Vitronectin specific cell receptor derived from mammalian mesenchymal tissue', US Patent No: 4,789,734.
74. *Integra Lifesciences Ltd v Merck KgaA* 1999 WL 398180 (1999).
75. *Integra Lifesciences Ltd v Merck KgaA* 331 F. 3d 860 at 864 (2003).
76. *Integra Lifesciences Ltd v Merck KgaA* 331 F. 3d 860 at 872 (2003).
77. *Integra Lifesciences Ltd v Merck KgaA* 331 F. 3d 860 at 875 (2003).
78. *Integra Lifesciences Ltd v Merck KgaA* 331 F. 3d 860 at 873 (2003).
79. *Integra Lifesciences Ltd v Merck KgaA* 331 F. 3d 860 at 872–873 (2003).
80. *Integra Lifesciences Ltd v Merck KgaA* 331 F. 3d 860 at 876 (2003).
81. *Integra Lifesciences Ltd v Merck KgaA* 331 F. 3d 860 at 873 (2003).
82. *Integra Lifesciences Ltd v Merck KgaA* 331 F. 3d 860 at 875 (2003).
83. *Integra Lifesciences Ltd v Merck KgaA* 331 F. 3d 860 at 876 (2003).
84. *Integra Lifesciences Ltd v Merck KgaA* 331 F. 3d 860 at 876 (2003).
85. *Integra Lifesciences Ltd v Merck KgaA* 331 F. 3d 860 at 876 (2003).
86. *Integra Lifesciences Ltd v Merck KgaA* 331 F. 3d 860 at 876 (2003).
87. *Integra Lifesciences Ltd v Merck KgaA* 331 F. 3d 860 at 876 (2003).
88. Consumer Project on Technology (2005), 'Brief of amici curiae Consumer Project on Technology, Electronic Frontier Foundation and Public Knowledge in support of petitioner in *Merck KgaA v Integra Lifesciences I, Ltd*', 2005 WL 435894, 22 February.
89. Ibid., 2–3.
90. Ibid., 29.
91. Ibid., 27.
92. Ibid., 28.
93. Intellectual Property Professors (2005), 'Brief of intellectual property professors as amici curiae in support of neither party in *Merck KgaA v Integra Lifesciences I, Ltd*', 2005 WL 435892, 22 February.
94. Ibid., 2.
95. Ibid., 11.
96. *Merck KGaA v Integra Lifesciences I, Ltd.* 545 U.S. 193, 125 S.Ct. 2372 U.S. (2005).
97. *Merck KGaA v Integra Lifesciences I, Ltd.* 545 U.S. 193, 125 S.Ct. 2372 U.S. (2005).
98. *Merck KGaA v Integra Lifesciences I, Ltd.* 545 U.S. 193, 125 S.Ct. 2372 U.S. (2005).
99. *Eli Lilly & Co v Medtronic* 496 US 661 (1990).
100. *Merck KGaA v Integra Lifesciences I, Ltd.* 545 U.S. 193 at 204 and 205 (2005).

101. *Merck KGaA v Integra Lifesciences I, Ltd.* 545 U.S. 193 (2005).
102. *Merck KGaA v Integra Lifesciences I, Ltd.* 545 U.S. 193 (2005).
103. *Merck KGaA v Integra Lifesciences I, Ltd.* 545 U.S. 193 (2005).
104. *Merck KGaA v Integra Lifesciences I, Ltd.* 545 U.S. 193 at 207 and 208 (2005).
105. *Merck KGaA v Integra Lifesciences I, Ltd.* 545 U.S. 193 at 207 and 208 (2005).
106. *Merck KGaA v Integra Lifesciences I, Ltd.* 545 U.S. 193 (2005).
107. *Merck KGaA v Integra Lifesciences I, Ltd.* 545 U.S. 193 (2005).
108. *Merck KGaA v Integra Lifesciences I, Ltd.* 545 U.S. 193 (2005).
109. *Integra Lifesciences I, Ltd. v Merck KGaA* 421 F.3d 1289 C.A.Fed., (2005).
110. Mota, S. (2006), '*Merck v Integra Lifesciences* – the Supreme Court protects the use of patented compounds in preclinical studies', *Hamline Law Review*, **29**, 53–62; and O'Connor, D. and T. Valoir (2006), 'The Supreme Court tilts toward drug developers' drug discovery after *Merck v Integra*', *Chicago-Kent Journal of Intellectual Property*, **5**, 124–41.
111. Hareid, J. (2006), 'Testing drugs and testing limits: *Merck KGAA* v. *Integra Lifesciences I, Ltd.* and the scope of the Hatch–Waxman safe harbor provision', *Minnesota Journal of Law, Science & Technology*, **7**, 713–56.
112. Wegner, H. (2005), 'Post-Merck experimental use and the "Safe Harbor",' *Federal Circuit Bar Journal*, **15**, 1–36 at 36.
113. Ibid.
114. Helm, K. (2006), 'Outsourcing the fire of genius: the effects of patent infringement jurisprudence on pharmaceutical drug development', *Fordham Intellectual Property, Media and Entertainment Law Journal*, **17**, 153–206 at 180–81; Williams, T. (2006), '*Merck KGAA v Integra Lifesciences I, Ltd*: does the breadth of safe harbor protection toll the death knell for biotech research companies?', *Mercer Law Review*, **57**, 917–31; and McPherson, J. (2006), 'The impact of the Hatch–Waxman Act's safe harbor provision on biomedical research tools after *Merck KGAA. v Integra Lifesciences I. Ltd*', *Michigan State University Journal of Medicine & Law*, **10**, 369–83.
115. Gajarsa, A. and L. Cogswell (2006), 'A review of recent decisions of the United States Court of Appeals for the Federal Circuit', *American University Law Review*, **55**, 821–44 at 843–4.
116. Lynn, R. (2006), '*Merck KgAA v Integra Lifesciences I. Ltd*.: judicial expansion of §271 (*e*)(1) signals a need for a broad statutory experimental use exemption in patent law', *Berkeley Technology Law Journal*, **21**, 79–100 at 94.
117. Federal Trade Commission (2003), *To Promote Innovation: The Proper Balance of Competition and Patent Law and Policy*, Washington, DC: Federal Trade Commission, http://www.ftc.gov/os/2003/10/innovationrpt.pdf.
118. Sampson, T. (2004), 'Madey, Integra and the wealth of nations', *European Intellectual Property Review*, **26**(1), 1–6 at 6.
119. *Canada: Patent Protection of Pharmaceutical Products: Complaint by the European Communities and their Member States*, 17 March 2000, WT/DS114/R.
120. *Canada: Patent Protection of Pharmaceutical Products: Complaint by the European Communities and their Member States*, 17 March 2000, WT/DS114/R, pp. 55–6.
121. *Canada: Patent Protection of Pharmaceutical Products: Complaint by the European Communities and their Member States*, 17 March 2000, WT/DS114/R, p. 56.
122. *Canada: Patent Protection of Pharmaceutical Products: Complaint by the European Communities and their Member States*, 17 March 2000, WT/DS114/R, p. 56.

7. The Utah saints: patent law and genetic testing

In March 2000, a Washington production of Margaret Edson's Pulitzer-prize winning play *Wit*,[1] told the story of English literature professor, Vivian Bearing, and her struggle with ovarian cancer.[2] Tucked into the Playbill was an advertisement for Myriad Genetics Inc. It pictured an earnest woman with her left hand held against her right breast and asked: 'If you could discover your risk for a second breast cancer or for ovarian cancer, would you? There is no stronger antidote to fear than information.'[3]

The emergence of Myriad Genetics is instructive about the history of the biotechnology industry. Dr Mark Skolnick and Nobel Laureate Walter Gilbert founded the company in 1993.[4] Its focus was upon developing diagnostics, rather than pharmaceutical drugs. The company's press release from April 1994 stated:

> Myriad is establishing a genetic information business based on testing for genes which predispose individuals to major common diseases. The genetic information business represents a multi-billion dollar market opportunity for the Company just for the testing of individuals affected with disease and their family members. As genetic disease testing moves toward a general population screen, the market size increases dramatically.[5]

Myriad's official corporate mission statement proclaimed that it was 'building a worldwide business based on the discovery and commercialization of genes linked to major disorders such as cancer and heart disease'.[6] Although concentrating on breast and skin cancer, Myriad also vowed to find genes for prostate, lung and colon cancer, obesity and hypertension. The company states that it would 'capitalize on its discoveries, by providing testing and genetic information services' and 'develop human therapeutic products independently and in conjunction with commercial partners'.[7]

In 1994, Dr Mark Skolnick and the private biotechnology company, Myriad Genetics, were able to isolate the location of the genes BRCA1 and BRCA2 with the help of the genealogical records of the Utah Mormons, and the processing power of super-computers. Myriad Genetics applied for patents on the discovery, including a 'composition-of-matter' patent on the gene itself and a 'method-of-use' patent for the application of BRCA1 and

BRCA2 in the diagnostic and therapeutic arenas. Skolnick believed that such patents were vital in order to encourage private investment and entrepreneurship, which are playing a large role in fuelling genetic discoveries. He declared: 'If it's not patented you won't get some group to spend money to develop it, and you won't get a high-quality, inexpensive test.'[8]

Myriad Genetics has also lodged a number of other patents in the United States.[9] In the 2001 Annual Report of Myriad Genetics, the chairman Hugh D'Andrade and the chief executive officer Peter Meldrum stressed to their shareholders:

> Intellectual Property is vital to Myriad. Accordingly, we have stepped up the pace of patent submissions to match the rapid discovery rate of potential therapeutic targets. To date, the company has submitted patent applications on over 1,800 genes, proteins, protein interactions, potential drug candidates, and predictive medicine opportunities.[10]

Myriad Genetics, Inc. has also obtained a number of foreign patents covering the BRCA1 and BRCA2 breast and ovarian cancer genes and their use in the development of therapeutic and predictive medicine products. Meldrum of Myriad Genetics commented: 'International patents provide important protection for us and our marketing partners as we expand the availability of our predictive medicine products globally.'[11] The company has obtained patents in Europe,[12] Canada,[13] Australia,[14] New Zealand and Japan.

Myriad Genetics has maintained that its commercial genetic predisposition testing is a boon for patient care and the delivery of health care. Greg Critchfield, the President of the company, observed:

> The real question for Australia and many other countries is this – do we want to continue with the old health care system or do we want to use the power of the new genetics? Do we want to give women the ability to choose interventions that will make a difference and prevent cancer and overall save a lot more money for the health care system by spending a little bit more on the up-front technologies?[15]

Myriad Genetics has consistently denied that its control of patents in respect of genetic testing is, in any way, an impediment to scientific research. The Vice President of Marketing, William Risconi, reassured a sceptical audience at the National Academy of Sciences, that the Utah company had no intention to interfere with scientific research.[16]

A number of public researchers and scientists have expressed doubts about the validity of the patents held by Myriad Genetics, particularly in respect of BRCA1 and BRCA2. Public health officials in the European Union, North America and Australasia have been fearful that the patents may have an adverse impact upon patient access to treatment, research into breast cancer and the administration of health care.[17] Most notably, the French research institution, the Institut Curie, was concerned about the breadth of the patent granted to Myriad Genetics.[18] The Institut Curie argued that the approach of Myriad Genetics undermined the public health system in France and other European countries:

> Myriad's monopoly position has given rise to a market for genetics in the United States which tends to dissociate actual testing from genetic counselling and high risk patient care and follow-up. This approach goes very much against the way we view public health care, in France and in most other European countries, where clinicians work within a model which integrates biological research, clinical investigation, and patient care, taking into account the medical and psychological aspects of diagnosis as well as the clinical history of high risk patients and their families.[19]

The Institut Curie argued that the costs of commercial testing would be a barrier to patient care. It observed that the initial family mutation searches performed by Myriad are billed 2400 United States dollars (18 000 francs – 2744 euros),[20] as against an estimated cost of 5000 francs (762 euros) for testing in French laboratories, which makes Myriad three and a half times more expensive. The increase in the cost of genetic screening would be directly borne by hospital budgets or by medical insurance schemes, and would thus require considerable additional expenditure on the part of the French social and medical system.

The Institut Curie claimed that the monopoly on patent exploitation will lead to a loss of expertise and information among physicians and research scientists in Europe, as they would no longer be allowed to improve diagnostic technologies and methods, and would therefore not be in a position to further their research under acceptable circumstances.[21] The 'compulsory' sending to Myriad of DNA samples obtained from high-risk individuals would help the United States corporation build up its copyright-protected genetic data bank. This in turn would grant it unchallenged control over the main research materials concerning genes coding for breast and ovarian cancer predisposition, thereby allowing it to make further discoveries and ultimately filing further patent applications as a result of such discoveries.

The controversy over Myriad Genetics crystallized a long-standing debate over whether medical diagnostics should be patentable subject matter.[22] This chapter argues that the patents held by Myriad Genetics should be overturned because of problems with their validity and their harmful effects upon medical research and health care. This argument considers the competing points of view of researchers and scientists who are working in the field of genetic testing. This chapter is divided into three parts. Section one examines opposition to the patents held by Myriad Genetics in relation to genetic testing in respect of BRCA1. It focuses squarely upon the challenge by the Institut Curie and European governments to the validity of the patents. Section two considers the competing patent claims of Myriad Genetics and Cancer Research Campaign Technology in relation to BRCA2. It examines the controversy about genetic discrimination,[23] which has arisen after the European Patent Office limited Myriad Genetics' patent claims to Ashkenazi Jewish women. Section three examines legislative responses to the problems posed by Myriad Genetics in the context of the *European Union Directive on the Legal Protection of Biotechnological Inventions* 1998 (EU). Possible solutions in respect of an expanded defence of experimental use, and compulsory licensing, are canvassed.

I   BRCA1

In 1990, leading geneticist Mary Claire-King mapped the position of the gene BRCA1.[24] She narrowed down its position to a small region on chromosome 17 containing about 1000 genes. This scientific breakthrough launched a race between a number of rival scientific teams to sequence the gene.

King was determined to isolate the relevant gene for BRCA1. She collaborated with two of the leading geneticists in the United States: Anne Bowcock, an expert in family studies and molecular biology, and Francis Collins, who had been successful in cloning the gene for cystic fibrosis.

The European Breast Cancer consortium brought together the British Institute of Cancer Research, with researchers from France, the Netherlands, Germany, Switzerland and Scandinavia. Professor Rodney Scott from the John Hunter Hospital was part of this team of scientists searching for BRCA1. One member of the team had managed to find the gene for BRCA1, but had not managed to prove that it was the breast cancer culprit in time.

Dr Mark Skolnick and Myriad Genetics were the first to locate the gene responsible for BRCA1. The research group relied upon the immense genealogical resources that existed in Utah, and the extensive registry of tumours in the State. The Mormon Church made it a priority to research and preserve the family histories of its members. Skolnick created a vast computer database of genealogies, and linked that information with the Utah Cancer Registry. This allowed him to identify those families that were most likely to help the study of various cancers.

In the United States, the National Institutes of Health (NIH) decided that it was not in its best interests for Myriad to be awarded a potentially exclusive patent on the BRCA1 gene.[25] The Myriad patent application did not include the names of NIH co-discoverers – Roger Wiseman and Andrew Futreal – because the inclusion of government-supported scientists would prevent the company from being awarded exclusive rights.[26] The NIH decided to file their own patent application on the gene, including the names of Wiseman and Futreal, and seven other key scientists from Utah. Harold Varmus said: 'We have taken all necessary measures to ensure that the government's contribution is recognized and to maximize the public benefit.'[27] Skolnick responded: 'If Wiseman and Futreal should be on the patent, we definitely want them on it, because we don't want to invalidate the patent.'[28] In February 1995, Myriad and the NIH finally agreed that Wiseman and Futreal should be added to the patent application. This settlement seemed to resolve the dispute in the United States. However, there remained opposition to the patents held by Myriad Genetics from research organizations and governments elsewhere in the world.

In the European Union, there have been a number of opposition proceedings against patents granted to Myriad Genetics Inc. in respect of diagnostic genetic testing relating to BRCA1.

A *The Institut Curie v Myriad Genetics Inc.* (EP 699754)

Myriad Genetics applied for patents on the discovery of BRCA1, including a 'composition-of-matter' patent on the gene itself and a 'method-of-use' patent for the application of BRCA1 in the diagnostic and therapeutic arena.[29] The abstract of the patent application gives a sense of the sheer breadth and scope of the claims: 'Specifically, the present invention relates to methods and materials used to isolate and detect a human breast and ovarian cancer predisposing gene (BRCA1), some mutant alleles of which cause susceptibility to cancer, in particular, breast and ovarian cancer.'[30]

The patent application stipulates a number of claims related to methods for diagnosing predisposition for breast cancer and ovarian cancer in a human subject. The patent application envisages a number of methods of use, ranging from the preparation of recombinant or chemically synthesized nucleic acids, to nucleic acid and peptide diagnosis and diagnostic kits, to drug screening, rational drug design and gene as well as peptide therapy. For therapies relying on BRCA1, Myriad sold those rights to Indianapolis-based Eli Lilly and Co. but retained rights for sequencing BRCA1.[31] The Utah biotechnology company has granted an exclusive licence to Genetic Technologies Limited to use and exploit its medical diagnostics in Australia.

There are a number of foreign equivalents to this European patent for a method for diagnosing a predisposition for breast and ovarian cancer.[32] In France, the Institut Curie – the famous medical and research institution named after dual Nobel prize winner, Marie Curie – initiated an opposition procedure against patent EP 699754 granted to Myriad Genetics for a method for diagnosing a predisposition for breast and ovarian cancer associated with the BRCA1 gene.[33] The Institut Curie challenged the patent granted by the European Patent Office on three grounds: lack of novelty, lack of inventive step and insufficient description. The action was also supported by the Hôpitaux de Paris and the Institut Gustave Roussy of France, the Belgian Society of Human Genetics, and the Associazone Angelasserra per la Ricerca sul Cancro of Italy. The opponents highlighted discrepancies in ten DNA letters between Myriad's original 1994 patent application, and the BRCA1 gene sequence described in Myriad's patent, issued in 2001. The opponents stressed that it was not until 1995 that Myriad submitted an updated sequence matching exactly the one in the issued patent. By that time, the crucial sequence had already been published openly on the scientific database, GenBank – so-called 'prior art'. After holding a hearing, the Opposition Division of the European Patent Office revoked the patent, EP 699754.[34]

First, the Institut Curie argued that the patent held by Myriad Genetics was invalid because it suffered from a lack of novelty arising from the availability of a number of predisposition tests based on indirect methods.

The Opposition Division of the European Patent Office considered whether the patent application complied under Article 54 of the *European Patent Convention* 1973 (EU) with the requirement for novelty.[35] The Opposition Division noted that there was great debate as to whether a full and complete version of the genetic sequence was first disclosed by Myriad Genetics Inc. or published on the genetic database, GenBank.[36] The Opposition Division observed that 'there arises some doubt as to the exact nature of the BRCA1 sequences which were available from GenBank at the time when D1 and D2 were published'.[37] The Opposition Division had to weigh 'the probability that the present inventors submitted a BRCA1 sequence to GenBank that was not only different to the one they were using for their own priority filings at the time but also different to the one they themselves published in D2'.[38] In the end, the Opposition Division gave the benefit of the doubt to the patent applicant on the question of novelty: 'Therefore, due to the absence of sufficient evidence from the Opponents to substantiate their allegations, the Opposition Division finds itself unable to reach a decision on this point based on the Opponents' arguments alone or to establish the facts of its own motion.'[39]

Second, the Institut Curie argued that the patent application lacked an inventive step: '[T]he patent application, as granted, has an excessively broad scope which does not correspond to the significance of Myriad's contribution to the public domain, at the date the patent was filed.'[40] The

Opposition Division of the European Patent Office held that the patent application failed to comply with Article 56 of the *European Patent Convention* 1973 (EU), which requires an inventive step:

> In light of D2 the skilled person would be aware that he could use any BRCA1 reference sequence to carry out the claimed methods, provided it allowed him to determine whether or not the three predisposing mutations were present, i.e. it would have to be unchanged (or 'wild-type') with respect to the D2 reference sequence at these positions.[41]

The Opposition Division noted that the requirement of an 'inventive step' had a higher threshold than mere 'novelty' because it addressed the question of whether or not an invention was inventive as compared to the prior art.

Finally, Institut Curie contended that the patent held by Myriad Genetics was insufficiently descriptive, lacking in industrial character and utility. The Oncological Genetics Unit at the Institut Curie, under Dr Dominique Stoppa-Lyonnet, contended that industrial methods focusing on the detection of point or small-sized abnormalities, and in particular methods such as the direct sequencing technology used by Myriad Genetics, failed to detect 10 to 20 per cent of all expected mutations.[42] The Opposition Division, though, rejected such arguments, claiming that the requirements of Article 83 of the *European Patent Convention* 1973 (EU) had been satisfied. The Opposition Division did not agree with the allegations of the Institut Curie that 'a number of BRCA1 mutations are still unclassified with regard to their diagnostic significance, such that the claims of the request are not enabled across their whole breadth'.[43]

The Institut Curie was jubilant at the decision of the Opposition Division of the European Patent Office.[44] The organization observed: 'This is a victory for all those whose conception of public health puts the principle of equal access to care before commercial interests.'[45] The Institut Curie also declared: 'It is also a victory for advocates of basic and clinical research founded on public–private partnerships in the drive for innovation in Europe and around the world.'[46] The Institut Curie also trumpeted: 'It is a victory too for ethics and rights over monopolistic abuses leading to unjustified appropriation of key know-how likely to lead to health improvements for people around the world.'[47]

Myriad Genetics Inc. sought to play down the significance of the decision. Spokesman William A. Hockett observed the European decision would be appealed, calling it just 'another step in a long administrative review process'.[48] He maintained that the decision would have no impact on Myriad's patent position or business in the United States or even on its overall business. The *New York Times* reported, though, that the stock of Myriad Genetics dropped 1.7 per cent in value after the European decision.[49]

Myriad Genetics Inc. quietly divested itself of the ownership of the patent, assigning its interest to the University of Utah Research Foundation. In January 2005, the University of Utah Research Foundation and the United States of America, henceforth sole owners of patent EP 699754, appealed against the May 2004 decision to revoke patent EP 699754.

B Social Democrat Party of Switzerland and the Institut Curie v The University of Utah Research Foundation (EP 705902 and EP 705903)

The Institut Curie and a number of parties filed an opposition procedure against patent EP 705902.[50] Again, the grounds for opposition were lack of priority and novelty, lack of inventive step and insufficient description. The opponents included the Social Democratic Party of Switzerland, Greenpeace, the Institut Curie, the Hopitaux de Paris, the Institut Gustave Roussy, the Belgian Society of Human Genetics and similar associations throughout the European Union, Dr Wilhelms of Germany and the State of Netherlands.

The Opposition Division of the European Patent Office upheld the validity of the patent in an amended form.[51] The Opposition Division ruled that EP 705902 had demonstrated an inventive step over the prior art:

> In summary, when considering all the options the skilled person would have been confronted with and the uncertain outcome of standard techniques proposed for BRCA1 cloning he could have selected from, the OD considers that the skilled person would not have obtained the BRCA1 gene and accordingly the probes therefore with a reasonable expectation of success when starting from D11. There is therefore no doubt that the cloning of the BRCA1 gene and the provision of probes suitable to clone said gene had to involve an inventive activity.[52]

The Opposition Division concluded that 'the BRCA1 sequence errors reported in the priority documents were caused by technical difficulties that had been overcome by repetitive sequencing'. The Division maintained: 'These errors were not the result of the use of probes which were unsuitable to isolate the BRCA1 gene.'[53]

The Opponents also argued that the patent application offended public morality and notions of human rights, because it would hinder research and laboratory testing, and it would lead to the serious obstruction of public health systems. The Opposition Division rejected such arguments:

> None of the objections of the Os demonstrated that the publication or exploitation of the invention in suit is contrary to 'ordre public' or morality. The objections were aiming at the negative effects of the patenting itself of the invention, at the financial and economic drawbacks and at the dependencies and negative consequences for the national health systems.[54]

The reasoning here is somewhat odd – the Opposition Division dismisses the ethical and human rights objections of the opponents by characterizing them as merely economic arguments. The decision highlights how the European Patent Office narrowly construes exclusions from patentable subject matter on the basis of 'ordre public or morality.' There is an appeal on foot against the decision of the Opposition Division.

In addition, the Institut Curie and its supporters have initiated an opposition procedure against patent EP 705903 B1 granted on 23 May 2001 to Myriad Genetics by the European Patent Office for 'Mutations in the 17q-linked breast and ovarian cancer susceptibility gene.'[55] They mount similar arguments about novelty, inventive step and lack of industrial applicability. The opponents also included the Hôpitaux de Paris and Institut Gustave Roussy of France, Vereniging van Stichtingen Klinische Genetica Leiden of the Netherlands, the State of Netherlands and Greenpeace.

The Opposition Division upheld the patent, albeit in an amended form.[56] The Opposition Division stressed that the patent claims were related to specific probes, vectors and cells, and not to genes. The Opposition Division denied that such an invention could be contrary to 'public ordre' or morality because of 'possible negative effects of the patenting of the invention in suit and pointed to financial and economic drawbacks or dependences or negative consequences for the health system'.[57] There is an appeal on foot against the decision of the Opposition Division.

II BRCA2

A group at the Institute of Cancer Research in Sutton, Surrey, led by Professor Michael Stratton, discovered the location of a second gene that predisposes some women to breast cancer: BRCA2 on Chromosome 13. They had been collaborating with Skolnick at the University of Utah. Stratton, the Professor of Cancer Genetics at the Institute of Cancer Research comments:

> The identification of a gene like BRCA 1 and 2 is the culmination of many years of work requiring the input of activities and information from many, many different groups. It is like building a wall – the final brick in the wall is the identification of the gene. And awarding the patent, in other words the full rewards to those individuals, to that group that has put that last brick in the wall, is unfair because it doesn't recognize the contribution of others.[58]

Stratton observed: 'We do not believe pieces of the human genome are inventions: we feel it is a form of colonization to patent them.'[59] He added: 'I don't think it is appropriate for BRCA1 to be owned by a commercial company because there is inevitably a demand for profit.'[60] After Myriad Genetics lodged patents in respect of BRCA1, Stratton was concerned by the prospect that Myriad Genetics would patent BRCA2.[61] He was moved to publish the discovery of BRCA2 in *Nature*, while keeping it secret from his collaborators until the last possible moment, so that it was not leaked to the researchers in Utah. Enough information reached Skolnick to enable him to locate the gene himself and submit a patent application.

In response, Stratton engaged in defensive patenting in an effort to protect his team's discovery from commercial exploitation. The Institute of Cancer Research took out one patent on the first mutation as soon as it was discovered, and another later covering more mutations. Meanwhile Myriad's patent applications claimed rights to the whole gene.

There have been a number of decisions in the European Patent Office on the validity of the patent claims of both Myriad Genetics and the Cancer Research Campaign to inventions relating to BRCA2.

A *The Belgian Society of Human Genetics and The Institut Curie v The University of Utah Research Division* **(EP 785216)**

In 1996, Myriad Genetics Inc. filed a patent application, EP 785216, in respect of 'Chromosome 13-linked breast cancer susceptibility gene BRCA2'.[62] The original patent application covered a wide range of diagnostic applications. The European Patent Office granted the patent in January 2003. The Belgian Society of Human Genetics and the Institut Curie filed opposition proceedings against the patent later in December 2003. Myriad Genetics Inc. transferred ownership of the patent to the University of Utah Research Foundation in 2004.

After holding hearings, the Opposition Division of the European Patent Office (EPO) decided in 2005 that European patent EP 785216 was to be maintained in an amended form.[63] The patent claims were narrowed to the use of a particular nucleic acid carrying a mutation of the BRCA 2-gene which is associated with a predisposition to breast cancer for in vitro diagnosing of such a predisposition in Ashkenazi Jewish women. There was much debate about the validity of patent claims focusing on identification of one particular mutation 'for diagnosing a predisposition to breast cancer in Ashkenazi Jewish women.'

First, the opponents argued that the term 'Ashkenazi-Jewish women' lacked the clarity required by Article 84 of the *European Patent Convention* 1973 (EU). The opponents cited evidence that 'The term "Ashkenazi", which usually refers to European origin, is imprecise and immeasurable for historic and scientific reasons.'[64] Third parties pointed out technical uncertainties and the lack of precision associated with the definition of racial, ethnic or otherwise hereditary determined groups. In response, the patent applicants argued that a number of documents, published before and after the relevant date of the claim, use the term 'Ashkenazi-Jewish'. Thus, the University of Utah Research Foundation argued that 'the obvious conclusion is that said term is clear to the skilled person involved in genetics and dealing with assigning a certain mutation to a specific population'.[65]

The Opposition Division of the European Patent Office ruled in favour of the patent applicant on this matter, finding that the term 'Ashkenazi-Jewish' was a legitimate classification.[66] The Opposition Division ruled that the requirements of Article 84 of the *European Patent Convention* 1973 (EU) has been satisfied: 'Therefore, cited documents published before the above-mentioned date establish that the term "Ashkenazi-Jewish" has been used to characterize ethnic/religious background but also to characterize a group of individuals in terms of inherited genetic material.'[67] This reasoning seems a particularly clumsy way of resolving issues about individual and group identity.

Second, the opponents argued that the method of claim 1 was not sufficiently disclosed, under Article 83 of the *European Patent Convention* 1973 (EU), because 'it is not evident how a skilled person may use a fragment of more than 10 000 nucleotides with a mutation on one particular nucleotide position for diagnosing a predisposition to breast cancer'.[68] The patent applicant responded that the skilled person had a host of standard methods to identify whether a mutation was present in a DNA sample. The Opposition Division ruled that 'the application indeed discloses in Example 10 a method whereby use of standard technique of PCR leads to a readable measurement which is indicative of the presence of the mutation in a genetic sample and, thus, one may diagnose predisposition to breast cancer'.[69]

Third, the opponents argued, under Article 54 of the *European Patent Convention* 1973 (EU), that 'since an Ashkenazi woman can hardly be considered to be part of an in vitro system, the limitation to Ashkenazi-Jewish women is not part of the claimed method'.[70] The patent applicant argued that the claim represented a novel selection of a subgroup 'Ashkenazi-Jewish women' from the wider group of women. The Opposition Division took the view 'that document 20 discloses the materials and methods for detecting a mutation in BRCA2 but not a method for diagnosing predisposition to breast cancer, let alone predisposition to breast cancer in Ashkenazi Jewish women'.[71] Accordingly, the Opposition Division held that 'the selection is considered novel and renders the claimed use novel'.[72]

Fourth, the opponents argued that documents disclosed a mutation, which was present in the BRCA2 gene, in women with breast cancer. They maintained that a person skilled in the art would use such documents to identify the presence of the said BRCA2 mutation in the Ashkenazi-Jewish population. The patent applicant argued that 'none of the documents reveals a high prevalence of 6174deIT mutation in Ashkenazi-Jewish women which is, thus, considered a surprising technical effect'.[73] The Opposition Division held the requirements of Article 56 of the *European Patent Convention* 1973 (EU) had been fulfilled: 'In view of the uncertainty surrounding the task and the great number of already known mutations, the skilled person is not expected to be able to arrive at the "correct" mutation without the exercise of inventiveness.'[74]

Fifth, the opponents argued that the patent application failed to meet the requirements of Article 53(a) of the *European Patent Convention* 1973 (EU). The opponents observed that 'the limitation of the method of Claim 1 to one particular group results in the test to be carried out for free for women who are not or who do not identify themselves as Ashkenazi-Jewish'.[75] The opponents concluded that 'the method amounts to nothing else than discrimination among women, that is, between the group of Ashkenazi-women who have to pay for the present test, and the rest of the women, who do not have to pay for the same test'.[76] The opponents suggested that the exploitation of the invention may be seen by some people as being contrary to morality because it is limited to the minority group of Ashkenazi-Jewish women. The patent applicants countered that, 'since the opponents, themselves, are practising genetic testing in Ashkenazi Jewish population or other ethnic populations, they are not entitled to plead that the claimed technology is unethical'.[77]

Taking the view that it was axiomatic that patent rights provided a public benefit, the Opposition Division dismissed the ethical complaints of the opponents as illogical and irrational: 'As a matter of fact, the field of medical diagnosis and therapy is striving to achieve as specialized diagnostic tests and therapies as possible, meaning that diagnostic tests and medicaments are preferentially designed to suit small and specific groups where the effect is expected to be direct and of utmost benefit.'[78] Rather unconvincingly, the Opposition Division wards off allegations of genetic discrimination:

> 'Genetic discrimination' is usually associated with disadvantaging a person or group of persons on the basis of their genetic identity. In the present method, the person or group of persons is offered the possibility of early diagnosing breast cancer which cannot be considered as disadvantaging the person or group of persons. Therefore, the term 'discrimination' is not appropriately used.[79]

Thus, the Opposition Division asserted that 'there is no reason to consider the present diagnostic method as claimed to be excluded from patentability since it could be used for the highly desirable goal of early diagnosis of predisposition to breast cancer'.[80]

The Opposition Division also nervously addressed third-party allegations that the patent is based on 'a racist idea which may cause discrimination in Israel and around the world' against Ashkenazi women.[81] The Opposition Division responded: 'The diagnostic method cannot be considered as racist or discriminatory if it has been adopted, apparently widely and successfully, in Israel.'[82]

The curious and peculiar decision of the Opposition Division of the European Patent Office to limit the patent claims to Ashkenazi Jewish women has raised larger ethical questions about linking ethnic identity to genetic disease.[83] Gert Matthijs[84] of the Department of Human Genetics and the Catholic University in Leuven and the Belgian Society of Human Genetics stated 'there is something fundamentally wrong if one ethnic group can be singled out by patenting'.[85] He commented:

> We still believe that there is something fundamentally wrong if one ethnic group can be singled out by patenting. Women coming to be tested for breast cancer will have to be asked whether they are Ashkenazi Jewish or not. If they are, the healthcare providers will only be able to offer the test if they paid for a license, or they will have to send the women's samples abroad. Women who are not Ashkenazi Jewish – or who just don't know that they have Ashkenazi Jewish ancestors – will be entitled to a test which is free. This is the first time that this kind of situation has arisen in genetic testing, and we find it very worrying.[86]

As a result of the decision, the European Patent Office has been criticized for countenancing racial and genetic discrimination. Dorit Lev, head of the Israel Association of Medical Geneticists, complained that the situation was unacceptable: 'It's not right that they should be discriminated against.'[87] Dominique Stoppa-Lyonnet of the Institut Curie worried that it would be discriminatory to compel a doctor to ask a woman about her ancestry before offering a consultation.[88] The pathbreaking researcher, Mary-Claire King, observed of the controversy: 'Is that fair? Of course not. [But] it's controlled by one company, and they set prices.'[89]

B *Myriad Genetics Inc. v Cancer Research Campaign Technology Inc.* (EP 0858467)

The British patent for the BRCA2 gene was awarded to a consortium of the Cancer Research Campaign Technology and Duke University,[90] and this consortium has in turn granted an exclusive worldwide licence to the patent for diagnostic services and products to the company OncorMed.[91] OncorMed had to meet certain strict conditions in exercising its licence: broad sub-licensing of diagnostic tests to other concerns, a requirement for pre- and post-test counselling for women tested, a ban on direct advertising to the public for screening tests, and no charge for use of the techniques by the UK National Health Service. OncorMed and Myriad Genetics sued one another for patent violations.[92] However, in a financial settlement, Myriad Genetics obtained exclusive rights to OncorMed's patents for BRCA1 and BRCA2 breast and ovarian cancer genetic testing, in exchange for undisclosed fees.[93] OncorMed agreed to discontinue offering BRCA1 and BRCA2 genetic testing services. Subsequently, OncorMed was bought out by Gene Logic.

Cancer Research Campaign Technology has continued to battle Myriad Genetics over the ownership of patents related to BRCA2.[94] The charity has blocked a bid by Myriad Genetics to patent the gene in Britain on the basis that it funded the British research that led to BRCA's 2 identification. The charity decided it would give Britain's National Health Service free access to the gene, known as BRCA2, for use in tests for women with a strong family history of breast cancer. The charity's financial manager and managing director of its technology transfer company, James Davidson, gave in-principle agreement to do the same for Australia. 'Our aim would be to make it available at non-commercial rates. We're not there to make money.'[95] The Cancer Research Campaign Technology has already had its application for an Australian patent covering BRCA2 approved. But Myriad Genetics is fighting this decision.

In January 2004, the European Patent Office awarded Cancer Research Technology Limited a patent, EP0858467 on 'materials and methods relating to the identification and sequencing of the BRCA2 cancer susceptibility gene and uses thereof'. The named inventors included Andrew Futreal, Richard Wooster, Alan Ashcroft and Michael Stratton. There are a number of overseas equivalents for this patent application.[96]

In 2004, Myriad Genetics Inc. filed opposition proceedings in the European Patent Office, requesting that the patent be revoked in its entirety.[97] The Utah company argues that the subject-matter is not patentable within the terms of Articles 52 to 57, Article 83 and Article 123(2) of the *European Patent Convention* 1973 (EU). Myriad Genetics Inc. contends that the 'patentee's attempt to complete an incomplete invention or to add an essential, but missing, element by defining the nucleic acid claimed on a detour is bound to fail'.[98] The company stressed that the landmark work of its scientists should take priority over Stratton and his colleagues. Myriad Genetics Inc. has requested that the patent be revoked for lack of novelty, inventive step and enablement.

In 2006, the Opposition Division of the European Patent office requested an oral hearing over the opposition proceedings. The matter is still on foot as of the beginning of 2007.

III   THE EUROPEAN REBELLION

The opposition proceedings by the Institut Curie against Myriad Genetics sparked a number of policy inquiries around the world in respect of patent law and genetic testing. Tim Caulfield, Robert Cook-Deegan, F. Scott Kieff and John Walsh observed that the Myriad Genetics controversy had a galvanizing impact:

> [T]he Myriad case ... emerged as emblematic of the fear that patents on human genetic material would have an adverse impact on access to useful technologies, both for research and for clinical use. This is most likely because the controversy, more than any other, resonated so well with the theoretical concerns that existed in the literature. In addition, the clinical consequences were easy to understand and highly visible breast cancer constituencies were engaged.[99]

The team of academics question whether the controversy was representative of wider trends in respect of gene patenting: 'Although the available evidence suggests that the concerns associated with the Myriad case have merit in the context of diagnostic tests, the data are hardly definitive, and empirical research suggests that data about diagnostics cannot be generalized to other uses.'[100] The authors speculate why there have been few similar controversies: 'One possibility is that the Myriad story has become a cautionary tale for the holders of similar gene patents, guiding them toward more constructive patent enforcement strategies.'[101] The controversy over patent law and genetic testing has attracted the attention of a number of legislators and policy makers in the European Union,[102] the United Kingdom,[103] the United States,[104] Canada[105] and Australia.[106] Such discussion papers have canvassed a range of initiatives to reform patent law. However, national governments have been slow to respond to such recommendations.

A   The European Union Biotechnology Directive

After 13 years of great controversy and public debate,[107] the European Parliament voted to adopt the *European Union Directive on the Legal Protection of Biotechnological Inventions* 1998 (EU).[108]

Article 1 of the Directive provides that 'Member States shall protect biotechnological inventions under national patent law' and 'shall, if necessary, adjust their national patent law to take account of the provisions of this Directive'. The Recitals recognize that 'biotechnology and genetic engineering are playing an increasingly important role in a broad range of industries' and that 'research and development require a considerable amount of high-risk investment and therefore only adequate legal protection can make them profitable'. The Recitals stress that 'effective and harmonised protection throughout the Member States is essential in order to maintain and encourage investment in the field of biotechnology'. The Recitals fear that differences in the legal protection of biotechnological inventions 'could create barriers to trade and hence impede the proper functioning of the internal market'.

In 2001, the Netherlands Government sought to annul the Directive in the European Court of Justice.[109] It argued that the Directive interfered with the internal markets of European Union countries and breached the principle of subsidiarity, and the principle of legal certainty. The Netherlands insisted that the Directive was incompatible with international obligations – in particular, the *TRIPS Agreement* 1994, the *Rio Convention on Biological Diversity* 1992 and the *Agreement on Technical Barriers to Trade* 1994. Further, it argued that the Directive breached fundamental rights by providing for the patentability of body parts, and that the Directive undermined human dignity and integrity. The European Court of Justice rather tersely rejected such arguments, and refused the application to annul the Directive.

The opposition proceedings taken against Myriad Genetics provided an opportunity to reopen the debate over the controversial *European Union Directive on the Legal Protection of Biotechnological Inventions* 1998 (EU). The legal action is a means to broach broader policy questions about bioethics in regard to such concepts as the human body, self and human dignity.

In 2001, the European Parliament passed a resolution on the patenting of the BRCA1 and BRCA2 genes.[110] It 'reiterates its call on the Council, the Commission and the Member States to adopt the measures required to ensure that the human genetic code is freely available for research throughout the world and that medical applications of certain human genes are not impeded by means of monopolies based on patents'.[111] The European Parliament was concerned that the granting of patents by the European Patent Office could create a monopoly for Myriad Genetics, which could seriously impede the further use of existing genetic tests for breast cancer. It stressed that 'this development could have an unacceptable detrimental effect on the women concerned and constitute a serious drain on the funds of public health services; whereas moreover it could seriously impede the development of and research into new methods of diagnosis'.[112]

This challenge to the validity of the patents held by Myriad Genetics represents a fundamental ambivalence in Europe: on the one hand, France was at the forefront of pushing forward the *European Union Directive on the Legal Protection of Biotechnological Inventions* 1998 (EU) and, on the other, it has also led the mutiny against the patenting of genes and gene sequences. Such a paradox is worthy of explanation and explication. The European Patent Office remains frustrated with the contrary approach of the European Parliament.[113] In its statement to the administrative council, the European Patent Office declares:

> The European Patent office is aware of the fact that the patenting of genes is and will continue to be a controversial issue in society. However, the Office is not the legislature which has to balance conflicting interests and lay down legal rules. The EPO is an administrative agency which applies and interprets the rules laid down by the legislature.[114]

The European Patent Office refers to the directive and how it allows the patenting of isolated human genes in certain instances. It also cited with approval the decision of the European Court of Justice to reject the challenge by the Dutch Government to the *European Union Directive on the Legal Protection of Biotechnological Inventions* 1998 (EU).[115]

There remain great divisions amongst European Union members over the *European Union Directive on the Legal Protection of Biotechnological Inventions* 1998 (EU). Margaret Llewelyn has observed of this mutinous state of affairs:

> Any semblance of cohesion and conformity within and across the EU is merely an illusion. Scratch the surface of the current provision and it would seem that Europe is as separated in its approach to protecting genetic material as it was before 1998. Clearly, the implementation of the Directive chosen by member states, such as France and Germany, reflects national priorities but these provisions remain untested in the courts and by the European Court of Justice in particular.[116]

She concludes: 'There is a very real sense of rebellion across the EU, not merely by member states, but also by the Commission itself, giving rise to the impression that that which is agreed, whether at the EU or international level, is open for interpretation and reinterpretation according to the agenda to be followed.'[117]

B Experimental Use

The debate over Myriad Genetics highlights the dearth of defences in respect of patent law. It demonstrates the need for the expansion of the defence of experimental use in patent law, along the lines of a defence of fair dealing or fair use in copyright law.[118]

The European Union has sought to encourage harmonization amongst its member states in respect of the research exemption under patent law. Article 27(b) of the *Community Patent Convention* (CPC) 1989 (EU) has provided the basis for an experimental use exception which exempts 'acts done for experimental purposes relating to the subject mater of the patented invention'. The European Court of Justice has affirmed that patent laws may specify that certain acts do not constitute infringement: 'Experimental use is one such exception: experiments aimed at perfecting, improving or further developing protected inventions do not infringe the patent.'[119] However, there remain variations in the interpretation of the defence of experimental use in different jurisdictions. William Cornish comments that, in some European countries, the defence was confined to the private and personal use of a scientific experimenter.[120] However, the changing nature of research has led to a step-wise expansion of the experimental use exception: 'No longer is any exception confined to the strictly non-commercial, because frequently scientific curiosity operates in conjunction with the desire to turn successful work to account.'[121]

In the United Kingdom, section 60(5) of the *Patents Act* 1977 (UK) provides: 'An act which, apart from this subsection, would constitute an infringement of a patent for an invention shall not do so if (a) it is done privately and for purposes which are not commercial; (b) it is done for experimental purposes relating to the subject-matter of the invention.' The courts have taken a liberal interpretation of the research exemption in a range of circumstances, dealing with agricultural herbicides,[122] pharmaceutical drugs,[123] virucidal compositions,[124] proteins[125] and clinical testing.[126]

By contrast, in the Netherlands, the courts have taken a conservative reading of the defence of experimental use, limiting its operation to acts solely related to research of the patented invention.[127] Article 53(3) of Dutch Patent Law provides: 'The exclusive right shall not extend to acts solely serving for research on the patented subject-matter, including the product obtained directly as a result of using the patented process.' The qualifier, 'solely', provides a significant limitation on the scope of experimental use in this country. Consequently, it appears that the Dutch courts have not interpreted the notion of experimental use as broadly as other jurisdictions in the European Union.

In *Wellcome Foundation v Parexel International & Flamel*, the Tribunal de Grande Instance de Paris emphasized that, under French law, it is a defence to patent infringement that the use of the patent was necessary to carry out experimental work 'relating to the subject matter of the invention'.[128] However, this exception covers only three types of activity: use of a patented invention for purely academic purposes; trials carried out to assess what the patent teaches, and its validity; and the use of the patented invention for technological development. In *E.R. Squibb & Sons Inc. v Giovannia Aguggini*, the Court of Milan in Italy held that a patent holder could not prevent a generic manufacturer from experimental activity in connection with an application for regulatory review during the term of the patent.[129]

In Germany, the Supreme Court has taken an expansive view of the defence of experimental use, maintaining that the doctrine can apply to both commercial and non-commercial uses. In *Klinische Versuche I* (Interferon Gamma), the German Supreme Court considered the scope of the defence of experimental use.[130] The Supreme Court thus takes a broad reading of the defence of experimental use, observing that it could potentially cover both non-commercial and commercial uses: '§11 No. 2 of the Patents Act in principle exempts all experimental acts as long as they serve to gain information and thus to carry out scientific research into the subject-matter of the invention, including its use.'[131] In *Klinische Versuche II* (Erythropoietin), the German Supreme Court provided a rearticulation of the defence of experimental use: 'As section 11 No. 2 of the Patent Act neither qualitatively nor quantitatively limits the experimental activities, we are given to understand that the examinations and tests can range from purely scientific experiments to commercially-oriented tests.'[132] In May 2000, the German Constitutional Court affirmed that the decision in *Klinische Versuche I* (Interferon Gamma) was in full conformity with the right of property under Article 14, Section 1 of the German Constitution.[133]

There has also been significant policy discussion in Europe about the scope and operation of the general defence of experimental use, particularly in the context of biotechnology and pharmacology. There has been some call for the further refinement of the defence in the European Union.

As part of its inquiry into patent law and scientific research, the Royal Society of the United Kingdom observed: 'We recommend that governments consider clarifying and harmonising the existing exceptions for "private and non-commercial" and "experimental" use.'[134] The Nuffield Council on Bioethics stressed that there was a need for law reform to clarify the defence of experimental use in Europe and the United States: 'Even in Europe, where there is a statutory basis for the research exemption, the scope of the exemption is not clear.'[135] The United Kingdom Department of Health commissioned a report on intellectual property rights and genetic technologies.[136] The authors of the report included William Cornish, from Cambridge University, and Margaret Llewelyn and Mike Adcock from Sheffield University. The report suggested that the present meaning of experimental use needed further clarification in three key contexts: research tools, clinical work and genetic testing.[137] It argued that there was greater scope for drawing reasonable boundaries over what constituted experimentation.

Reviewing European case law, Trevor Cook comments upon the legal developments in the European Union concerning the interpretation of

Article 27(b) of the *Community Patent Convention* 1989 (EU), and its national co-ordinates: 'As to the appropriate response to such issues in the United Kingdom and the rest of Europe, new legislation to more finely craft an experimental-use defence is tempting, but there is the risk that to the extent that this involves abandoning existing wording it will achieve no more than to exacerbate existing uncertainties and add new ones.'[138]

C Compulsory Licensing

The Institut Curie hopes to anticipate the implementation of a legal framework, actually mapped out by French authorities, better suited to the specificity of genetic testing.

The bill about the partial transposition of the 98/44 directive extends the product field at present subjected to drug ex officio licensing to medical and in vitro diagnostic devices and related therapeutic products, as well as ex vivo diagnostic methods. It facilitates as well the activating of this legal process and at the same time of compulsory licensing. This bill plans to strictly define patent applicants' claims. As France's Minister in charge of Research, Roger-Gérard Schwartzenberg, pointed out:

> [W]here gene sequence function is patent-protected, subsidiary patents will be available. Work must be done at the European level to develop a system of compulsory and ex officio licensing in the interest of society at large and of public health promotion in cases where subsidiary patents protect new therapeutic or diagnostic applications of previously patented sequences.[139]

There is also scope for the introduction of compulsory licensing in relation to patents for genetic testing in the United Kingdom,[140] the United States,[141] Canada[142] and Australia.[143] Such regulation would help limit excessive profits (the cost of developing a test kit for mutations in a gene is not great and this should be reflected in the price of the product).[144] There is widespread concern that patents will reduce access to genetic testing because of higher cost: government will be less able to fund testing and, if this occurs, access to clinically indicated genetic tests will be determined, for many people, by capacity to pay.[145] It provides no incentive for the technological improvement and price reduction that comes with competition.

Implementing the *European Union Directive on the Legal Protection of Biotechnological Inventions* 1998, the Belgium Government introduced an enlarged defence of experimental use and compulsory licences in respect of public health, as part of its amendments to the Belgian patent laws in 2005.[146] Geertrui van Overwalle comments on this initiative: 'The Minister specified that the newly designed compulsory licence particularly aims at securing a delicate balance between different stakeholders and to prevent ending up in American situations, like the Myriad case.'[147] She notes: 'The new compulsory licence for domestic public health will hopefully address undesirable effects and unreasonable behaviour from patent holders in an adequate manner, thanks to its preventive and dissuading effect towards patent holders applying (extremely) restricting licensing policies.'[148]

There has been opposition to compulsory licensing, particularly within the biotechnology and pharmaceutical industries. Some are uncertain whether the monopoly of Myriad Genetics will prevail. An economist from Boston University, Iain Cockburn, questioned whether there was a need for government regulation: 'If Myriad were making lots of money that's one thing, but they aren't.'[149] A member of the Chicago school of law and economics, Richard Epstein, maintained that the marketplace should be left to solve problems with respect to patents in the field of biotechnology: 'Compulsory licenses cannot replicate the complex provisions that regulate the scope of the permitted use, the creation of sub-licensees, the sharing of information between the two parties, the extension of the license term and the host of other provisions included for mutual advantage in voluntary licenses.'[150]

In the United Kingdom, the Nuffield Council on Bioethics convened a group of experts to discuss the ethics of gene patenting.[151] After its deliberations, the Nuffield Council on Bioethics concluded that the criteria for the granting of patents, particularly the criterion of inventiveness, should be strongly applied to patent applications in respect of genetic testing. Furthermore, it suggested that compulsory licensing may be required to ensure reasonable licensing terms are available to enable alternative tests to be developed.

In the Report on Intellectual Property Rights and Genetics, William Cornish, Margaret Llewelyn and Mike Adcock submit that the United Kingdom Department of Health needs to play a more active role in relation to gene patents: 'The Department needs to develop a coherent policy for both the receipt and the provision of patented material.'[152] The report recommended that the Department of Health should instigate a robust central policy for 'licensing in' designed to moderate excessive demands by licensors by considering, as possible options, the use of compulsory licensing, competition law and Crown use.

In the context of access to essential medicines, the European Union had developed a regulation to deal with the export of patented pharmaceutical drugs to developing countries.[153] Perhaps a similar Community-wide approach to compulsory licensing in respect of public health concerns in respect of genetic diagnostic testing would be desirable.

CONCLUSION

In light of the controversy over Myriad Genetics, there is a need for a comprehensive review of the *European Union Directive on the Legal Protection of Biotechnological Inventions* 1998 in respect of the protection of biotechnological inventions. There is a concern that broad patents on genetic material and medical treatments will have a deleterious effect on patient care, research and the administration of public health care. The European Community needs to remedy the paucity of defences in the field of patent law. A broad, modernized experimental use exemption would ensure that medical researchers could contemplate follow-on innovation, without the fear of litigation. Furthermore, medical practitioners should be exempted from patent infringement suits in respect of utilizing surgical procedures and medical diagnostics.

The European Community should also consider the provision of compulsory licensing of patents relating to the provision of medical diagnostics and genetic tests. This measure would ensure that the private rights of patent holders do not impinge upon the wider public interest. The European Community also needs to take further measures to protect research participants and patients from genetic discrimination.

The controversy over Myriad Genetics has much in common with the controversy over the access to essential medicines, such as AIDS drugs, and the Cipro drugs for anthrax. Seth Shulman suggests that the dispute over Myriad Genetics needs to be resolved at an international level:

> The Curie Institute's legal action is an important protest. But more proactive work is needed to clarify appropriate limits on similar health-care claims. Ideally, a panel of stakeholders under the auspices of an international body like the World Health Organization ought to tackle the job. Without such a group, we will likely see many divisive fights like this one – needlessly restricting medical knowledge and potentially undermining the Hippocratic oath.[154]

Another possible international forum would be the World Trade Organisation.[155] The signatories to the *TRIPS Agreement* 1994 may exclude from patentability 'diagnostic, therapeutic and surgical methods for the treatment of humans or animals'.[156] Furthermore, the *Doha Declaration on the TRIPS Agreement and Public Health* 2001 and the *WTO General Council Decision* 2003 has affirmed the use of compulsory licences.[157]

NOTES

1. Edson, Margaret (1998), *Wit: A Play*, New York: Farrar, Straus and Giroux.
2. Ibid.

3. Blanton, K. (2002), 'Corporate takeover', *Boston Globe*, 24 February.
4. Davies, Kevin and Michael White (1995), *Breakthrough: The Quest to Isolate the Gene for Hereditary Breast Cancer*, London: Macmillan Books, pp. 259–60.
5. Ibid., 223.
6. Ibid., 260.
7. Ibid., 261.
8. Ibid., 289.
9. A search for 'myriad genetics' under 'assignee' (AN) on the Patent Database in March 2007 yielded 64 US patents. Patents relating to breast and ovarian cancer totalled 14: http://www.uspto.gov.
10. Myriad Genetics (2001), *2001 Annual Report*, Salt Lake City: Myriad Genetics.
11. Myriad Genetics (2001), 'Patents issued in Europe, Canada, Australia and New Zealand encourage broad availability of predictive medicine', Salt Lake City, 15 May.
12. A company name search as at March 2007 for 'Myriad Genetics' on the European Patent Database yielded 391 results. Patents relating to breast cancer in the abstract amounted to 32 entries. European Patents Database (Database for World patents): http://ep.espacenet.com.
13. A search for 'myriad genetics' as at March 2007 on the Patent Database yielded 19 Canadian Patents. Patents relating to breast and ovarian cancer totalled 4: Canadian Patents Database: http://cipo.gc.ca.
14. A name search for 'myriad genetics' as at March 2007 on the Patent Database yielded 57 Australian Patents (not including one held by Myriad Genetics Corporation). Patents relating to breast cancer totalled 6: Australian Patents Database: http://ipaustralia.gov.au.
15. Australian Broadcasting Corporation (2001), 'Fears genetic testing may be confined to the Rich', *7:30 Report* (Transcript), 14 March.
16. Rusconi, W. (2005), 'The National Academies Fifth Meeting of the Committee on Intellectual Property Rights in Genomic and Protein-Related Inventions', http://www7.nationalacademies.org/step/Genomics_Committee_Meeting_6_transcript.pdf, 11 February.
17. Walpole, I., H. Dawkins, P. Sinden and P. O'Leary (2003), 'Human gene patents: the possible impacts on genetic services healthcare', *Medical Journal of Australia*, **179**, 203–5; Williams-Jones, B. (2002), 'History of a gene patent: tracing the development and application of commercial BRCA testing', *Health Law Journal*, **10**, 123–46; Merz, J. and M. Cho (2005), 'What are gene patents and why are people worried about them?', *Community Genetics*, **8**, 203–8; Nicol, D. (2005), 'Balancing innovation and access to healthcare through the patent system – an Australian perspective', *Community Genetics*, **8**, 228–34; Parthasarathy, S. (2005), 'The patent is political: the consequences of patenting the BRCA genes in Britain', *Community Genetics*, **8**, 235–42; Garforth, K. (2005), 'Health care and access to patented technologies', *Health Law Journal*, **13**, 77–97; and Weck, E. (2005), 'Exclusive licensing of DNA diagnostics: is there a negative effect on quantity and quality of healthcare delivery that compels NIH rulemaking?', *William Mitchell Law Review*, **31**, 1057–91.
18. Sevilla, C., C. Julian-Reynier, F. Eisinger, D. Stoppa-Lyonnet, B. Bressac-de Paillerts, H. Sobel and J.-P. Moatti (2003), 'The impact of gene patents on the cost-effective delivery of care: the case of BRCA1 testing', *International Journal of Technology Assessment in Health Care*, **19**(2), 287–300.
19. Institut Curie (2000), 'Opposition procedure with the European Patent Office', http://www.curie.net/actualities/myriad/declaration_e.htm, 12 September.
20. Assuming an exchange rate of 7.50 FF to the dollar.
21. Institut Curie (2001), 'Opposition procedure with the European Patent Office', http://www.curie.net/actualities/myriad/declaration_e.htm, 12 September.
22. Loughlan, P. (1995), 'Of patents and patients: new monopolies in medical methods', *Australian Intellectual Property Journal*, **6**, 5–15; Shulman, Seth (1999), 'The new medical

licenses', *Owning The Future*, Boston: Houghton Mifflin Company, pp. 33–59; and Martin, T. (2000), 'Patentability of methods of medical treatment: a comparative study', *Journal of the Patent and Trademark Office Society*, **82**(6), 381–423.
23. For a wider discussion of genetic privacy and genetic discrimination, see Australian Law Reform Commission (2003), *Essentially Yours: The Protection of Human Genetic Information*, Sydney: Australian Commonwealth, http://www.austlii.edu.au/au/other/alrc/publications/reports/96/.
24. Hall, J.M., M. Lee, B. Newman, J. Morrow, L. Anderson, B. Huey and M.-C. King (1990), 'Linkage of early-onset familial breast cancer to chromosome 17q21', *Science*, **250**, 1684–9.
25. Davies, Kevin and Michael White (1995), *Breakthrough: The Quest To Isolate The Gene For Hereditary Breast Cancer*, London: Macmillan Books, pp. 289–90.
26. Nowak, R. (1994), 'NIH in danger of losing out on BRCA1 patent', *Science*, **266**, 209; and Elliot, M. (1995), 'NIH gets a share of BRCA1 patent', *Science*, **267**, 1086.
27. Davies, Kevin and Michael White (1995), *Breakthrough: The Quest to Isolate the Gene for Hereditary Breast Cancer*, London: Macmillan Books, pp. 289–90.
28. Ibid.
29. Skolnick, M. and D. Goldgar (1995), 'Method for diagnosing a predisposition for breast and ovarian cancer', European Patent No: EP699754.
30. Skolnick, M. and D. Goldgar (1995), 'Method for diagnosing a predisposition for breast and ovarian cancer', European Patent No: EP699754.
31. Blanton, K. (2002), 'Corporate Takeover', *Boston Globe*, 24 February.
32. AU3242895, AU3321695, AU691331, AU691958, CA2196790, CA2196795, CN1159829, DE69519834D, DE69519834T, EP0705902, FI970514, FI970515, NO970625, NO970626, NZ291624, PT699754T, *WO9605307*, *WO9605308*.
33. For a press release on the initial action, see Institut Curie (2001), 'The Institut Curie is initiating an opposition procedure with the European Patent Office', Press Release, 12 September.
34. *Institut Curie v Myriad Genetics Inc.*, European Patent Office Opposition Division, Division Revoking the European Patent EP0699754 (3 November 2004).
35. *Institut Curie v Myriad Genetics Inc.*, European Patent Office Opposition Division, Division Revoking the European Patent EP0699754 (3 November 2004).
36. *Institut Curie v Myriad Genetics Inc.*, European Patent Office Opposition Division, Division Revoking the European Patent EP0699754 (3 November 2004).
37. *Institut Curie v Myriad Genetics Inc.*, European Patent Office Opposition Division, Division Revoking the European Patent EP0699754 (3 November 2004).
38. *Institut Curie v Myriad Genetics Inc.*, European Patent Office Opposition Division, Division Revoking the European Patent EP0699754 (3 November 2004).
39. *Institut Curie v Myriad Genetics Inc.*, European Patent Office Opposition Division, Division Revoking the European Patent EP0699754 (3 November 2004).
40. Institut Curie (2001), 'Opposition procedure with the European Patent Office', http://www.curie.net/actualities/myriad/declaration_e.htm, 12 September.
41. *Institut Curie v Myriad Genetics Inc.*, European Patent Office Opposition Division, Division Revoking the European Patent EP0699754 (3 November 2004).
42. Gad, S., M. Scheuner, S. Pages-Berhouet, V. Caux Moncoutier, A. Bensiman, A. Aurias, M. Pinto and D. Stoppa-Lyonnet (2001), 'Identification of a large rearrangement of the BRCA1 gene using colour bar code on combed DNA in an American breast/ovarian cancer family previously studied by direct sequencing', *Journal of Medical Genetics*, **38**(6), 388.
43. *Institut Curie v Myriad Genetics Inc.*, European Patent Office Opposition Division, Division Revoking the European Patent EP0699754 (3 November 2004).
44. Institut Curie (2004), 'The European Patent Office has revoked the Myriad patent', Press Release, http://www.curie.fr/upload/presse/190504_gb.pdf, 21 May.
45. Ibid.
46. Ibid.
47. Ibid.

48. Pollack, A. (2004), 'Patent on test for cancer is revoked by Europe', *The New York Times*, 19 May.
49. Ibid.
50. Skolnick, M. et al. (1995), '17q linked breast and ovarian cancer susceptibility gene', European Patent No: EP705902; and Institut Curie and others (2002), 'Against Myriad Genetic's monopoly on tests for predisposition to breast and ovarian cancer associated with the BRCA1 gene: third French opposition', Press Release, 26 September.
51. *Sozialdemokratische Partei der Schweiz and the Institut Curie v The University of Utah Research Foundation*, European Patent Office Opposition Division, Interlocutory Decision in Opposition Proceedings Against EP705902 (19 September 2005).
52. *Sozialdemokratische Partei der Schweiz and the Institut Curie v The University of Utah Research Foundation*, European Patent Office Opposition Division, Interlocutory Decision in Opposition Proceedings Against EP705902 (19 September 2005).
53. *Sozialdemokratische Partei der Schweiz and the Institut Curie v The University of Utah Research Foundation*, European Patent Office Opposition Division, Interlocutory Decision in Opposition Proceedings Against EP705902 (19 September 2005).
54. *Sozialdemokratische Partei der Schweiz and the Institut Curie v The University of Utah Research Foundation*, European Patent Office Opposition Division, Interlocutory Decision in Opposition Proceedings Against EP705902 (19 September 2005).
55. Shattuck-Eidens, D., J. Simard, E. Mitsuru, Y. Nakamura and F. Durocher (1995), 'Mutations in the 17q-linked breast and ovarian cancer susceptibility gene', European Patent No: EP 705903; and The Institut Curie and others (2002), 'BRCA1 gene-linked forms of breast and/or ovarian cancer: The Institut Curie, the Assistance Publique-Hôpitaux de Paris and the Institut Gustave-Roussy file a joint opposition to a second Myriad Genetics patent', Press Release, 22 February.
56. *Institut Curie v The University of Utah Research Foundation*, European Patent Office Opposition Division, Interlocutory Decision in Opposition Proceedings Against EP705903 (9 June 2005).
57. *Institut Curie v The University of Utah Research Foundation*, European Patent Office Opposition Division, Interlocutory Decision in Opposition Proceedings Against EP705903 (9 June 2005).
58. Canadian Broadcasting Corporation (2000), 'The impact of gene patents on health care and medical research: the case of breast cancer genetic screening', 21 March.
59. Ibid.
60. Ibid.
61. Sulston, John and Georgina Ferry (2002), *The Common Thread: A Story of Science, Politics, Ethics and the Human Genome*, London: Bantam Press, p. 142.
62. Tavtigian, S., A. Kamb, J. Simard, F. Couch, J. Rommens and B. Weber (1996), 'Chromosome 13-linked breast cancer susceptibility gene BRCA2', European Patent No: EP 785216.
63. *The Belgian Society of Human Genetics and the Institut Curie v The University of Utah Research Foundation*, European Patent Office Opposition Division, Interlocutory Decision in Opposition Proceedings Against EP785216, (29 June 2005).
64. Ibid.
65. Ibid.
66. Ibid.
67. Ibid.
68. Ibid.
69. Ibid.
70. Ibid.
71. Ibid.
72. Ibid.
73. Ibid.
74. Ibid.
75. Ibid.
76. Ibid.

77. Ibid.
78. Ibid.
79. Ibid.
80. Ibid.
81. Ibid.
82. Ibid.
83. Brandt-Rauf, S., V. Raveis, N. Drummond, J. Conte and S. Rothman (2006), 'Ashkenazi Jews and breast cancer: the consequences of linking ethnic identity to genetic disease', *American Journal of Public Health*, **96**(11), 1979.
84. Matthijs, G. (2006), 'The European opposition against the BRCA gene patents', *Familial Cancer*, **5**, 95–102.
85. Kienzlen, G. (2005), 'BRCA2 patent upheld', *The Scientist*, **6**(1), http://www.the-scientist.com/news/20050701/01/.
86. European Society of Human Genetics (2005), 'EPO upholds limited patent on BRCA2 gene: singling out an ethnic group is a "dangerous precedent" says European Society of Human Genetics', http://www.eshg.org/ESHGPressRelease01July2005.pdf, 1 July.
87. Abbott, A. (2005), 'Genetic patent singles out Jewish women', *Nature*, **426**, 12, http://www.nature.com/nature/journal/v436/n7047/full/436012a.html, 7 July.
88. Marshall, E. (2005), 'BRCA2 patent faces new challenge', *Science*, **308**(5730), 1851, 24 June.
89. Keim, B. (2006), 'Breast cancer research neglects non-Jewish groups, experts charge: patent monopolies have skewed research on breast cancer genetics', *Nature Medicine*, http://www.nature.com/news/2006/061127/full/nm1206-1335a.html, 29 November.
90. Futreal, P., R. Wooster, A. Ashworth and M. Stratton (1995), 'Materials and methods relating to the identification and sequencing of the BRCA2 cancer susceptibility gene and uses thereof', European Patent No: EP 0858 467.
91. Editorial (1997), 'Constraints imposed in breast cancer gene patents', *Nature*, 7 November.
92. Marshall, E. (1997), 'The battle over BRCA1 goes to court; BRCA2 may be next', *Science*, **278**, 1874.
93. Myriad Genetics (1998), 'Myriad Genetics obtains OncorMed's BRCA1/BRCA2 genetic testing program in patent settlement: OncorMed and Myriad settle BRCA patent disputes', 18 May.
94. Smith, D. (2001), 'Patent battle looms over cancer gene', *The Sydney Morning Herald*, 15 March.
95. Ibid.
96. WO9719110 (A1); US6045997 (A1); EP0858467 (A0); EP0858467 (B1); ES2217328T (T3); DE69631540T (T2); and AU707636B (B2).
97. *Myriad Genetics Inc. v Cancer Research Campaign Technology Inc.*, European Patent Office, Opposition to EP 0858 467 (11 November 2004).
98. *Myriad Genetics Inc. v Cancer Research Campaign Technology Inc.*, European Patent Office, Opposition to EP 0858 467 (11 November 2004).
99. Caulfield, T., R.C. Cook-Deegan, F.C. Kieff and J. Walsh (2006), 'Evidence and anecdotes: an analysis of human gene patenting controversies', *Nature Biotechnology*, **24**(9), 1091–4 at 1093.
100. Ibid.
101. Ibid.
102. European Parliament (2001), 'European Parliament resolution on the patenting of BRCA1 and BRCA2 ("breast cancer") genes', Texts Adopted by Parliament, Provisional Edition: 04/10/2001, B5-0633, 0641, 0651 and 0663/2001.
103. Nuffield Council on Bioethics (2002), *The Ethics of Patenting DNA, A Discussion Paper*, London: Nuffield Council on Bioethics, http://www.nuffieldbioethics.org/go/ourwork/patentingdna/publication_310.html.
104. Rivers, L. (2002), 'Introduction of *The Genomic Research And Diagnostic Accessibility Act* of 2002 H.R. 3967 and *The Genomic Science And Technology Innovation Act* of 2002 H.R. 3966', Congressional Record, 14 March, E353; and X. Becerra and D. Weldon

(2007), 'Representatives Becerra and Weldon introduce bill to ban the practice of gene patenting', United States Congress, http://weldon.house.gov/News/DocumentSingle.aspx?DocumentID=57930, 9 February.
105. Ontario State Government (2002), 'Genetics, testing and gene patenting: charting new territory in healthcare', *Draft Report To The Provinces and The Territories*, Toronto: Ontario State Government; and Canadian Biotechnology Advisory Committee (2006), *Human Genetic Materials, Intellectual Property, and the Health Sector*, Ottawa: Canadian Biotechnology Advisory Committee, http://cbac-cccb.ca/epic/internet/incbac-cccb.nsf/en/ah00578e.html.
106. Australian Law Reform Commission (2003), *Gene Patenting and Human Health, Issue Paper 27*, Sydney: Australian Commonwealth, http://www.austlii.edu.au/au/other/alrc/publications/issues/27/, July; Australian Law Reform Commission (2004), *Gene Patenting and Human Health, Discussion Paper 68*, Sydney: Australian Commonwealth, http://www.austlii.edu.au/au/other/alrc/publications/dp/68/, February; and Australian Law Reform Commission (2004), *Genes and Ingenuity: Gene Patenting and Human Health, Report 99*, Sydney: Australian Commonwealth, http://www.austlii.edu.au/au/other/alrc/publications/reports/99/, June.
107. For an account of this history, see Llewelyn, M. (2006), 'European bio-protection laws: rebels with a cause', in M. Rimmer (ed.), *Patent Law and Biological Inventions, Law in Context*, **24**(1), 11–33; and Llewelyn, Margaret and Mike Adcock (2006), *European Plant Intellectual Property*, Oxford: Hart Publishing, pp. 341–94.
108. *Directive 98/44/EC of the European Parliament and of the Council* of 6 July 1998 on the Legal Protection of Biotechnological Inventions, <http://europa.eu.int/eur-lex/pri/en/oj/dat/1998/l_213/l_21319980730en00130021.pdf.
109. *Netherlands v European Parliament* (2001) 3 CMLR 49; Scott, A. (1999), 'The Dutch challenge to the bio-patenting directive', *European Intellectual Property Review*, **21**(4), 212–15; Moore, S. (2002), 'Challenge to the biotechnology directive', *European Intellectual Property Review*, **24**(3), 149–54; and Beyleveld, D. and R. Brownsword (2002), 'Is patent law part of the EC legal order? A critical commentary on the interpretation of Article 6(1) of Directive 98/44/EC in Case C-377/98', *Intellectual Property Quarterly*, **1**, 97–110.
110. European Parliament (2001), 'European Parliament resolution on the patenting of BRCA1 and BRCA2 ("breast cancer") Genes', Texts Adopted by Parliament, Provisional Edition: 04/10/2001, B5-0633, 0641, 0651 and 0663/2001.
111. Ibid.
112. Ibid.
113. Moore, S. (2002), 'Challenge to the biotechnology directive', *European Intellectual Property Review*, **24**(3), 149–54.
114. Ibid., 154.
115. *Netherlands v European Parliament* (2001) 3 CMLR 49.
116. Llewelyn, M. (2006), 'European bio-protection laws: rebels with a cause', in M. Rimmer (ed.), *Patent Law and Biological Inventions, Law in Context*, **24**(1), 11–33 at 31–2.
117. Ibid.
118. O'Rourke, M. (2000), 'Toward a doctrine of fair use in Patent law', *Columbia Law Review* **100**(5), 1177–250; and Gitter, D. (2001), 'International conflicts over patenting human DNA sequences in the United States and the European Union: an argument for compulsory licensing and a fair use exemption', *New York University Law Review*, **76**, 1623–91.
119. *Netherlands v European Parliament* (C377/98), 2001 WL 758973, [2001] ECR I-7079, Celex No. 698C0377, EU: Case C-377/98, ECJ, June 14, 2001.
120. Cornish, W. (1998), 'Experimental use of patented inventions in European States', *International Review Of Industrial Property And Copyright Law*, **29**, 735–53.
121. Ibid., 752.
122. *Monsanto v Stauffer Chemical Co (UK)* [1985] RPC 515.
123. *Smith Kline & French Laboratories Ltd v Evans Medical Ltd* [1989] F.S.R. 513.

124. *Auchincloss v Agricultural & Veterinary Supplies Ltd* [1997] R.P.C. 649; on appeal [1999] R.P.C. 397.
125. *Kirin-Amgen Inc. v Transkaryotic Therapies Inc.* (No. 2) [2002] RPC 3.
126. *Inhale Therapeutic Systems v Quadrant Healthcare Plc* [2002] RPC 21.
127. *ICI/Pharbia and Medicopharma* (Atenolol) [1993] NJ 735 (1993) GRUR Int. 887 (Dutch Supreme Court); *Applied Research Systems/Organon* (Follicle-Stimulating Hormone) [1996] NJ 463, 28 IIC 558 (1997); (Dutch Supreme Court, affirming a more extensive judgment of The Hague Court of Appeal, 29 IIC 702 (1998)); *Kirin Amgen/Boehringer Mannheim* (Erythropoietin), Judgment of 3 February 1994 (docket No. 93/960) (The Hague Court of Appeal). Affirmed on other grounds by the Dutch Supreme Court: [1996] NJ 462; and *Generics BV v Smith Kline & French Laboratories Ltd.* (1997) R.P.C. 801, 803 (European Ct. of J. 1997); Binns, R. and B. Driscoll (1999), 'Are the generic companies winning the battle?', *Managing Intellectual Property*, **86**, 36.
128. *Wellcome Foundation v Parexel International & Flamel*, Tribunal de Grande Instance de Paris, 20 February 2001.
129. *E.R. Squibb & Sons Inc. v Giovannia Aguggini*, 12 June 1995, T. Milan.
130. *Klinische Versuche I* (Interferon Gamma) [1997] RPC 623.
131. Ibid., 639.
132. *Klinische Versuche II* (Erythropoietin) [1998] RPC 423 at 432–3.
133. *Klinische Versuche I* (Interferon Gamma) 2001 GRUR 43; 1 BvR 1864/95, http://www.bundesverfassungsgericht.de/cgi-bin/link.pl?entscheidungen (in German); Goddard, H. (2002), 'The experimental use exception: a European perspective', Center for Advanced Studies and Research on Intellectual Property, University of Washington, Seattle, 7, http://www.law.washington.edu/casrip/Symposium/Number 7/1-Goddar.pdf.
134. Royal Society of the United Kingdom (2003), *Keeping Science Open: The Effects of Intellectual Property Policy on the Conduct of Science*, London: The Royal Society, April, http://www.royalsoc.ac.uk/files/statfiles/document-221.pdf, p. 11.
135. Nuffield Council on Bioethics (2002), *The Ethics of Patenting DNA, A Discussion Paper*, London: Nuffield Council on Bioethics, http://www.nuffieldbioethics.org/go/ourwork/patentingdna/publication_310.html.
136. Cornish, William, Margaret Llewelyn and Mike Adcock (2003), *Intellectual Property Rights and Genetics: A Study into the Impact and Management of Intellectual Property Rights within the Healthcare Sector*, Cambridge: Public Health Genetics Unit, http://www.phgu.org.uk/about_phgu/intellect_prop_rights.html.
137. Ibid., 26.
138. Cook, T. (2006), 'Responding to concerns about the scope of the defence from patent infringement for acts done for experimental purposes relating to the subject matter of the invention', *Intellectual Property Quarterly*, **3**, 193–222 at 220.
139. Schwartzenberg, R.-G. (2001), 'First large scale analysis of the human genome sequence', Minister of Research, France, http://www.recherche.gouv.fr/english/ministre/discours.htm, 12 February.
140. Nuffield Council on Bioethics (2002), *The Ethics of Patenting DNA, A Discussion Paper*, London: Nuffield Council on Bioethics, http://www.nuffieldbioethics.org/go/ourwork/patentingdna/publication_310.html.
141. Federal Trade Commission (2003), *To Promote Innovation: The Proper Balance of Competition and Patent Law and Policy*, Washington, DC: Federal Trade Commission, http://www.ftc.gov/os/2003/10/innovationrpt.pdfl; National Academy of Sciences (2004), *A Patent System for the 21st Century*, Washington, DC: National Academy of Sciences, http://books.nap.edu/catalog.php?record_id=10976; and National Research Council Committee on Intellectual Property Rights in Genomic and Protein Research and Innovation (2005), *Reaping the Benefits of Genomic and Proteomic Research: Intellectual Property Rights, Innovation and Public Health*, Washington, DC: National Academies Press.
142. Canadian Biotechnology Advisory Committee (2006), *Human Genetic Materials, Intellectual Property, and the Health Sector*, Ottawa: Canadian Biotechnology

Advisory Committee, http://cbac-cccb.ca/epic/internet/incbac-cccb.nsf/en/ah00578e.html.
143. Australian Law Reform Commission (2004), *Gene Patenting and Human Health, Discussion Paper 68*, Sydney: Australian Commonwealth, http://www.austlii.edu.au/au/other/alrc/publications/dp/68/, February; and see also Lawson, C. (2002), 'Patenting genes and gene sequences and competition: patenting at the expense of competition', *Federal Law Review*, **30**, 97–133.
144. Gitter, D. (2001), 'International conflicts over patenting human DNA sequences in the United States and the European Union: an argument for compulsory licensing and a fair use exemption', *New York University Law Review*, **76**, 1623–91.
145. Human Genetics Society of Australasia (2001), 'HGSA position paper on the patenting of genes', 3.6.
146. van Overwalle, G. (2006), 'The implementation of the biotechnology directive in Belgium and its after-effects: the introduction of a new research exemption and a compulsory licence for public health', *International Review of Intellectual Property and Competition Law*, **37**(8), 889–920.
147. Ibid., 908.
148. Ibid., 919.
149. Westphal, S.P. (2002), 'Your money or your life', *New Scientist*, **175**, 29 at 33.
150. Epstein, R. (2002), 'If it ain't Broke' *FT.Com*, 2 July http://www.law.uchicago.edu/news/epstein-genome.html; see also: Epstein, Richard (2003), 'Steady the course: property rights in genetic material', in F. Scott Kieff (ed.), *Perspectives on Properties of the Human Genome Project*, Amsterdam: Academic Press, Elsevier, p. 159.
151. Nuffield Council on Bioethics (2002), *The Ethics of Patenting DNA, A Discussion Paper*, London: Nuffield Council on Bioethics, http://www.nuffieldbioethics.org/go/ourwork/patentingdna/publication_310.html.
152. Cornish, William, Margaret Llewelyn and Mike Adcock (2003), *Intellectual Property Rights and Genetics: A Study into the Impact and Management of Intellectual Property Rights within the Healthcare Sector*, Cambridge: Public Health Unit.
153. European Union, Regulation (EC) No 816/2006 of the European Parliament and the Council of 17 May 2006 on compulsory licensing of patents relating to the manufacture of pharmaceutical products for export to countries with public health problems.
154. Shulman, S. (2001), 'Doctors without patents', Owning the Future, *Technology Review*, December.
155. Faunce, T. and P. Drahos (1998), 'Trade related aspects of intellectual property rights (TRIPS) and the threat to patients: a plea for doctors to respond internationally', *Medicine And Law*, **17**, 299–310.
156. Article 27.3(*a*) of the *TRIPS Agreement*, 1994.
157. Nielsen, J. and D. Nicol (2002), 'Pharmaceutical patents and developing countries: the conundrum of access and incentive', *Australian Intellectual Property Journal*, **13**, 289–308.

8. The alchemy of junk: patent law and non-coding DNA

> Genius of Junk is the story of how Malcolm Simons turned Junk into gold, enflaming one of the greatest controversies of our time – the control and ownership of our genetic material. ('Genius of Junk', *Catalyst*, Australian Broadcasting Corporation)[1]

GeneType was founded in 1989 by immunologist Dr Malcolm Simons and medical practitioner Dr Mervyn Jacobson. Their website provides this foundation story:

> [Simons and Jacobson] resolved to prove the non-coding ('junk' DNA) region of the human HLA gene complex [the human leukocyte antigen system] on Chromosome 6 is in reality not 'junk' but in fact a valuable and highly ordered reservoir of useful genetic information, largely overlooked by the rest of the world. The commercial mission then evolved that GeneType would seek exclusive ownership over access to this important genetic information and, ultimately, to exploit it globally for profit.[2]

Genetic Technologies Limited (GTG) was the result of a merger in 2000 between the original holding company, the private Swiss-owned GeneType AG, and a publicly listed Australian company, Duketon Goldfields Limited. After the corporate restructuring, GTG set a new goal of conversion to a biotechnology company.

GTG was able to obtain broad patents on a range of scientific inventions arising out of the work of Malcolm Simons. Most significantly, the United States Patent and Trademark Office (USPTO) awarded US Patent No. 5 612 179 to GTG for an invention entitled 'Intron sequence analysis method for detection of adjacent and remote locus alleles as haplotypes.'[3] Furthermore, the USPTO also issued US Patent No. 5 851 762 to GTG for an invention entitled 'Genomic mapping method by direct haplotyping using intron sequence analysis.'[4] The company has also applied for patents in respect of foetal cell recovery, retroviral-immuno therapy, and an ACTN3 genotype screen for athletic performance.[5]

A wide spectrum of the community could be affected by the patents related to non-coding DNA. GTG asserts that its genomic mapping methods can deal with monogenic diseases such as cystic fibrosis, sickle-cell anaemia and beta-thalassemia. Furthermore, it suggests that its markers can help identify multigenic diseases such as diabetes, colon cancer, and breast and ovarian cancer. In addition to identifying individuals at risk for genetic diseases, GTG argues that its patented inventions could be used in respect of forensics and paternity testing. The company also asserts that the patents have wider implications for agriculture, because they are relevant to the genetic testing of plants and animals.

Long-term, GTG aspires to become a comprehensive centre for genetic testing in the Asia-Pacific region.[6] With entrepreneurial bravado, Jacobson predicted, 'Our mission in relation to service testing now is to become the leading genetic testing facility in the Asia-Pacific region – the biggest and the best.'[7] Since the 1990s, the company has provided genetic testing in the field of disputed paternity. It currently provides paternity testing services to Queensland Legal Aid. AgGenomics Pty Ltd, the joint venture with Agriculture Victoria Services Pty Ltd, provides genetic testing and genomic services at the Plant Biotechnology Centre at La Trobe University, focusing mainly on plant and agricultural opportunities. In October 2002, GTG joined with Myriad to announce a strategic alliance in comprehensive cancer susceptibility testing. GTG would offer such testing in Australia, New Zealand and South East Asia, and Myriad would offer such testing in the rest of the world. In 2003, GTG announced that it would provide forensic testing at its laboratory services.[8] In 2005, GTG was listed on the NASDAQ, a move which the company hopes will give it access to capital markets in the US.[9] The company has also enlisted the former Australian treasurer, John Dawkins, as a director.[10]

Controversially, GTG has demanded large licence fees from private companies for access to its patents in respect of non-coding and genomic mapping. As at June 2007, the Australian firm had issued 32 commercial licences to biotechnology companies, and 5 research licences to universities and research institutions. The licences cover a range of activities, including plant genomics; animal testing; human genetic diagnostics; paternity and pathology testing; and the development of reagents. The licensees originate from the United States, Canada, the European Union, Australia and New Zealand. However, this aggressive licensing strategy has met with some resistance. Several United States companies, including Myriad Genetics,[11] Nuvelo Inc.,[12] Covance Inc.,[13] LabCorp,[14] Applera,[15] Monsanto and GE Healthcare Bio-Sciences Corp., initially brought legal action against GTG before reaching a settlement. There has also been legal action over the validity of the patents of GTG in New Zealand. After being asked to pay considerable patent royalties, the Auckland District Health Board brought an action against GTG, alleging groundless threats of legal proceedings. The matter was withdrawn after mediation, with the parties bearing their respective costs. However, a number of Crown Research Institutes in New Zealand agreed to pay commercial licensee fees to GTG.[16] GTG has encouraged universities and research institutions to take out research licences to use its patents on non-coding DNA and genomic mapping. In response, public researchers have raised doubts about the inventiveness of the patents held by GTG. The leaders of the public consortium of the Human Genome Project, Dr Francis Collins and Sir John Sulston, were particularly vocal in their concerns.[17]

After initial litigation,[18] Myriad Genetics paid US$1 million plus for licences to GTG's key non-coding DNA analysis patents and granted an exclusive licence to GTG to use and exploit its medical diagnostics in Australia.[19] The press release observed:

> Under the terms of the agreement, Myriad will receive a broad, non-exclusive license to Genetic Technologies' non-coding DNA analysis and mapping patents for all applications in human therapeutics and diagnostics. Genetic Technologies will become Myriad's exclusive marketing agent in Australia and New Zealand for its world-leading predictive medicine products for a range of important diseases, including breast cancer, ovarian cancer, colon cancer, melanoma and hypertension.[20]

After the announcement of this cross-licensing deal, GTG stressed that it was the only lawful provider of genetic testing held by Myriad in Australia. It emphasized that other service providers were guilty of patent infringement. After adverse publicity,[21] GTG announced that it would not enforce the patents on BRCA1 and BRCA2 in Australia. A press release said: 'GTG also announced that the intellectual property rights it had obtained from Myriad for breast cancer susceptibility testing will not be enforced by GTG against other service providers in Australia and New Zealand – and were a gift from GTG to the people of Australia and New Zealand.'[22] It was suggested by the ALRC that the company has now been estopped from taking legal action because of its public statements.[23] However, there remains some ambiguity about the status of this 'gift'. Although the Melbourne company can certainly decide not to enforce the patents, it cannot bind the patent owner, Myriad Genetics. Moreover, the *Four Corners* report revealed that GTG will still be seeking royalties in respect of the use of its non-coding DNA patents.[24]

This chapter[25] considers the recent international controversy over the patents held by GTG in respect of non-coding DNA and genomic mapping.[26] It questions whether Simons is a 'genius of junk', and whether his patents have the requisite novelty and inventiveness, according to the scientific knowledge at that time. Section one focuses upon the litigation between GTG and Applera. In the United States, GTG brought a legal action for patent infringement against Applera and its subsidiaries. Applera denied such allegations and counter-claimed that the patents of GTG were invalid because they failed to comply with the requirements of US patent law such as novelty, inventive step and written specifications. The matter was eventually settled. Section two examines parallel litigation in New Zealand. In New Zealand, the Auckland District Health Board brought an action against GTG in the Auckland High Court (in which it questioned the validity of the patents, claiming that it did not infringe such patents) and the Ministry of Health and the Ministry of Economic Development have reported to cabinet on the issues relating to the patenting of genetic material.[27] Section three considers the policy developments resulting from the controversy over patent law and non-coding DNA in Australia. The Australian Law Reform Commission (ALRC) has undertaken an inquiry into gene patents and human health,[28] and the Advisory Council on Intellectual Property (ACIP) is considering whether there should be a new defence in respect of experimental use and research.[29]

I JUNKYARD DOGS: *GENETIC TECHNOLOGIES LIMITED v APPLERA CORPORATION*

On 26 March 2003, GTG filed a patent infringement action in the US District Court for the Northern District of California against the major life sciences company, Applera, and its subsidiaries: Applied Biosystems Group, the Celera Genomics Group and Celera Diagnostics.[30] The complaint alleged that the companies were infringing US Patent No: 5,612,179, entitled 'Intron Sequence Analysis Method for Detection of Adjacent and Remote Locus Alleles as Haplotypes.'[31] The allegedly infringing products were cystic fibrosis reagent kits sold through Celera Diagnostics, and products 'relating to methods of analysis of non-coding sequence variants.'[32] The complaint also alleged that the companies in the Applera Group were infringing US Patent No: 5 851 762.[33] GTG sought 'monetary damages, costs, expenses, injunctive relief, and other relief as the court deems proper.'[34]

In its complaint, GTG alleged that a number of activities of the Applera Corporation infringed its patents on non-coding DNA and genomic mapping. First, the Applera Genome Initiative is engaged in the identification and selection of approximately 200 000 SNPs located in genes or gene-regulatory regions, and the validation of approximately 90 sets of human DNA, by generating individual genotypes and allele frequency data.[35] It also provided a framework for the haplotype map of the human genome. Second, Applied Biosystems develops and markets instrument-based systems, reagents, software and contract services such as

Assays-on-Demand™ SNP Genotyping kits, the Assays-by-Design[SM] SNP Genotyping kits, and the Celera Discovery System to the life science industry and research community. Third, Celera Diagnostics is involved in the discovery, development and commercialization of diagnostic products, including the cystic fibrosis ASR product. Fourth, Applera and its operating divisions are involved in commercial relationships to detect, identify and determine the chromosomal locations of various genes associated with one or more traits. Finally, Celera Genomics is engaged in integrated advanced technologies to discover and develop new therapeutics by leveraging its capabilities in proteomics, bioinformatics and genomics to identify and validate novel drug targets and to discover novel therapeutic candidates. GTG alleged that all of such activities fell within the scope of its patents.

A   Patent Infringement

GTG claimed that Applera has engaged in a number of activities which fall within the scope of US patent 5 612 179: 'Upon information and belief, Applera has designed, tested, manufactured, marketed, offered to sell, and sold its products and/or services, including, but not limited to the Cystic Fibrosis ASR kit, the Assays-on-Demand™ SNP Genotyping kits, the Assays-by-Design[SM] SNP Genotyping kits, and the Celera Discovery System.'[36] Furthermore, the company argued, 'Upon information and belief, Applera has engaged in activities within the scope of the Applera Genome Initiative with full knowledge of the claims of the US patent 5 612 179, and with full knowledge of GTG's rights therein.'[37]

Similarly, GTG argued that Applera had engaged in a number of activities that fall within the scope of US patent 5 851 762. It submitted that Applera had infringed the patent by undertaking activities in genomic haplotype analysis, including 'creating a haplotype map of the human genome; creating a haplotype map of the mouse genome; creating haplotype maps of genomic DNA of species other than mouse or human; and obtaining genomic DNA samples from individuals exhibiting a range of traits'.[38] Consequently, GTG argued that 'Applera has designed, developed, tested, manufactured, marketed and used certain products and/or services relating to haplotype analysis of genomic DNA with full knowledge of the claims of the US patent 5 851 762, and with full knowledge of GTG's rights therein.'[39]

In its first claim of relief, GTG alleged that Applera has engaged in direct infringement of its patent rights in both the 5 612 179 and the 5 851 762 patents.[40] The company observed: 'Applera's direct infringement of both the 5 612 179 and the 5 851 762 Patents has been with full knowledge of

GTG's rights therein.'[41] GTG claims that it is suffering 'irreparable harm' as a result of Applera's continuing direct infringement of both the 5 612 179 and the 5 851 762 patents.[42]

In its second claim of relief, GTG asserts that Applera has induced third parties to infringe both patents,[43] in that it 'has offered training, instruction or other advice to its customers, licensees or others in the use of its products, services and technology, including, but not limited to, the Cystic Fibrosis ASR kit, the Assays-on-Demand™ SNP Genotyping kits, the Assays-by-Design^{SM} SNP Genotyping kits, and the Celera Discovery System'.[44] It maintains that the alleged inducement of patent infringement was both 'wilful and malicious'.[45]

In its third claim of relief, GTG further claims that Applera has been engaged in contributory patent infringement[46] in having 'manufactured, offered for sale, and sold certain products and/or services' that 'constitute a material part of the inventions claimed in the '179 Patent'.[47] It argued, 'Applera's products and/or services relating to non-coding sequence variation detection and haplotype analysis and the methods of their use are not staples of industry, and they are not commodities suitable for substantial non-infringing use.'[48] It concluded, 'As a result of Applera's development, testing, manufacture and/or sale of its products and/or services, and as a result of its instruction, training or advice in the use of such products and/or services, Applera has contributorily infringed GTG's rights in the '179 Patent.'[49] It insisted that such contributory patent infringement was both 'wilful and malicious'.[50]

GTG has sought preliminary and permanent injunctions barring Applera from engaging in all of these allegedly infringing activities. In addition to costs, GTG sought damages 'in no event less than a reasonable royalty to GTG for the rights secured in both the '179 and the '762 Patents' and 'treble damages in compensation for the exceptional circumstances of Applera's infringement'.[51]

B   Answer and Counterclaims

In response, Applera put forward a number of affirmative defences to the claims of GTG.[52] Foremost, it declared that Applera had not infringed, and was not infringing, either patent. Further, it alleged that both patents failed to comply with the requirements of the US Patent Laws, including lack of novelty, inventive step and problems with the written specifications.[53]

Applera argued that the patents of GTG were invalid because they were anticipated and obvious in light of prior art. The company identified a range of prior art, which it alleged anticipated the patent applications filed by GTG.[54] The company observed the following:

At least as early as 1978, researchers began publishing RFLP studies showing correlations between non-coding DNA sequence variations and disease-related coding sequence variations. With the advent of PCR in the mid-1980's, many more researchers published such correlations and described them in patents . . . By the time GTG filed its first patent application, other researchers had applied non-coding sequence analyses to the 13-thalassemia, apolipoprotien B, and phenylalanine hydroxylase genes. Indeed, at least one researcher suggested that 'any coding gene, defective or normal, will be surrounded by a unique set of DNA polymorphisms,' and another observed that if non-coding gene regions are not conserved, '[i]t should be thus possible to derive from the non-coding and flanking regions of the HLA-DQa gene probes capable of distinguishing a specific allogenotype in the presence of another'.[55]

In addition, Applera asserted that, by reason of the proceedings in the USPTO during prosecution of the applications that resulted in the issuance of both patents, 'GTG is estopped from asserting any construction of the claims of the patent which would cover or include any of the purported acts of infringement of which GTG complains'.[56] Finally, Applera argued that the claims of GTG were barred by the defence of laches. The company sought declaratory judgments of non-infringement and invalidity of both patents.

There has been great debate over the validity and the scope of the patents held by GTG, especially in terms of their novelty and inventiveness. Mervyn Jacobson of GTG contends that the patents are novel, non-obvious and useful.[57] He maintains that this view is supported by a number of testimonials from famous scientists, such as Professor Leroy Hood, Professor Peter Gresshoff, Professor Pablo Rubinstein, and Dr Brian Tait.[58] Jacobson claimed that the research on non-coding DNA by Malcolm Simons was breakthrough science. He argued that geneticists failed to appreciate the significance of the work performed by immunologists: 'the HLA experts had their own conferences and don't particularly meet with other human geneticists, who don't meet with the other animal geneticists, who don't meet at conferences with plant geneticists – in fact they even use different language, the nomenclature, they use words differently, it's almost like they're on different planets'.[59]

GTG has only quoted selectively from such expert testimony in its public relations.[60] In July 1993, Professor Leroy Hood provided a testimonial in respect of the patents held by Malcolm Simons: 'This observation was a great surprise to me.'[61] His own data indicated the presence of informative polymorphisms in non-coding regions of these 'vastly different types of genes'[62] in a range of species. He concluded that these informative polymorphisms, which are indicative of haplotypes and alleles, are 'present throughout the eukaryotic genome'.[63] In February 1992, Professor Peter Gresshoff commented that Simons' work was unexpected: 'Simons' data may indicate that recombination within functional regions is suppressed, and occurs only at "hotspots" at the end of a transcriptional unit, a genetic focus.'[64] Professor Pablo Rubinstein described Simons' work as 'totally unexpected'.[65] He said, 'There was no reason, *a priori*, to believe that introns would be informative in this regard.' He said that there was no reason to expect that Simons' data relating to the HLA system would be 'an isolated curiosity of nature'.[66] He believed that it could be reasonably expected that the same principles would govern the evolution of alleles in other multi-allelic gene families. Dr Brian Tait wrote in February 1992: 'The state of the art at the time of Malcolm Simons' invention was that introns and other non-coding regions were thought to be fairly random DNA arrangements.'[67] He observed, 'However, we did not recognise that the non-coding region polymorphisms generally were sufficiently conserved to provide a complete typing system.'[68] He described Simons' methods as 'a significant advance' that could be used on a daily basis in the laboratory to type alleles.[69] The company was reluctant to release publicly such expert testimony in full in the course of the legal proceedings with Applera.[70] The quality and credibility of such expert evidence would have an important bearing on the determination of the patents' validity.

Malcolm Simons and his friend the patent attorney, Carol Nottenburg, argue that the patents are indeed novel and inventive, but have expressed reservations about the scope of the claims asserted by GTG.[71] The original inventor has issued a clarification concerning his view of the non-coding DNA patents.[72] He sought to dispel some of the misconceptions about his research: 'It has become obvious to me that the disbelief in, and reactions to, the patents mainly arise from imprecision in the wording of the patent claims, and from a failure to clearly distinguish my discoveries and inventions from prior art.'[73] Simons comments on the relationship of his research to the prior art:

> I am unaware of any current DNA molecular laboratory test for disease-associated gene diagnostics, or for HLA typing, that utilizes the method of the Intron Diagnostic patent. By contrast, all uses of population-based LD / Allele association fine-mapping seem to me to be encompassed by the Genome Mapping patent.[74]

Thus Simons has also expressed reservations about the breadth of the patent claims made by GTG. Indeed, the morning after the *Four Corners* report, the scientist appeared on Nine's *Today* programme to protest GTG's behaviour. The immunologist told *Forbes* that he was willing to take the stand for either side to clarify the patent if the Applera lawsuit goes to trial.[75] The inventor, therefore, is something of a maverick. He is an unpredictable, uncertain element in the whole controversy.

By contrast, a number of researchers and scientists were sceptical of the inventiveness of the patents, finding the broad claims to be dubious. Members of the public Human Genome Project have expressed doubts about the validity and the scope of the patents. Francis Collins was surprised that the patent office had granted patents to GTG in respect of non-coding DNA given the requirements of the USPTO's novelty, non-obviousness and utility standards: 'After all, there were many prior published reports on the correlation of variation in noncoding regions with important mutations, going back at least to Kan and Dozy's *The Lancet* report on the sickle mutation back in 1978.'[76] Nobel Laureate Sir John Sulston was sceptical of the portrayal of Simons as 'the genius of junk'. He observed, 'the generality that there are very important sequences outside the narrow specific protein coating bits was well known throughout the 80s'.[77] Sulston argues that the scope of the patent is of particular concern: 'What he cannot possibly be justified in doing at that time is claiming all of the non-coding sequence in all organisms.'[78] Martin Bobrow of Cambridge University argued that the patents were a sign that biotechnology patents were too easy to obtain and that the rules governing patentability needed to be tightened: 'Broad patents that lead to extraordinarily large rewards for extraordinarily little inventive input are a wholly destructive trend.'[79]

A number of Australian and New Zealand geneticists entered the debate about the patents relating to non-coding DNA and genomic mapping. Professor John Mattick of the Institute of Molecular Biosciences was concerned that the *Catalyst* programme gave the misleading impression that his work vindicated Malcolm Simons' claims to inventive insights about non-coding sequences.[80] He expressed his views that the patents were lacking in novelty and inventive step on the *Four Corners* programme.[81] Professor Peter Little of the University of New South Wales doubted GTG's claim that its principals made important discoveries concerning the nature of 'junk' DNA that were not appreciated in 1989 by the scientific community.[82] He observed the following:

> It is unclear to me why, in 1989, it was necessary to prove the idea that linked polymorphisms could be used to analyse functional variation: the fundamental principles and practice had been widely published, and these could be simply applied to any gene, including the HLA complex. Importantly, the concepts of haplotypes, linkage disequilibrium, and linkage had all been identified as directly relevant to the DNA-based analyses then available. GTG's contention that its principals had discovered something that was 'largely overlooked' is not supported by the scientific literature. The comment that non-genic DNA is 'a valuable and highly ordered reservoir of useful genetic information' is simply a restatement of what was first demonstrated in 1978 and applied widely. In this strict sense, such DNA can never be truly 'junk' by virtue of its linkage to genes and must always be of potential utility.[83]

Professor Joe Sambrook of the Peter MacCallum Cancer Institute observed that the idea of using stretches of junk DNA to track genes was well-established by the mid-1980s.[84] Similarly, Dr Graeme Suthers, a clinical geneticist, commented, '[t]he usefulness of non-coding DNA in biomedical research has been recognised for decades'.[85] Associate Professor Paul Waring has contended that Simons' invention should be limited to the use of non-coding haplotypes to determine coding region alleles in individuals without the need for information about other family members. He doubts whether the patent claims can be expanded from the HLA locus to non-coding DNA in a wide range of organisms.[86]

C Settlement

Applera was initially reluctant to yield to the demands to pay licence fees in respect of non-coding DNA. It was very proud that it was leader of the private efforts to sequence the human genome. The company was unwilling to relinquish the status and kudos associated with the achievement of that 'Big Science' project. Ironically, given its large patent holdings in the field of biotechnology, Applera became the white knight for universities and public research institutions. The company drew upon public research to help challenge the validity of the patents held by GTG.

In the meantime, GTG struggled to generate revenue and provide dividends to its shareholders. Since its formation as a public company, GTG has accumulated significant losses, in spite of its occasional licensing windfalls. The company protested that its revenues had been affected by ongoing litigation. The *2005 Annual Report* noted, 'It is unfortunate to note that, in recent times, the legal action against Applera Corporation has absorbed management resources that could have otherwise been spent pursuing licenses from other companies.'[87] To be fair, GTG is unexceptional in this respect: many biotechnology firms struggle to generate profits.

After court-ordered mediation, GTG and Applera reached a final settlement of the patent dispute in December 2005.[88] The two companies executed a number of binding agreements, including a final settlement agreement, licence agreements and a supply agreement. The terms of the settlement remained confidential: 'The commercial terms of the settlement reached between GTG and Applera are subject to confidentiality requirements, but it can be disclosed that the settlement also includes a license to the GTG non-coding patents.'[89] The Northern California District Court formally dismissed the law suit between GTG and Applera on 30 December 2005.

After this announcement, the Australian Stock Exchange demanded additional details regarding the material terms of the agreement. In response, GTG reported that the final settlement included Applera taking a licence to the GTG non-coding patents, and making payments to GTG in the form of cash, equipment, reagents and intellectual property.[90] GTG estimated that the value of such agreements was $AU15 million. The company stressed that the settlement would have further strategic benefits: 'GTG believes that its settlement with Applera is its most strategic and therefore valuable deal to date.'[91] For the most part, the marketplace appeared to be disappointed with the size of GTG's settlement with Applera. Financial pundits noted that 'the market was obviously expecting something bigger. It swiped more than 20 per cent off the market value of the company'.[92]

Jacobson was triumphant about the settlement over the Applera litigation, suggesting that the share price of GTG should be re-valued in light of the outcome: 'For a little Australian company to file a lawsuit against Applera, and have the resources to see it through, and bear the associated legal costs for three years, is obviously very significant'.[93] Applera was disgruntled by Jacobson's characterization of the settlement. Company representative Peter Dworkin said that such statements were misleading: 'The facts are that Applera has never conceded the validity or infringement of GTG's patents, and settled the case on very favourable terms for Applera in order to spare it and its customers further distraction by the litigation.'[94]

II TRANS-TASMAN RIVALRY: *AUCKLAND DISTRICT HEALTH BOARD v GENETIC TECHNOLOGIES LIMITED*

In 2000, the New Zealand Government held a Royal Commission into Genetic Modification.[95] Chapter 10 of the final report provided a brief overview of some of the issues arising in respect of intellectual property and biotechnology.[96] The report briefly canvassed some of the ethical issues arising from patenting living organisms.

In July 2003, Cabinet directed officials of the Ministry of Health and the Ministry of Economic Development to report to Cabinet on the issues relating to the patenting of genetic material and, in particular, whether there was a need for further public consultation on these issues.[97]

In November 2003, the Minister for Health, Annette King, and the Associate Minister of Commerce, Judith Tizard, released a report, *Implications of Granting of Patents over Genetic Material*.[98] The report mentioned the litigation over the patents held by Myriad Genetics in respect of BRCA1 and BRCA2. It discussed at length particular concerns about the patents held by GTG: 'A number of the organisations approached by GTG expressed concern at the relatively high licence fees being charged, the excessive breadth of the patents and have questioned whether these patents in fact presented any novel information at the time of issue.'[99] The report recommended that the committee 'note the health and life science sectors are currently in discussion with an Australian company, Genetic Technologies Ltd, regarding their patents on non-coding DNA'.[100] It observed: 'These patents are of concern because of their breadth and the excessively high licensing fees being asked.'[101] It also stressed the need for a response: 'The Ministry of Health is concerned to limit the risk from such patents being granted or exercised in the future.'[102]

The report noted that the *Patents Act* 1953 (NZ) was currently under review. A number of reforms could have an important impact upon gene patents:

> The introduction of examination for obviousness will reduce the likelihood that patents will be granted over genetic material, or applications of genetic material that are no more than obvious variations on what is already known. Changes to the utility requirement will mean that patents will not be granted unless the invention is shown to have a 'substantial, credible and specific' use. This makes it less likely that patents will be granted over genetic material that has no demonstrated 'real world' use.[103]

Furthermore, the report notes, '[t]he amended Patents Act will expand on the current exclusion from patent protection, of inventions whose commercial exploitation would be contrary to morality or '*ordre public*', to include 'where the prevention of such exploitation is necessary to protect human, animal or plant life or health, or to avoid serious prejudice to the environment'.[104] Most significantly, it will be possible to revoke a granted patent on this ground.

In June 2004, a second Cabinet paper was released, dealing with options to address genetic patents.[105] First, the New Zealand Government was reluctant to prohibit patents on genetic material per se because of the difficulties in defining types of genetic material (given the rapid advancements in the field of biotechnology), the large amount of genetic material already patented, and New Zealand's international treaty obligations.[106] Second, the New Zealand Government emphasized the need to limit the breadth of claims made in biotechnology patents through the strict application of the criteria for patent validity. It noted that 'the way in which the Intellectual Property Office of New Zealand (IPONZ) applies the new criteria on applications involving biological material will be crucial in determining the breadth of patents on genetic material'.[107] Third, the New Zealand Government recommended a review of the *Patents Act* 1953 (NZ): 'This review would examine how the criteria of the Act have been implemented and how the new criteria have been applied.'[108]

A Licensing

In a letter dated May 2003, GTG proposed two alternative proposals to Dr David Sage, the Chief Medical Officer of the Auckland District Health Board (AHDB) for use of its patented inventions. In its preferred proposal, GTG mooted a public–private partnership with the ADHB and the Ministry of Health in New Zealand.[109] GTG proposed that '[c]onsideration for such a national licence would involve a signing fee to address past infringement nationwide and an ongoing annuity for the life of the patents which, in the case of New Zealand, is until May 10, 2011'.[110] It suggested that '[w]e propose the fee for signing and waiving of past infringement be NZD10 million and the ongoing annuity be NZD2.0 million per year.'[111] The public–private partnership would consist of $NZ 5 million being offered back to New Zealand universities for research.

Alternatively, GTG proposed granting a licence to each testing laboratory individually. Such a licence would include the following:

(a) Unrestricted use of the GTG non-coding DNA patents in human diagnostic testing by that particular laboratory.
(b) Free access to the breast cancer susceptibility testing IP of Myriad Genetics that have been exclusively licensed to GTG for New Zealand.
(c) Co-operation in potential subcontracting of laboratory testing between the particular laboratory and GTG on a commercial basis.[112]

The company observed, 'We believe that volume-related licence conditions would be impractical to administer and instead propose a simple arrangement, applicable to each laboratory, that would comprise a signing fee of NZD2.0 million and an ongoing annuity of NZD200 000 p.a. to May 10, 2011.'[113]

Jonathon Holmes questioned Jacobson as to whether the New Zealand Government should pay such fees for the use of the non-coding DNA patents. The journalist noted that, under the proposed agreement, '[I]f they don't pay for the non-coding patents, they won't get the Myriad patent either.'[114] Jacobson responded that 'the New Zealanders should be pleased that the test that they've been getting which is subject to a New Zealand government patent and performed by government agencies illegally will be offered lawfully.'[115] He elaborated that the quantum of the amount should not be given too much attention: 'Some of those organisations themselves don't think it's an awful lot of money but it's rather difficult to set a balanced value when you don't know how many tests are being done and you don't have all the information on which to base an intelligent assessment.'[116]

The journalist asked whether GTG would retain intellectual property rights in research that it invested in. Jacobson observed, 'That's normally the process when private or for-profit companies fund research in public institutes.'[117] He nonetheless maintained that there would be some benefit to the New Zealand taxpayer arising from the commercialization of any resulting intellectual property.[118]

Jonathan Holmes observed that at least a dozen Crown Research Institutes in New Zealand, and a number of privately funded companies, had received letters from GTG suggesting that they needed licences. He asked, 'is GTG using New Zealand as some kind of test bed for Australia, in terms of licensing non-coding patents?'.[119] Jacobson denied any such intentions: 'We see Australia and New Zealand as our home territory and we are trying to adopt a very supportive and helpful view in bringing our technology to Australia and New Zealand.'[120] Jacobson concluded, 'These are flexible ongoing negotiations, the final quantum always has to fit in or it doesn't happen or it doesn't work.'[121]

In 2004, GTG representatives made several visits to New Zealand to talk to the Government, Crown Research Institutes and private research laboratories about licences they say they must buy in order to carry out gene technology research legally.

The ADHB acted on behalf of all the district health boards in New Zealand.[122] Jacobson said that the company revised its licence fee down to the one-off figure of $NZ 560 000. ADHB's lawyer, Bruce Northey, had corrected the original figures and provided more accurate information about the level of use of the genetic tests by the 20 health boards, which was much lower than originally estimated. He observed that the time it would take to settle the matter depended on 'how much he [Jacobson] wants to arm wrestle'.[123]

B Litigation

In August 2004, ADHB announced that it was taking pre-emptive legal action against GTG in the High Court in New Zealand.[124] Northey said ADHB decided to challenge GTG following an extensive analysis of the scientific and legal basis of GTG's patents, involving members and advisers to the New Zealand health services community. ADHB was satisfied that its DNA testing did not infringe the patents and doubted that the patent claims were valid. ADHB has accused GTG of using implied threats of legal action to force the ADHB to pay access and royalty fees for the GTG patents.

Northey confirmed that GTG's co-founder and inventor of the company's non-coding DNA gene-testing techniques, Dr Malcolm Simons, had been advising the ADHB and its legal counsel during its analysis. It had yielded a 'consensus understanding' of the claims and their relevance to all the genetic tests used by the ADHB that could be subject to licence fees if GTG's interpretation of its patent rights was upheld.[125] Northey said, 'Malcolm's summary was that, while the patents can be interpreted to relate to what he invented, nobody employing the methodology would be infringing the patents.'[126] Northey said that the ADHB was prepared for the court case to happen as quickly as possible. The ADHB issued proceedings in the High Court in Auckland alleging 'groundless threats of infringement' by GTG.[127]

In response, Jacobson was critical of ADHB's account (in the media release) of negotiations with GTG over the past 18 months. He questioned why the New Zealand Government would issue a patent and then challenge the validity of such an invention: 'By failing to respect a valid patent, the ADHB – itself a government agency – is undermining the integrity of the New Zealand patent system.'[128]

Such comments, of course, overlook the obvious point: the New Zealand Intellectual Property Office operates independently from the Government. The health boards are perfectly entitled to take legal action to challenge the validity of granted patents.

C   Settlement

In June 2005, the New Zealand High Court appointed a retired judge, Justice Barry Pattinson, to oversee mediation between GTG and ADHB.[129] In July 2005, the parties reached a settlement. As a consequence of this settlement, the current High Court proceedings between the parties will now be withdrawn, without payment by either party to the other. In addition, both parties have agreed not to pursue the other in future, in relation to these patents.[130] In addition, as part of the same settlement, GTG is now granting commercial licences to the non-coding patents to four commercial New Zealand entities – AgResearch, HortResearch, Forest Research and Livestock Improvement Corporation – who will together pay $NZ 450 000 to GTG. The parties agreed to keep the precise terms of the settlement confidential, while allowing each to comment on its circumstances.

The *New Zealand Herald* reported the result as a victory for the ADHB: 'In the first case of its kind against Government-funded organisations, GTG has been sent packing empty-handed after mediation which saw High Court proceedings withdrawn without costs and both parties agreeing not to pursue each other again in future over the patents.'[131] Likewise, the *Independent Business Weekly* also suggested, 'New Zealand's publicly funded health boards continue to defy the licence-holder of a breast-cancer gene test they use, despite settling a two-year legal dispute last week.'[132] ADHB lawyer Bruce Northey said that the mediation process was 'interesting'. He was satisfied with the outcome and with the fact that the District Health Boards would not have to pay GTG fees. 'You come along to a country and demand considerable sums of money and you walk away with nothing.'[133] GTG's decision not to claim royalties 'vindicated' the ADHB's decision to challenge the validity of GTG's patent claims. He said that GTG had declined to produce evidence for 'robust scientific and commercial scrutiny' of the patents and their relationship to the health boards' activities.[134]

GTG executive chairman Mervyn Jacobson, was philosophical about the settlement, seeking to play down its significance, suggesting that New Zealand was a special case:

> The overall message is that drawn-out disputes are not good for either side. We needed to reach a settlement and get on with life. For whatever reason, they initially chose to negotiate, and then fight, but finally, we've reached a settlement that is very acceptable to both parties. From GTG's perspective, it addresses one small group, in one small country. We can now get back to the big picture, including our Nasdaq listing and our global licensing strategy.[135]

Jacobson said that he was also happy with the outcome or the company would not have settled. GTG was receiving royalties from other patents in New Zealand and the decision would have no impact on those or on future business opportunities. However, he said Northey was being 'a little self-serving talking about vindication'.[136] Jacobson concluded that, at the very best, the ADHB had only gained a pyrrhic victory: 'The reality is they spent between $700 000 and $1m to prove a point which they could have achieved from working with us for half that amount.'[137]

III GENE PATENTING AND HUMAN HEALTH: THE AUSTRALIAN LAW REFORM COMMISSION INQUIRY

In December 2002, the Federal Attorney General, The Honourable Daryl Williams, commissioned the Australian Law Reform Commission (ALRC) to undertake a review of intellectual property rights over genes and genetic and related technologies, with a particular focus on human health issues.[138] This inquiry was prompted in particular by concerns about the impact of patents held on genetic testing by biotechnology companies such as Myriad Genetics and GTG.[139] In this context, the ALRC considered the impact of gene patents upon research (and its subsequent application and commercialization), the Australian biotechnology sector, and the cost-effective provision of health care in Australia. There was much political debate over the impact of gene patents upon research, health care and competition.

Along with biotechnology and pharmaceutical companies, GTG argued that the current regime of patent law should not be changed, lest investment in the marketplace be affected. Jacobson stated:

> I see the patent process as a very wholesome process. It's been around for 400 years. It started in Britain, but most countries in the world have adopted it, modified it. It works very well. You interfere with that process, you interfere with invention, you interfere with innovation, with risk taking. And, in fact, in Australia, if you drastically interfere with an established process, you run the risk of damaging or destroying biotechnology in Australia, which not only harms biotechnology companies but it's negative for Australian health care.[140]

GTG argued that isolated genetic materials and genetic products should be regarded as 'inventions' rather than 'discoveries,' for the purposes of Australian patent law. It argued against prohibitions in respect of the patenting of genes, methods of human treatment, or medical diagnostics. GTG was hostile to the idea that ethical and social concerns about patents on genetic materials and technologies should be addressed through the patent system. It denied that there was any need to make special provision for individuals or groups whose genetic samples are used to make a patented invention, to benefit from any profits from the patent. GTG argued that changes to the requirements for patentability under Australian law for inventions involving genetic materials and technologies would hinder Australia's quest to be at the cutting edge of global research and conflict with Australia's obligations under the *TRIPS Agreement* 1994.[141]

By contrast, health care providers, scientific groups and academics argued that there was a need to reform the patent system in order to accommodate genetic technologies.[142] Some considered solutions that regulated the grant of patents, such as refining the standards of novelty, inventive step and utility. Others considered whether there should be new defences to claims of infringement of gene patents, such as where patents are used for research, for private non-commercial purposes or for medical treatment. Some supported solutions that regulated the exploitation and abuse of patents. They investigated the circumstances in which Crown use,[143] compulsory acquisition[144] and the compulsory licensing provisions[145] of the *Patents Act* 1990 (Cth) could be used.

Finally, a number of idealists expressed per se objections to the patenting of genes and gene sequences. For instance, Graeme Suthers of the South Australian Clinical Genetics Service argued that biological inventions should not be allowed to be patentable subject matter: 'The law should be changed so that human genetic information cannot be placed under private control.'[146] Such complaints were grounded in ethical concerns about the commercialization of scientific discoveries.

A   Final Report

The ALRC released an issue paper in July 2003,[147] a discussion paper in January 2004,[148] and a final report in June 2004, which was tabled in Parliament in August 2004.[149] The leviathan 678-page final report contains 50 recommendations. The ALRC took a decidedly tentative approach in its final report: 'In view of the equivocal nature of evidence about adverse impacts on research and healthcare, the ALRC considers that it should adopt a cautious approach towards recommending radical changes in patent law and practice.'[150] The ALRC offered various recommendations for reform to government, independent agencies, industry and funding agencies. It noted, 'in a more complex environment in which authority is more diffused, modern law reform efforts usually involve a mix of strategies, including legislation, guidelines, principles, education programs, and changed practices'.[151] As a result, only a few policy recommendations required legislative action.

First, the ALRC recommended that the Australian Government reform the *Patents Act* 1990 (Cth) to provide that an invention will satisfy the requirement of usefulness in section 18(1)(c) of the *Patents Act* 1990 (Cth) 'only if the patent application discloses a specific, substantial, and credible use'.[152] The law reform body observed the following: 'The ALRC considers that reform is needed to the way in which the usefulness of an invention is addressed in the requirements for patentability.'[153] Article 17.9.12 of the *Australia–United States Free Trade Agreement* 2004[154] declares, 'Each Party shall provide that a claimed invention is useful if it has a specific, substantial, and credible utility.'[155] As a result of this Article, the Australian Government will be obliged to adopt USPTO utility standards. Recently, *In re Fisher*, the United States Court of Appeals for the Federal Circuit held that Monsanto could not patent express sequence tags in respect of maize because there was a lack of utility and a lack of disclosure.[156] No doubt this will also be a persuasive authority in Australia.

Second, the ALRC proposed that the Australian Government should recognize a defence under patent law for experimental use:

> The ALRC believes it is desirable to remove uncertainty about the existence and scope of an experimental use exemption in Australian law. This approach received broad support in submissions. The existing uncertainty is unhelpful to the research community and commercial organisations. It has the potential to result in under-investment in basic research; and to hinder innovation if researchers become concerned that their activities may lead to legal action by patent holders.[157]

The ALRC rejected the narrow view of the research exemption adopted by the US Court of Appeals for the Federal Circuit in *Madey*, which affirmed that the defence was limited to actions performed 'for amusement, to satisfy idle curiosity, or for strictly philosophical inquiry'.[158] It instead supported the approach in the European Union, in which the defence of experimental use extended to research on a patented invention, but not with an invention. The ALRC concluded, 'Moreover, basing a new provision on the European Union model would promote harmonization of Australian patent law with the law of a major trading bloc, and would give Australian courts the benefit of considering European case law in applying the new provisions.'[159]

Third, the ALRC proposed a number of amendments to the existing compulsory licensing regime in chapter 12 of the *Patents Act* 1990 (Cth). 'Given the unique nature of many biotechnology inventions, and hence their possible lack of substitutability, the anti-competitive exploitation of a patent could have significant implications for downstream research or access to certain healthcare services.'[160] The ALRC recommended that the federal government should amend the *Patents Act* 1990 (Cth) to insert a competition-based test as an additional ground for the grant of a compulsory licence. It believed that such a test would address those circumstances in which there is a public interest in enhanced competition in a market, and the patent holder has not met reasonable requirements for access to the patented invention. The provision of compulsory licensing would be useful in the future, if a company abuses its dominant market position. Such measures would also be relevant in circumstances where the patent holder blocked access to inventions for research, treatments and diagnosis.

Fourth, the ALRC promoted the use or acquisition of patented technologies pursuant to the Crown use provisions in chapter 17 of the *Patents Act* 1990 (Cth).[161] The report observed: 'These [Crown use] provisions may be seen as a "safety valve" in particular cases, preventing the public interest from being subverted by the patent system.'[162] The ALRC advised that the Commonwealth should amend the *Patents Act* 1990 (Cth) to clarify that, for the purposes of the Crown use provisions, an invention is exploited 'for the services of the Commonwealth or the State' if the exploitation of the invention is for the provision of health care services or products to members of the public. Furthermore, it suggested that the Commonwealth should amend the *Patents Act* 1990 (Cth) to provide that, when a patent is exploited or acquired under the Crown use or Crown acquisition provisions in chapter 17 of the *Patents Act* 1990 (Cth), the Crown must pay remuneration or compensation.[163]

Finally, the ALRC considered the relationship between intellectual property law and competition law in the context of biotechnology.[164] The

ALRC recommended that the Commonwealth should amend section 51(3) of the *Trade Practices Act* 1974 (Cth) (*TPA*) to clarify the relationship between Part IV of the *TPA* and intellectual property rights. The ALRC also advised that the Australian Competition and Consumer Commission (ACCC) should develop guidelines to clarify the relationship between Part IV of the *TPA* and intellectual property rights. The ALRC envisaged that such guidelines should extend to the exploitation of intellectual property rights in genetic materials and technologies, including patent pools and cross-licensing. As the need arises, the ACCC should review the conduct of firms dealing with genetic materials and technologies protected by intellectual property rights, to determine whether their conduct is anti-competitive within the meaning of Part IV of the *TPA*. The ALRC also recommended that Commonwealth, State and Territory health departments, and other stakeholders, should make use of existing complaint procedures under the *TPA* where evidence arises of conduct that may have an adverse impact on medical research or the cost-effective provision of health care.

Arguably, this reform agenda is a modest one. The solutions proffered by the ALRC are aimed mainly at the exploitation of gene patents. The fundamental weakness of the final report was its failure to address the initial grant of gene patents by the Patent Office. The ALRC failed to address patent criteria dealing with the threshold of inventiveness, such as 'novelty' and 'inventive step'. As a result, there remains a danger of the Patent Office granting broad patents of dubious inventiveness under the *Patents Act* 1990 (Cth). Arguably, the ALRC should have sought to amend the *Patents Act* 1990 (Cth) to raise the standards of 'novelty' and 'inventive step' in section 18(1)(b).

In his dissenting judgment in *Aktiebolaget Hässle v Alphapharm*, Kirby J observed that patents should only be granted if there is sufficient ingenuity: 'It is not diligence and determination or the input of time, labour, skill and effort or the expenditure of resources that meet the criteria in the Act.'[165] His Honour argued that there was a need for the test of 'novelty' and 'inventive step' of the *Patents Act* 1990 (Cth) to reflect the complexities of contemporary science:

> But the Act talks to science and invention at different stages. Its origins lie in earlier centuries and nowadays science, in the field of nuclear physics and the field of biology and in the field of informatics, has gone beyond the scope, immediate Eureka-type exclamations, it is more complex, and therefore, if the Act is to speak with relevance to science and technology as they exist today, the ultimate question that has to be addressed is whether in that moving context what is obvious moves with that change and therefore that with the advance of the availability of information.[166]

The ALRC should have also directed judges to attribute greater creativity and problem-solving skills to a person 'skilled in the art'. McGill University academics Richard Gold and Karen Durrell have argued that 'the skilled reader permits the courts to introduce flexibility into patent law so that the context in which inventions are made and used is considered'.[167] These Canadian authors suggested that this device is particularly useful in dealing with biotechnological inventions: 'The use of the skilled reader permits the adaptation of patent law in a flexible yet transparent manner that at once ensures the continued relevance and functioning of patent law with the ability to adapt the law to take into account the particularities of new technologies such as biotechnology and information technology.'[168] Perhaps the GTG patents on non-coding DNA and genomic mapping would have never been granted if such rigorous standards had been in place.

Furthermore, the ALRC was remiss in its failure to canvass a wider range of patent infringement and exceptions to patent infringement. The introduction of a defence of experimental use alone is insufficient. Given the expansion of the scope of patentable subject matter, there is a need in turn to broaden the range of exceptions to patent infringement. The ALRC should have sought to take advantage of the flexibilities available for patent exceptions under the *TRIPS Agreement* 1994. In addition, there should be a defence in respect of personal, non-commercial use of patented inventions. The federal government should also introduce a limited liability for medical practitioners in respect of patent infringement. This position is a necessary response given that methods of human treatment have been treated as patentable subject matter.[169] The report was also somewhat hollow because it was reluctant to engage with the ethical concerns held by many of the public in respect of gene patents. It is recommended that the *Patents Act* 1990 (Cth) be amended to provide the requirement that a patent can only be granted if there is evidence of informed consent and benefit sharing.[170] This requirement is a general one. It could arise in the context of genetic testing and biomedical research.

B Government Responses

Some commercial entities sought to portray the report as showing no empirical evidence of impacts on research and health care. GTG Director, Deon Venter, argued, 'The report points out that many of the fears expressed about the much-touted negative impacts on research and healthcare were groundless.'[171] In response, GTG made the following controversial comments about the verdict of the final report delivered by the ALRC:

GTG wishes to reaffirm it fully supports the work of ALRC. Indeed, GTG was pleased to be invited twice, and to have appeared twice, before ALRC, to offer its expertise and views to ALRC. In summary, ALRC has confirmed to GTG that it saw no problem with GTG's patents or licensing practices, and GTG is pleased with the findings of ALRC.[172]

Such a gloss on the final report is inaccurate. The ALRC has given no such public or private undertakings that it had no problem with GTG's patents or licensing practices. Indeed, the law reform body studiously avoided dwelling upon any particular controversy in the belief that such an approach would be unrepresentative. Instead, the ALRC offered general recommendations on reforms to patent law and genetic technologies. It is not the role of the law reform body to rule one way or the other upon the legitimacy of GTG's inventions. The validity of the patents and the legitimacy of the licensing practices is ultimately a matter for the Patent Office and the courts.

It is fair to say that the ALRC did not favour the opponents of gene patents. The law reform body was unwilling to tamper with the broad, open-ended definition of 'manner of manufacture'. Graeme Suthers argued that patents should not be granted in respect of genes and gene sequences on the grounds that they were mere scientific discoveries.[173] By contrast, GTG argued that isolated genetic materials and genetic products should be regarded as 'inventions', rather than 'discoveries', for the purposes of Australian patent law. The company argued against prohibitions in respect of the patenting of genes and gene sequences. In the end, the ALRC held that there should be no absolute prohibitions upon the patenting of genes and gene sequences.

Ultimately, the ALRC did not accept the submissions of GTG that there should be no reforms to the Australian patent system, and that the status quo should be preserved. The Commission made a number of modest recommendations for reform of the *Patents Act* 1990 (Cth). The ALRC recommended the establishment of a defence of experimental use, despite the protests of GTG that there was no need for such an exemption. The ALRC also called for reforms to compulsory licensing, and Crown use, to enable such mechanisms to be more accessible. By contrast, GTG was hostile to government intervention in relation to patents that had been granted by the Patent Office. In its view, the marketplace should resolve issues concerning access to patented inventions.

The Federal Government showed some initial reluctance in tabling the final report of the ALRC on gene patents and human health. In August 2004, there was a debate raging over whether the proposed *Australia–United States Free Trade Agreement* 2004 would result in the evergreening of pharmaceutical drug patents.[174] The Liberal Federal Government argued that the patent system was working well, and there was no need to introduce reforms to the implementing legislation. By contrast, the Opposition maintained that there was a need for amendments to discourage the evergreening of pharmaceutical drug patents. In the end, the major parties agreed to pass the *Australia–United States Free Trade Agreement* 2004 subject to the amendments proposed by the Opposition. Once this controversy subsided, the Federal Government released the final report of the ALRC on gene patents and human health, in September 2004. There was little political comment upon the findings of the final report, which was unsurprising given that the federal election took place in October 2004.

Since its re-election, the Liberal Federal Government has yet to respond to the recommendations of the law reform body in respect either of gene patenting and human health, or of the protection of human genetic information. There remain doubts as to whether the topic will be a legislative priority. Particular issues have been hived off for further consideration by other law reform bodies. The Advisory Council on Intellectual Property has undertaken inquiries into both the defence of experimental use and Crown Use.[175] The government has yet to table its response to such investigations.

The Federal Government added a competition test to the compulsory licensing regime and expanded the scope of springboarding under the *Intellectual Property Laws Amendment Act* 2006 (Cth). However, such amendments are minor, and will do little to address the policy issues raised by the Australian Law Reform Commission inquiry.

CONCLUSION

This chapter has considered the public debate over the patents held by GTG in respect of non-coding DNA, haplotyping and genomic mapping. It has suggested that such patents are deserving of particular attention, because of both the exceptional breadth of their claims, and their potential impact upon industry, science and health care. Professor John Mattick of the Institute for Molecular Biosciences comments upon the scope of these patents: 'I think the chances they'll be challenged somewhere are very, very high – simply because unlike other patents, this one . . . claims provenance over 98% of the human genome, and not just the human genome – the bovine genome, the eucalyptus genome – any genome.'[176] The important point is that the non-coding DNA patents have a general application, and have been used in research in the fields of agriculture, health and the environment. There has been much animated discussion as to whether such far-reaching and extravagant patent claims are sustainable in light of the scientific prior art.

The robust licensing strategies of GTG have caused some consternation amongst private biotechnology companies, health care providers and public researchers. GTG has sought significant commercial licence fees from an array of biotechnology companies who are engaged in genetic testing of plants, animals and humans. The firm has hoped that companies would be willing to pay royalties, rather than endure the expensive, risky and time-consuming process of patent litigation. GTG has also sought research licence fees from universities and research institutions. The company has also sought to claim licence fees from public hospitals which are conducting clinical genetic tests. Perhaps the licensing tactics of GTG are symptomatic of a wider phenomenon. In response to its licensing strategy, GTG has faced concerted opposition. Several private biotechnology companies, most notably Applera, have questioned the validity of its patents, in light of the scientific prior art. The ADHB took pre-emptive action against GTG for groundless threats of legal proceedings. Some private companies have voiced complaints about GTG in the media. The Chief Scientist of Sequenom Inc complained that its tactics involved 'blackmail'.[177] Scientists from universities and public researchers have been busy documenting scientific publications and research as prior art that could undermine the validity of GTG's patents. It is uncertain whether the business model of GTG will be viable in the long term. In spite of its robust approach, the company has struggled to generate profits from its licence fees. The company has been investigated by the Australian Securities and Investments Commission in respect of illicit share trading activity.[178]

Since the settlement of the legal actions with Applera and ADHB, GTG has become embroiled in new conflicts. In August 2005, GTG and the agricultural biotechnology firm, Monsanto, entered into negotiations surrounding the use of the non-coding DNA and genomic mapping patents.[179] In February 2006, Dr Jacobson of GTG alleged that Monsanto had infringed its non-coding DNA patents through its DNA marker mapping and marker-assisted breeding.[180] In June 2006, Monsanto brought a legal action in the United States District Court for the Eastern District of Missouri.[181] The company noted that 'GTG's aggressive actions have created a reasonable apprehension on the part of Monsanto that it will face an infringement suit under the non-coding DNA patents . . . based upon Monsanto's activities in using DNA markers.'[182] Calling for a jury trial, Monsanto has sought a declaration that a patent held by an Australian company, GTG, was invalid. Alternatively, Monsanto sought a declaration that it was not infringing the patent held by GTG. In June 2007, Monsanto and GTG announced a settlement of the dispute, with the agricultural biotechnology company agreeing to pay the Melbourne company a fee of $US 5 million. There is also emerging patent litigation in the United States District Court for the

Southern District of New York between GE Healthcare Bio-Sciences Corp. and GTG.[183] The two companies are negotiating a resolution of this conflict.

NOTES

1. Australian Broadcasting Corporation (2003), 'Genius of Junk (DNA)', *Catalyst*, http://www.abc.net.au/catalyst/stories/s 898887.htm, 10 July.
2. Genetic Technologies Limited (2003), 'Corporate details', http://web.archive.org/web/20030207163028/www.gtg.com.au/CorpDetails.html, February.
3. Simons, M. (1992), 'Intron sequence analysis method for detection of adjacent and remote locus alleles as haplotypes', US Patent No: 5 612 179.
4. Simons, M. (1994), 'Genomic mapping method by direct haplotyping using intron sequence analysis', US Patent No: 5 851 762.
5. The principal applications in these patent families include: Rarecell project, US Patent Nos: 5153117 and 5447842, and US Patent Applications 10/516,430 and 10/547,721 (Pending), Immunaid project, Australia Patent Applications, 2003905858 and 2004905118; and ACTN3 genotype screen for athletic performance. Earliest priority date 16-Sep-02 US Patent Application, 2003258390 (Pending). For a full list of the company's patent holdings, see Genetic Technologies Limited, 'Intellectual Property', http://www.gtg.com.au/index.asp?menuid=060.070.190.010.
6. Genetic Technologies Limited (2004), 'GTG reports growth in revenue from genetic testing', http://www.gtg.com.au/index.asp?menuid=060.070.130.010&artid=155, 23 July.
7. Trudinger, M. (2004), 'BIO profile: Mervyn Jacobson, genetic technologies', *Australian Biotechnology News*, http://www.biotechnews.com.au/index.php?id=218823635, 1 June.
8. Genetic Technologies Limited (2004), 'GTG reports growth in revenue from genetic testing', http://www.gtg.com.au/index.asp?menuid=060.070.130.010&artid=155, 23 July.
9. Genetic Technologies Limited (2005), 'GTG celebrates NASDAQ listing with the ringing of the Stock Market Closing Bell', http://www.gtg.com.au/index.asp?menuid=060.070.130&artid=292&function=NewsArticle, 7 November.
10. Genetic Technologies Limited (2004), 'John Dawkins Joins GTG Board of Directors'. http://www.gtg.com.au/index.asp?menuid=060.070.130.010&artid=174, 25 November.
11. *Myriad Genetics Inc. v Genetic Technologies Limited* (2002, United States District Court for the District of Utah, Central Division), 2-02CV-0964 BSJ.
12. *Genetic Technologies Ltd. v Nuvelo Inc.* (2003, United States District Court, Northern District of California).
13. *Genetic Technologies Ltd. v Covance Inc.* (2003, United States District Court, Northern District of California) 3:2003cv01315.
14. *Laboratory Corp. of America Holdings v Genetic Technologies Ltd.* (2003, United States District Court, District of New Jersey) 3:2003cv06067.
15. *Genetic Technologies Limited v Applera Corporation* (2003, United States District Court, Northern District of California), No. c-03-1316 PJH, 2003 WL 23796524.
16. Genetic Technologies Limited (2005), 'Mediation in New Zealand Results in Final Settlement of Legal Action', 7 July.
17. Holmes, J. (2003), 'Patently a problem', *Four Corners*, Australian Broadcasting Corporation, transcript, <http://www.abc.net.au/4corners/content/2003/transcripts/s922059.htm, 11 August.
18. *Myriad Genetics Inc. v Genetic Technologies Limited* (2002, United States District Court for the District of Utah, Central Division), 2-02CV-0964 BSJ.
19. Genetic Technologies Limited (2002), 'Genetic Technologies and Myriad Genetics announce strategic licensing agreement', 28 October.

20. Ibid.
21. Hayes, L. and N. Greenaway (2003), 'Profit motive', 60 Minutes, Channel Nine, http://sixtyminutes.ninemsn.com.au/sixtyminutes/stories/2003_04_20/story_806.asp,20 April.
22. Genetic Technologies Limited (2003), 'Genetic susceptibility testing – third progress report', Australian Stock Exchange Company Announcements, 22 May.
23. Australian Law Reform Commission (2004), *Gene Patenting and Human Health: Discussion Paper 68* (Sydney: Australian Commonwealth, February 2004, http://www.austlii.edu.au/au/other/alrc/publications/dp/68/, pp. 260 and 578.
24. Holmes, J. (2003), 'Patently a problem', Four Corners, ABC, http://www.abc.net.au/4corners/content/2003/20030811_patent/default.htm, 11 August.
25. For an extended version of this paper, see Rimmer, M. (2006), 'The alchemy of junk: patent law and non-coding DNA', *The University of Ottawa Law and Technology Journal*, 3(2), 539–99.
26. This case study is part of the larger policy debate over gene patents. There is a large literature on this topic. See Heller, M. and R. Eisenberg (1998), 'Can patents deter innovation? The anticommons in biomedical research', *Science* **280**(5364), 698, http://www.sciencemag.org/cgi/content/full/280/5364/698; Nuffield Council on Bioethics (2002), *The Ethics of Patenting DNA, A Discussion Paper*, London: Nuffield Council on Bioethics, http://www.nuffieldbioethics.org/go/ourwork/patentingdna/publication_310.html; Nicol, D. and J. Nielsen (2003), 'Patents and medical biotechnology: an empirical analysis of issues facing the Australian industry', Occasional Paper No. 6, Centre for Law and Genetics, University of Tasmania, http://www.ipria.org/ publications/workingpapers/BiotechReportFinal.pdf; Centre for Intellectual Property Policy (2005), *Genetic Patents and Health Care in Canada: An International Comparison of Patent Regimes of Canada and its Major Trading Partners*, Montreal: McGill Centre for Intellectual Property Policy, http://www.cipp.mcgill.ca/data/publications/00000015.pdf.
27. New Zealand, Ministry of Health and Ministry of Commerce (2004), *Memorandum to Cabinet Policy Committee: Report Back with Recommendations and Options for Addressing Genetic Material Patents*, http://www.med.govt.nz/templates/MultipageDocumentTOC____1148.aspx, May.
28. Australian Law Reform Commission (2003), *Gene Patenting and Human Health, Issue Paper 27*, Sydney: Australian Commonwealth, http://www.austlii.edu.au/au/other/alrc/ publications/issues/27/, July; Australian Law Reform Commission (2004), *Gene Patenting and Human Health, Discussion Paper 68*, Sydney: Australian Commonwealth, http://www.austlii.edu.au/au/other/alrc/publications/dp/68/, February; and Australian Law Reform Commission (2004), *Genes and Ingenuity: Gene Patenting and Human Health, Report 99*, Sydney: Australian Commonwealth, http://www.austlii.edu.au/au/other/alrc/publications/reports/99/, June.
29. Advisory Council on Intellectual Property (2004), *Patents and Experimental Use: Options Paper*, Canberra: Commonwealth Government, http://www.acip.gov.au/library/Experimental%20Use%20Options%20Paper%20A.pdf; and Advisory Council on Intellectual Property (2005), *Patents and Experimental Use: Final Report*, Canberra: Commonwealth Government, http://www.acip.gov.au/library/ACIP%20Patents%20&%20Experimental%20Use%20final%20report%20FINAL.pdf.
30. 'First amended complaint for patent infringement', *Genetic Technologies Limited v Applera Corporation* (2003, United States District Court, Northern District of California), No. c-03-1316 PJH, 2003 WL 23796524.
31. Simons, M. (1992), 'Intron sequence analysis method for detection of adjacent and remote locus alleles as haplotypes', US Patent No: 5,612,179.
32. 'First amended complaint for patent infringement', *Genetic Technologies Limited v Applera Corporation* (2003, United States District Court, Northern District of California), No. c-03-1316 PJH, 2003 WL 23796524.
33. Simons, M. (1994), 'Genomic mapping method by direct haplotyping using intron sequence analysis', US Patent No: 5,851,762.

34. Securities and Exchange Commission (2003), 'Quarterly report pursuant to Section 13 or 15(d) of the Securities Exchange Act of 1934 for the quarterly period ended March 31, 2003 for Applera Corporation', http://media.corporate-ir.net/media_files/NYS/ABI/reports/10q_q3_2003.pdf, March, p. 16.
35. 'First amended complaint for patent infringement', *Genetic Technologies Limited v Applera Corporation* (2003, United States District Court, Northern District of California), No. c-03-1316 PJH, 2003 WL 23796524.
36. Ibid.
37. Ibid.
38. Ibid.
39. Ibid.
40. Ibid.
41. Ibid.
42. Ibid.
43. 'Complaint for patent infringement', *Genetic Technologies Limited v Applera Corporation*, 2003 WL 23794369.
44. Ibid., 29.
45. Ibid. at para. 35.
46. 'First amended complaint for patent infringement', *Genetic Technologies Limited v Applera Corporation*, 2003 WL 23796523 at para. 55.
47. Ibid.
48. Ibid. at para. 57.
49. Ibid. at para. 58.
50. Ibid. at para. 61.
51. Ibid., 'Prayer for relief'.
52. 'Answer and counterclaims of defendant Applera Corporation', *Genetic Technologies Limited v Applera Corporation* (5 September 2003), 2003 WL 23794380.
53. SS 102, 103, 112 of the *Patent Act* 1952 (US).
54. Exhibit A refers to Kan, Y. and A. Dozy (1978), 'Polymorphism of DNA sequence adjacent to human beta-globin structural gene: relationship to sickle mutation', *Proceedings of the National Academy of Sciences*, **75**(11), 5631, <http://www.pnas.org/cgi/reprint/75/11/5631>; Exhibit B refers to Mullis, K. (1985), 'Process for amplifying nucleic acid sequences', US Patent No: 4,683,202; Exhibit C refers to Geisel, J., B. Weisshaar, K. Oette and W. Doerfler (1988), 'A new APA LI restriction fragment length polymorphism in the low 28 density lipoprotein receptor gene', *Journal of Clinical Chemistry and Clinical Biochemistry*, **26**(7), 429; Exhibit D concerns Little, P., G. Annison, S. Darling, R. Williamson, L. Camba and B. Modell (1980), 'Model for Antenatal Diagnosis of Beta-thalassaemia and other monogenic disorders by molecular analysis of linked DNA polymorphisms', *Nature*, **283**(5761), 144; and Exhibit E is Guardiola, J., A. Maffei, S. Carrel and R. Accolla (1988), 'Molecular genotyping of the HLA-DQ [α] Gene Region', *Immunogenetics*, **27**(1), 12.
55. 'Applera Corporation's claim construction brief', *Genetic Technologies Limited v Applera Corporation*, 2003 WL 23794399 at part B.
56. 'Answer and counterclaims of defendant Applera Corporation', *Genetic Technologies Limited v Applera Corporation* (5 September 2003), 2003 WL 23794380 at para. 66.
57. Holmes, J. (2003), 'Interview with Dr Mervyn Jacobson', http://www.abc.net.au/4corners/content/2003/20030811_patent/int_jacobson.htm, 11 August.
58. Ibid.
59. Ibid.
60. O'Neill, G. (2003), 'Prior art', *Australian Biotechnology News*, 22 August, p. 8.
61. Ibid.
62. Ibid.
63. Ibid.
64. Ibid.
65. Ibid.

66. Ibid.
67. Ibid.
68. Ibid.
69. Ibid.
70. *Genetic Technologies Limited v Applera Corporation* (2003, United States District Court, Northern District of California), No. c-03-1316 PJH, 2003 WL 23796524.
71. Simons, M. (2004), '"Junk DNA" non-coding DNA patents: the inventors' view', *Simons Haplomics*, http://www.simonsjunkdna.com/Full%20Article.htm, 18 May; and Nottenburg, C. and J. Sharples (2004), 'Analysis of "Junk DNA" Patents', *Simons Haplomics*, http://www.simonsjunkdna.com/junk%20dna%20analysis.pdf, July.
72. Simons, M. (2004), '"Junk DNA" non-coding DNA patents: the inventors' view', *Simons Haplomics*, http://www.simonsjunkdna.com/Full%20Article.htm, 18 May.
73. Ibid.
74. Ibid.
75. Moukheiber, Z. (2003), 'Junkyard dogs', *Forbes*, http://www.forbes.com/forbes/2003/0929/052_print.html, 29 September, p. 2.
76. Collins, F. (2003), 'A Patent's Place', *Bio-IT World*, http://www.bio-itworld.com/archive/081303/horizons_aussie_sidebar_1.html, 13 August.
77. Holmes, J. (2003), 'Interview with Sir John Sulston', http://www.abc.net.au/4corners/content/2003/20030811_patent/int_sulston.htm, 9 July.
78. Ibid.
79. Nowak, R. and D. Concar (2002), 'Footing the bill: should we all have to pay for one company's bright idea?' *New Scientist*, 4, 18 May.
80. Australian Broadcasting Corporation (2003), 'Genius of junk (DNA)', *Catalyst*, http://www.abc.net.au/catalyst/stories/s898887.htm, 10 July.
81. Holmes, J. (2003), 'Interview with Professor John Mattick', http://www.abc.net.au/4corners/content/2003/20030811_patent/int_mattick.htm, 10 July.
82. Little, P. (2003), 'GTG's inventions concerning "Junk" DNA', *Bio-IT World*, http://www.bio-itworld.com/archive/091103/letters.html, 11 September. Little cites a number of publications as prior art, including Kan, Y. and A. Dozy (1978), 'Antenatal diagnosis of sickle-cell anaemia by D.N.A. analysis of amniotic-fluid cells', *Lancet*, 2, 910; Kan, Y. and A. Dozy (1978), 'Polymorphism of DNA sequence adjacent to human beta-globin structural gene: relationship to sickle mutation', *Proceedings of the National Academy of Sciences*, **75**(11), 5631, <http://www.pnas.org/cgi/reprint/75/11/5631>; Little, P., G. Annison, S. Darling, R. Williamson, L. Camba and B. Modell (1980), 'Model for antenatal diagnosis of beta-thalassaemia and other monogenic disorders by molecular analysis of linked DNA polymorphisms', *Nature*, **283**(5761), 144; Phillips, J., S. Panny, H. Kazazian, C. Boehm, A. Scott and K. Smith (1980), 'Prenatal diagnosis of sickle cell anemia by restriction and endonuclease analysis: Hind III polymorphisms in gamma-globin genes extend test applicability', *Proceedings of the National Academy of Sciences*, **77**(5), 2853, http://www.pubmedcentral.nih.gov/articlerender.fcgi?artid=349503; and Botstein, D., R. White, M. Skolnick and R. Davis (1980), 'Construction of a genetic linkage map in man using restriction fragment length polymorphisms', *American Journal of Human Genetics*, **32**(3), 314.
83. Little, P. (2003), 'GTG's inventions concerning "Junk" DNA', *Bio-IT World*, http://www.bio-itworld.com/archive/091103/letters.html, 11 September.
84. Holmes, J. (2003), 'Patently a problem', *Four Corners*, Australian Broadcasting Corporation, transcript, http://www.abc.net.au/4corners/content/2003/transcripts/s922059.htm, 11 August.
85. Ibid.
86. Waring, P. (2003), 'Patenting genetic information', Mutation Detection, International Symposium on Mutations in the Genome, July.
87. Genetic Technologies Limited (2005), *Annual Report 2005*, http://esvc001057.wic005u.server-web.com/archives/1/070.130/822/Release%20of%202005%20Annual%20Report.pdf, p. 15.

88. Genetic Technologies Limited (2005), 'GTG reports final settlement of its patent dispute with Applera Corporation', http://www.gtg.com.au/index.asp?menuid= 060.070.130& artid=262&function=NewsArticle, 12 December.
89. Ibid.
90. Genetic Technologies Limited (2005), 'GTG provides further details of the settlement with Applera', http://www.gtg.com.au/index.asp?menuid=060.070.130&artid=264& function=NewsArticle, 15 December.
91. Genetic Technologies Limited (2006), 'GTG reports final dismissal of Applera suit, and plans for the future', http://www.gtg.com.au/index.asp?menuid=060.070.130&artid= 266&function=NewsArticle, 4 January.
92. Westerman, H. and R. Urban (2005), 'Genetic wins little fight over DNA work', *The Age*, 16 December.
93. O'Neill, G. (2005), 'GTG celebrates win over Applera in patent battle', *Australian Biotechnology News*, http://www.biotechnews.com.au/index.php?id=548699165, 12 December, p. 2.
94. Westerman, H. and R. Urban (2005), 'Genetic wins little fight over DNA work', *The Age*, 16 December.
95. New Zealand (2000), *Royal Commission on Genetic Modification*, http://www.mfe.govt.nz/issues/organisms/law-changes/commission/.
96. New Zealand (2002), *Report of the Royal Commission on Genetic Modification*, http://www.mfe.govt.nz/publications/organisms/royal-commission-gm/index.html, Chapter 10. http://www.mfe.govt.nz/publications/organisms/royal-commission-gm/chapter-10.pdf.
97. New Zealand, Ministry of Health and Ministry of Commerce (2004), *Memorandum to Cabinet Policy Committee: Report Back with Recommendations and Options for Addressing Genetic Material Patents*, http://www.med.govt.nz/templates/Multipage-DocumentTOC____1148.aspx, May, p. 1.
98. New Zealand – Minister of Health and Associate Minister of Commerce (2003), *Implications of the Granting of Patents Over Genetic Material*, Cabinet Policy Committee, available at *Internet Archive*, http://web.archive.org/web/*/http://www.med.govt.nz/buslt/int_prop/genetic-material/cabinet/implications/implications.pdf, November.
99. Ibid. at para. 18.
100. Ibid. at p. 7.
101. Ibid.
102. Ibid.
103. New Zealand – Minister of Health and Associate Minister of Commerce (2003), *Implications of the Granting of Patents Over Genetic Material*, Cabinet Policy Committee, available at *Internet Archive*, <http://web.archive.org/web/*/http://www.med.govt.nz/buslt/int_prop/genetic-material/cabinet/implications/implications.pdf, November, at para. 24.
104. Ibid. at para. 25.
105. New Zealand, Ministry of Health and Ministry of Commerce (2004), *Memorandum to Cabinet Policy Committee: Report Back with Recommendations and Options for Addressing Genetic Material Patents*, http://www.med.govt.nz/templates/Multipage-DocumentTOC____1148.aspx, May, 1.
106. Ibid., p. 2.
107. Ibid., p. 2.
108. Ibid., p. 2.
109. Christensen, I. (2003), 'A letter to David Sage, Chief Medical Officer of the Auckland District Health Board', available at Australian Broadcasting Corporation, http://www.abc.net.au/4corners/content/2003/20030811_patent/documents/page1.htm,23 May, p. 1.
110. Ibid.
111. Ibid.
112. Ibid.
113. Ibid.
114. Holmes, J. (2003), 'Interview with Dr Mervyn Jacobson', http://www.abc.net.au/4corners/content/2003/20030811_patent/int_jacobson.htm, 11 August.

115. Ibid.
116. Ibid.
117. Ibid.
118. Ibid.
119. Ibid.
120. Ibid.
121. Ibid.
122. Gorman, P. (2004), 'Board negotiates with research team: Kiwi funds money in junk', *Christchurch Press*, 16 February.
123. Ibid.
124. Gorman, P. (2004), 'Legal action against GTG', *Christchurch Press*, 19 August.
125. O'Neill, G. (2004), 'NZ Health Service takes on GTG over licence fees', *Australian Biotechnology News*, http://www.biotechnews.com.au/index.php?id=813052581, 24 August.
126. Ibid.
127. Gorman, P. (2004), 'NZ Health Boards in patent dispute', *Christchurch Press*, 21 August, p. 3.
128. Genetic Technologies Limited (2004), 'GTG supports patent Protection', http://www.gtg.com.au/index.asp?menuid=060.070.200.020.010&artid=205, 19 August.
129. Gorman, P. (2005), 'Mediation causes row', *Christchurch Press*, 26 March, p. 8.
130. Genetic Technologies Limited (2005), 'Mediation in New Zealand results in final settlement of legal action', 7 July.
131. New Zealand Herald (2005), 'GTG's bid for Health Board patent fees fails', *New Zealand Herald*, 8 July.
132. McNabb, D. (2005), 'Health boards defy gene test claims', *The Independent Business Weekly*, 13 July.
133. New Zealand Herald (2005), 'GTG's bid for Health Board patent fees fails', *New Zealand Herald*, 8 July.
134. Ibid.
135. O'Neill, G. (2005), 'GTG boss reflects on "unusual" license dispute', *Australian Biotechnology News*, http://www.biotechnews.com.au/index.php?id=1192139376, 8 July.
136. New Zealand Herald (2005), 'GTG's bid for health board patent fees fails', *New Zealand Herald*, 8 July.
137. Ibid.
138. Attorney General's Department (2002), 'Inquiry into human genetic property issues', press release, http://www.ag.gov.au/agd/WWW/attorneygeneralHome.nsf/Page/Media_Releases_2002_December_2002_Inquiry_into_human_genetic_property_issues_(17_December_2002), 17 December.
139. Australian Law Reform Commission (2004), *Genes and Ingenuity: Gene Patenting and Human Health*, Report 99, Sydney: Australian Commonwealth, http://www.austlii.edu.au/au/other/alrc/publications/reports/99/, June 2004.
140. Holmes, J. (2003), 'Patently a problem', *Four Corners*, Australian Broadcasting Corporation, transcript, http://www.abc.net.au/4corners/content/2003/transcripts/s922059.htm, 11 August.
141. *Agreement Establishing the World Trade Organization, Annex 1C: Agreement on Trade-Related Aspects of Intellectual Property Rights*, 15 April 1994, 1869 U.N.T.S. 299, <http://www.wto.org/english/docs_e/legal_e/27-trips_01_e.htm>, 33 I.L.M. 81 [TRIPS Agreement]. Article 27.1 of the TRIPS Agreement provides, 'patents shall be available and patent rights enjoyable without discrimination as to the [. . .] field of technology'. This is subject to Article 27.3 which allows Members to exclude from patentability, 'diagnostic, therapeutic and surgical methods for the treatment of humans or animals' and 'plants and animals other than micro-organisms, and essentially biological processes for the production of plants or animals other than non-biological and micro-biological processes'. Article 30 further provides, 'Members may provide limited exceptions to the exclusive rights conferred by a patent, provided that such exceptions do not

unreasonably conflict with a normal exploitation of the patent and do not unreasonably prejudice the legitimate interests of the patent owner, taking account of the legitimate interests of third parties.'
142. Australian Law Reform Commission (2004), *Genes and Ingenuity: Gene Patenting and Human Health, Report 99*, Sydney: Australian Commonwealth, http://www.austlii.edu.au/au/other/alrc/publications/reports/99/, June.
143. S. 163 of the *Patents Act* 1990 (Cth).
144. S. 171 of the *Patents Act* 1990 (Cth).
145. S. 133 of the *Patents Act* 1990 (Cth).
146. Suthers, G. (2004), 'Our genes: humanity's heritage or cash cow?', *Issues*, **67**, 23.
147. Australian Law Reform Commission (2003), *Gene Patenting and Human Health, Issue Paper 27*, Sydney: Australian Commonwealth, http://www.austlii.edu.au/au/other/alrc/publications/issues/27/, July.
148. Australian Law Reform Commission (2004), *Gene Patenting and Human Health, Discussion Paper 68*, Sydney: Australian Commonwealth, http://www.austlii.edu.au/au/other/alrc/publications/dp/68/, February.
149. Australian Law Reform Commission (2004), *Genes and Ingenuity: Gene Patenting and Human Health, Report 99*, Sydney: Australian Commonwealth, http://www.austlii.edu.au/au/other/alrc/publications/reports/99/, June.
150. Ibid., 79.
151. Ibid., 51.
152. Ibid.
153. Ibid., 155.
154. *Australia–United States Free Trade Agreement* (1 May 2004), 2005 A.T.S. 1, http://www.dfat.gov.au/trade/negotiations/us_fta/final-text/index.html.
155. Ibid., ch.17, art. 9.12.
156. *In re Fisher*, 421 F.3d 1365 (Fed. Circ. 2005).
157. Australian Law Reform Commission (2004), *Genes and Ingenuity: Gene Patenting and Human Health, Report 99*, Sydney: Australian Commonwealth, http://www.austlii.edu.au/au/other/alrc/publications/reports/99/, June, p. 366; for a summary of the wider debate over patent law and experimental use, see Rimmer, M. (2005), 'The freedom to tinker: patent law and experimental use', *Expert Opinion on Therapeutic Patents*, **15**(2), 167–200.
158. *Madey v Duke University* 307 F.3d 1351 (2002).
159. Australian Law Reform Commission (2004), *Genes and Ingenuity: Gene Patenting and Human Health, Report* 99, Sydney: Australian Commonwealth, http://www.austlii.edu.au/au/other/alrc/publications/reports/99/, June, p. 340
160. Ibid., 620.
161. Ibid., 593–610.
162. Ibid., 605.
163. Ibid.
164. Ibid., 555–80.
165. *Aktiebolaget Hässle v Alphapharm Pty Limited* (2002) 212 C.L.R. 411 at 468.
166. Kirby J in *Aktiebolaget Hässle v Alphapharm Pty* (2002), S287/2001, oral proceedings transcripts, http://www.austlii.edu.au/cgi-bin/disp.pl/au/other/hca/transcripts/2001/S287/1.html, 29 May.
167. Gold, E.R. and K. Durrell (2005), 'Innovating the skilled reader: tailoring patent law to new technologies', *Intellectual Property Journal*, **19**(1), 189–226 at 192.
168. Ibid., 224–5.
169. *Anaesthetic Supplies Pty Ltd v Rescare Ltd* (1994) 28 I.P.R. 383; *Bristol-Myers Squibb Company v FH Faulding & Co Limited* (2000) 46 I.P.R. 553; for the United States, see *Pallin v Singer* (1995, D. Vt.), 36 USPQ (2d) 1050; *Metabolite Laboratories Inc. v Laboratory Corporation of America Holdings (LabCorp)* 370 F.3d 1354 (2006); for Canada, see *Apotex Inc. v Wellcome Foundation Ltd.*, 2002 SCC 77, [2002] 4 S.C.R. 153; and for New Zealand, see *Pfizer v Commissioner of Patents* [2005] 1 N.Z.L.R. 364.

170. See *Greenberg v Miami Children's Hospital Research Institute, Inc.* 264 F.Supp.2d 1064 (S.D. Fla 2003); Gitter, D. (2004), 'Ownership of human tissue: a proposal for federal recognition of human research participants' property rights in their biological material', *Washington and Lee Law Review*, **61**(1), 257–346; and Rimmer, M. (2006), 'Miami heat: patent law, informed consent, and benefit-sharing', *Journal of International Biotechnology Law*, **3**, 177–92.
171. Trudinger, M. (2004), 'Gene patent system ain't broke, but needs fine tuning: ALRC', *Australian Biotechnology News*, http://www.biotechnews.com.au/index.php?id= 1801276438, 6 September.
172. Genetic Technologies Limited (2004), 'GTG reports on the current status of legal action to protect its patents', http://www.gtg.com.au/index.asp?menuid=060.070. 130.010&artid=158, 3 September.
173. Suthers, G. (2004), 'Our genes: humanity's heritage or cash cow?', *Issues*, **67**, 23.
174. For a discussion of the debate over patent law and evergreening, see Burton, K. and J. Varghese (2004), 'The PBS and the Australia–US Free Trade Agreement', Parliament of Australia Research Note no. 3 2004–2005, 22 July; Lawson, C. and C. Pickering (2004), ' "TRIPs-plus" patent privileges – an intellectual property "Cargo Cult" in Australia', *Prometheus*, **22**(4), 355–77; Drahos, P., B. Lokuge, T. Faunce, M. Goddard and D. Henry (2004), 'Pharmaceuticals, intellectual property and free trade: the case of the US–Australia Free Trade Agreement', *Prometheus*, **22**(3), 243–57; Sainsbury, P. (2004), 'Australia–United States Free Trade Agreement and the Australian Pharmaceutical Benefits Scheme', *Yale Journal of Health Policy, Law and Ethics*, **4**(2). 387–99; and Chalmers, R. (2006), 'Evergreen or deciduous? Australian trends in relation to the "evergreening" of patents', *Melbourne University Law Review*, **30**, 29–61.
175. Advisory Council on Intellectual Property (2005), *Patents and Experimental Use: Final Report*, Canberra: Commonwealth Government, http://www.acip.gov.au/library/ ACIP%20Patents%20&%20Experimental%20Use%20final%20report%20FINAL.pd; and Advisory Council on Intellectual Property (2005), *Review of Crown Use Provisions for Patents and Designs*, Canberra: Commonwealth Government, http://www.acip.gov. au/library/review_of_Crown_Use_provisions.pdf.
176. Holmes, J. (2003) 'Interview with Professor John Mattick', http://www.abc.net.au/ 4corners/content/2003/20030811_patent/int_mattick.htm, 10 July.
177. Holmes, J. (2003), 'Patently a problem', *Four Corners*, Australian Broadcasting Corporation, transcript, http://www.abc.net.au/4corners/content/2003/transcripts/ s922059.htm, 11 August.
178. Urban, R. (2007), 'Police seize executives' computers in genetic technologies raid', *The Sydney Morning Herald*, 8 March.
179. Complaint, *Monsanto Company v Genetic Technologies Limited* (2006, United States District Court, Eastern District of Missouri), No. 3:06-cv-00989-HEA, p. 3.
180. Ibid., p. 3.
181. Ibid., p. 3.
182. Ibid., p. 3.
183. *GE Healthcare Bio-Sciences Corp. v Genetic Technologies, Ltd.* (2006, United States District Court for the Southern District of New York, 1:06-cv-13172-WHP).

9. Still life with stem cells: patent law and human embryos

In a beguiling and sometimes disturbing collection of art works, the Australian artist Patricia Piccinini has explored social and ethical attitudes to scientific developments in stem cell research and therapeutic cloning.[1] Her artwork of a child playing with strange blobs of humanoid matter, *Still Life with Stem Cells*, was inspired by her first sight of stem cells pulsating in a petri dish:

> Last year I saw one of those extraordinary things, which reminds me that what I make is not so strange or far-fetched. As usual it was in a petri dish. This petri dish contained a small layer of cells, a thin skin of biological matter that was pulsating to rapid but steady rhythm. This was the first time that I had really seen stem cells. These ones had been differentiated into heart cells and they were doing what heart cells do; beating – flatly, geometrically, pointlessly. Stems cells are base cellular matter before it is differentiated into specific kinds of cells like skin, liver, bone or brain. Pure unexpressed potential, they contain the possibility for transformation into anything.[2]

The image, *Still Life with Stem Cells*, is an arresting one: it captures both a curiosity and an enthusiasm for scientific breakthroughs in the field of stem cell research and therapeutic cloning, together with a horror and revulsion at manipulating essential human biological matter.

There has been a strong push by stem cell researchers, biotechnology companies and pharmaceutical drug manufacturers to obtain intellectual property rights in respect of stem cell research. Back in 1999, the peak body, the Biotechnology Industry Organization (BIO), told Congress about the desirability of patent protection in respect of stem cell research:

> In regard to stem cell research, patents are vital and they should be freely transferable. These patents are essential to the continuation of stem cell research. No money has yet been made from selling stem cell products. It is unreasonable to expect any money to be made for many years to come from this research.[3]

BIO observed that patents would help scientific progress in this discipline: 'Once boundaries are established and claims granted in an exciting field such as stem cells, patents provide a powerful stimulus to competitive academic groups and companies to improve on technologies and/or find new routes to achieve the same effects.'[4] The peak body denied that patents would have an adverse impact upon stem cell researchers and scientists: 'Patents do not block university researchers from conducting research on patented inventions.'[5] The peak body insisted that 'patents increase the range of effective products available to treat intractable diseases and improve social welfare'.[6]

Policy makers have been concerned that stem cell patents may impede further research and development, particularly if the patent holders license such patents exclusively or on restricted terms. There has been a concern that the granting of broad patents to private companies would impair further research and development. There have been anxieties that patent holders could charge unreasonable fees for the use of their inventions, or block access altogether. Arti Rai comments: 'Control of embryonic stem cell research by the private sector may have significant justice-related consequences.'[7] Law reformers have proposed a number of general and specific mechanisms to facilitate access to stem cell technologies. Some have proposed a broad defence for experimental and research use to facilitate access to both genetic technologies and stem cell research. It has been suggested that patent pools, patent clearinghouses or collective rights organizations might also help address difficulties in obtaining access to patented stem cell lines and technologies. The provision of compulsory licensing and Crown use would be useful in the future, if a company abuses its dominant market position against rival research organizations in the field of stem cell research. Such measures would also be relevant in circumstances where access to treatments and diagnosis was blocked by misuse of patent rights. Some have contended that the development of stem cell banks would help facilitate and regulate access to stem cell lines by researchers. The United Kingdom Stem Cell Bank, for instance, seeks to promote wider use of cells, and to facilitate the possibility of new discoveries while protecting depositor intellectual property rights.[8] There has been scope for funding agencies to facilitate access to patented stem cell lines.[9]

A number of opponents have maintained that stem cell research should not be patentable subject matter, because human embryonic stems have the potential to develop into human beings. For instance, Dr Warwick Neville, an academic and a lawyer, representing the Australian Catholic Bishops Conference, submitted: 'The commodification of human life is inimical to the recognition and protection of human dignity.'[10] Such critics suggest that stem cell patents represent an inappropriate commodification of human biological material, and in particular human reproductive material. Moreover, it has been alleged that stem cell patents violate fundamental principles regarding the ownership of human beings and the free and informed consent of the donor.

This chapter considers the instrumental role played by patent law in respect of stem cell research and therapeutic cloning. It examines the relationship between commercialization, access to essential medicines and bioethics. Section one considers the debate over patent law and stem cell research in Australia. It examines the legislative history behind section 18(2) of the *Patents Act* 1990 (Cth), which provides that 'Human beings, and the biological processes for their generation, are not patentable inventions.' It considers the interpretation of section 18(2) of the *Patents Act* 1990 (Cth) in two key decisions by the Deputy Commissioner of Patents: *Fertilitescentrum AB and Luminis Pty Ltd* and *Woo-Suk Hwang*.[11] Section two examines the strong patent protection secured by the Wisconsin Alumni Research Foundation (WARF) and Geron Corporation in respect of stem cell research in the United States. It considers the challenge to the validity of such patents in the United States Patent and Trademark Office (USPTO) by the Californian-based Foundation for Taxpayer and Consumer Rights, and the New York-based Public Patent Foundation.[12] It considers a number of mechanisms to safeguard access to stem cell lines, and resulting drugs and therapies. Section three seeks to accommodate ethical concerns within the framework of the patent system. It examines the current indecision of the European Patent Office in respect of stem cell patent applications, such as the 'Edinburgh patent', a WARF patent application, and a California Institute of Technology (Caltech) patent application.[13] It also considers the inquiry of the European Group on Ethics in Science and New Technologies into 'The Ethical Aspects of Patenting involving Human Stem Cells'.[14]

I LIFE SCIENTISTS AND CLONING COWBOYS

In 1990, Independent Senator Brian Harradine introduced amendments into the Australian Parliament which became section 18(2) of the *Patents Act* 1990 (Cth): 'Human beings, and the biological processes for their generation are not patentable inventions.' He sought to illustrate the intent of his amendments:

> Let me give an extreme example of a process which my amendment would prohibit. I refer to the techniques that may well be developed for cloning a human embryo at the four-cell stage. That is an example, albeit an extreme one, of the type of technique or process which my amendment to this Bill would prohibit from being patentable.[15]

At the time, the amendments were criticized for a lack of clarity. Democrat Senator John Coulter queried: 'What are the biological processes for the generation of human beings?'[16] It is difficult to ascertain the scope of the excluded subject matter.

In its submission to the House of Representatives Standing Committee On Legal And Constitutional Affairs, IP Australia emphasized economic concerns related to the patenting of new technologies, and played down matters of ethics and social policy.[17] The organization maintained that section 18(2) of the *Patents Act* 1990 (Cth) prohibits human cloning, but not stem cell research:

> It is the understanding of IP Australia that its practice in granting patents for inventions involving human genes, cell lines and tissue is consistent with the provisions of subsection 18(2) of the Act. This is premised on a widely accepted view that human genes, cell lines and tissues are not regarded as human beings, as distinct from foetuses and embryos which are regarded as human beings and hence are not patentable.[18]

In 2004, the Australian Patent Office at IP Australia had the opportunity to interpret section 18(2) of the *Patents Act* 1990 (Cth) in two key decisions, *Fertilitescentrum AB and Luminis Pty Ltd* and *Woo-Suk Hwang*.[19] The Deputy Commissioner of Patents took a liberal interpretation of the clause, leaving room for embryonic stem cell research to be patentable subject matter in certain circumstances. Such a progressive view of the clause may run counter to the original intentions of Independent Senator Brian Harradine, an ardent opponent of embryonic stem cell research.

A *Fertilitescentrum AB and Luminis Pty Ltd*

In 2004, the Australian Patent Office made its first ruling on the meaning of section 18(2) of the Patents Act 1990 (Cth) in the matter of *Fertilitescentrum AB and Luminis Pty Ltd*.[20]

Fertilitescentrum AB and Luminis Pty Ltd filed an Australian Patent application No 44916/99 that disclosed and claimed an invention related to their discovery that a substance called 'granulocyte-macrophage colony-stimulating factor (GM-CSF)' was effective at substantially increasing the proportion of early embryos that develop to blastocyst and increasing the proportion of those embryos that continue to expanded blastocyst and then hatched blastocyst stages of development.[21] The substance is apparently present in the natural environment of the fallopian tube. The invention involved ensuring that it was present in an IVF environment. The filed patent application included claims to a culture medium, to a method of growing preblastocyst human embryos, and an IVF programme. The key claim was claim 10: 'A method of growing preblastocyst human embryos, the method including the step of incubating the embryos in vitro in a culture medium containing an effective amount of human GM-CSF to increase the chance of implantation of the embryos, the amount of the GM-CSF being sufficient to increase the proportion of blastocysts formed from the preblastocyst embryos when compared to embryos incubated in a medium lacking GM-CSF.' [22]

In the course of examination, the examiner at IP Australia objected to the claims to the method of growing preblastocyst human embryos, and the IVF programme, as being contrary to section 18(2) of the *Patents Act* 1990 (Cth). The examiner was of the view that the claimed method was a step along the path of generating a human being, and is therefore covered by the exclusion. In rejoinder, the applicant has deleted the claims to the IVF programme, but has maintained that the claims to the method of growing the human embryo did not offend against section 18(2) of the *Patents Act* 1990 (Cth). The applicant argued that the invention was merely a treatment of a human being, rather than a method of generating a whole human being.

In his decision, the Deputy Commissioner of Patents, Dave Herald, observed that section 18(2) of the *Patents Act* 1990 (Cth) was an enigmatic clause, which had not yet been elucidated by either the Australian Patent Office or the courts:

> This section, which superficially looks very simple, has the inherent difficulty of defining the exclusion by reference to 'human beings', without any definition of what constitutes a 'human being'. Reproductive technology exposes a range of fundamental issues concerning the nature of human life vis-à-vis human beings – issues that are essentially ethical or moral in nature, with no clear scientific answer.[23]

The Deputy Commissioner of Patents noted that the definition of a human being raised a host of ethical and social issues.

Favouring a developmental approach to the definition of a human being, the Deputy Commissioner provided this interpretation of section 18(2) of the *Patents Act* 1990 (Cth):

> The correct interpretation of section 18(2) is ascertained by recognizing a human being as being in the process of generation from the time of the processes that create a fertilized ovum (or other processes that give rise to an equivalent entity) up until the time of birth.
>
> The prohibition of 'human beings' in my view is a prohibition of patenting of any entity that might reasonably claim the status of a human being. Clearly a person that has been born is covered by this exclusion. But to the extent that there is a process of generation of a human being that lasts from fertilisation to birth, I consider that a fertilised ovum and all its subsequent manifestations are covered by this exclusion.
>
> The prohibition of 'biological processes for (the generation of human beings)' clearly covers all biological processes applied from fertilisation to birth – so long

as the process is indeed one that directly relates to the generation of the human being. I also consider the exclusion of biological processes includes the processes of generating the entity that can first claim a status of human being. For example, processes for fertilising an ovum; processes for cloning at the four–cell stage by division; processes for cloning by replacing nuclear DNA.[24]

It is somewhat debatable whether the 'developmental' interpretation adopted by the Australian Patent Office of section 18(2) of the *Patents Act* 1990 (Cth) would be compatible with the intention of its drafter, Senator Brian Harradine, a notable opponent of both stem cell research and therapeutic cloning. Applying this interpretation of section 18(2) of the *Patents Act* 1990 (Cth), Herald ruled that the patent application fell within the scope of the exclusion, because 'the process is one that directly relates to the generation of a human being'.[25] Accordingly, Herald directed the patent applicant to delete claims 10 to 23 from the patent application, as they were contrary to section 18(2) of the *Patents Act* 1990 (Cth).

B Woo-Suk Hwang

In 2004, Woo-Suk Hwang and his collaborators at the Seoul National University in South Korea claimed to have isolated embryonic stem cells through somatic cell nuclear transfer and generated the cell line (SCNT-hES 1).[26] The researchers claimed to have improved the efficiency of the technique in further research, using fewer human eggs to generate patient-specific embryonic stem cell lines.[27]

As part of this pioneering research, Hwang filed an Australian patent application in April 2003. During examination of this application, the examiner at IP Australia raised objections under section 18(2) of the *Patents Act* 1990 (Cth). That application lapsed through failure to gain acceptance within the 21-month period allowed.

Hwang filed a second patent application before the parent application lapsed.[28] The application claimed a method of producing a hybrid embryo using inter-species nuclear transplantation techniques. Specifically, the embryo was created by transferring the nucleus of a human cell into a bovine ovum, and activating the ovum. The specification contained 12 claims, all dependent on claim 1. Claim 1 is described as 'a method for producing chimeric embryos derived by nuclear transfer using human cells as nucleus donors and enucleated bovine oocytes as recipients'.[29] The method adopted by Hwang creates an embryo where the nuclear DNA is human, and the mitochondrial DNA is bovine.

On 28 January 2004, the examiner at IP Australia objected to the claims being in contravention of section 18(2) of the *Patents Act* 1990 (Cth) and also under s.50(1) of the *Patents Act* 1990 (Cth) asserting that the claimed invention was contrary to the *Prohibition of Human Cloning Act* 2002 (Cth). The examiner's report was also accompanied by a hearing notice, with advice that the Commissioner intended to refuse the application.

In the hearing, Herald considered whether the patent application violated section 18(2) of the *Patents Act* 1990 (Cth); and section 50(1) of the *Patents Act* 1990 (Cth).[30] He noted that the case was factually distinct from the previous decision in *Fertilitescentrum AB and Luminis Pty Ltd* in two important respects: first, there was no step of fertilization per se; and, second, the embryo was a hybrid involving both human and bovine DNA. Herald held that the patent application offended section 18(2) of the *Patents Act* 1990 (Cth):

> The embryo produced by the claimed process has both human and bovine DNA present. It is clear that the nuclear DNA is intended to be entirely human DNA. The mitochondrial DNA, which essentially is relevant to the energy use of the cell, is entirely bovine. The primary physical characteristics of mammals are governed by the nuclear DNA of the cells. In my view, the presence of the bovine mitochondrial DNA does not take away the essentially human characteristic of the embryo that is determined by the nuclear DNA. That is, the embryo that is produced by this method – while being hybrid – is properly described as human.[31]

Accordingly the Deputy Commissioner of Patents was satisfied that the claimed method fell within the exclusion of section 18(2) of the *Patents Act* 1990 (Cth), because it was a method for the generation of a human being.

Herald also considered whether the patent application was invalid under section 50(1) of the *Patents Act* 1990 (Cth), because it related to 'an invention the use of which would be contrary to law'. He noted that 'this provision is rarely invoked, and there is little precedent for its operation'.[32] The Australian Patent Office Manual of Practice and Procedure seems to envisage that the 'contrary to law' provision would deal with criminal acts.[33] The Deputy Commissioner of Patents observed:

> The Manual suggests that a relevant consideration is whether there are any uses of the invention that would not be contrary to law. While it might be thought that the method might have application in mammals other than humans, both the description and claims are specific in the application of the invention to humans only . . . The invention claimed in the present case is very clearly contrary to the *Prohibition of Human Cloning Act 2002*. Furthermore, in my view the prohibitions in s.20 of that Act are of a nature such that the prohibition is unlikely to be ephemeral.[34]

The Deputy Commissioner of Patents observed that section 20 of the *Prohibition of Human Cloning Act* 2002 (Cth) provided that it was an offence intentionally to create a chimeric or hybrid embryo, and was subject to a maximum penalty of ten years' imprisonment. He also noted that section 14(d) of the *Prohibition of Human Cloning Act* 2002 (Cth) provided that a hybrid embryo included 'an animal egg into which the nucleus of a human cell has been introduced'. Herald observed that 'the product of the present method is a hybrid embryo within the meaning of this Act'.[35] Accordingly, the Deputy Commissioner of Patents was satisfied that the patent application should be refused under section 50(1)(a) of the *Patents Act* 1990 (Cth) by reason of its being contrary to law.

It is surprising that this significant decision has received little academic analysis or commentary, especially given the prominence of Woo-Suk Hwang and the novelty of the technology. It is worth highlighting a number of notable features of this judgment. First, the decision is significant in that it reinforces the approach taken by the Australian Patent Office to section 18(2) of the *Patents Act* 1990 (Cth). However, it is worth noting that this interpretation of section 18(2) of the *Patents Act* 1990 (Cth) has yet to be confirmed by the Federal Court of Australia or the High Court of Australia. Second, the judgment demonstrates that the 'contrary to law' provisions are not dead letters in Australian patent law, as hitherto thought. The discretion to refuse a patent application claiming an invention whose use would be 'contrary to law' may be relevant in future patent applications in the field of stem cell research and therapeutic cloning.[36] Third, it is worth noting that the Australian Patent Office declined to invoke the 'generally inconvenient' proviso. In recent cases, the Federal Court has declined to rely upon the 'generally inconvenient' proviso, in relation to methods of human treatment and business methods.[37]

In 2005, Professor Woo-Suk Hwang was accused of scientific misconduct and fraud in relation to his work in respect of somatic cell nuclear transfer technology.[38] He was labelled a 'cloning cowboy' by bioethics expert Professor Peter Glasner.[39] In 2006, a Seoul National University investigation found that Hwang had fabricated scientific evidence for nine of 11 stem cell lines. The scientist retracted his publication in the prestigious journal *Science*, in light of such revelations. Hwang was indicted in 2006 on charges of fraud and embezzlement in respect of misuse of research funds. He was also accused of breach of bioethics regulations because he obtained human eggs from donors, without full and informed consent. Furthermore, he erred in seeking eggs from junior researchers for his research, and paying the donors of such eggs. Hwang has resigned in disgrace from Seoul National University. He is subject to ongoing inquiries and court proceedings.

In a soul-searching article, Lawrence Ebert has observed that the Hwang scandal has raised questions about maintaining quality control in assessing patent applications: 'The failure of editors and referees of the journal *Science* to detect the fraud in manuscripts of Woo-Suk Hwang prior to publication, and the widespread acceptance of the work after publication, illustrates some difficulties in relying on peer review to authenticate the validity of scientific work.'[40] Patent Offices retain the discretion to revoke patent applications in the case of fraud or misrepresentation. There remain unresolved disputes regarding the priority of patent applications in respect of somatic cell transfer technology between Hwang and the Seoul National University, and Gerald Schatten and the University of Pittsburgh.[41]

C Australian Law Reform Commission

There has been much parliamentary debate over patent law and stem cell research in Australia.[42] There has also been much policy discussion about the ethical regulation of stem cell research.[43] In Australia, the House of Representatives Standing Committee on Legal and Constitutional Affairs has released its report entitled *Human Cloning: Scientific, Ethical and Regulatory Aspects Of Human Cloning and Stem Cell Research*.[44] The Senate Committee on Community Affairs also produced a report on the *Research Involving Embryos Act* 2002 (Cth) and *The Prohibition of Human Cloning Act* 2002 (Cth).[45] There were a number of submissions which considered patent law and stem cell research. IP Australia put forward a submission which outlined its philosophy and practice in this particular area. There were a range of comments from Australian companies which were undertaking stem cell research. There were also a number of submissions which raised ethical objections to the patenting and commercialization of stem cell research. Lamentably, there was little sustained discussion of intellectual property in the final parliamentary reports. Such issues were considered to be secondary and ancillary to the regulation of stem cell research.

In April 2003, the United Kingdom Patent Office issued a Practice Note setting out its general approach to patent applications claiming stem cells derived from human embryos and processes involving human embryonic stem cells.[46] The Practice Note indicates that each patent application will be assessed on its merits, but goes on to provide as follows:

- processes for obtaining stem cells from human embryos are not patentable because the *Patents Act 1977* (UK) provides that uses of embryos for industrial or commercial purposes are not patentable inventions;
- 'human totipotent cells' are not patentable because they have the potential to develop into an entire human body, and the human body at its various stages of its formation and development is excluded from patentability under the *Patents Act 1977* (UK); and
- 'human embryonic pluripotent stem cells' will be patentable if such inventions satisfy the statutory criteria for patentability because such stem cells do not have the potential to develop into an entire human body.[47]

In addition, the United Kingdom Patent Office has concluded that the commercial exploitation of inventions involving human embryonic pluripotent stem cells is not, as a general matter, contrary to public policy or morality in the United Kingdom.

In 2004, the Australian Law Reform Commission provided a review of the issues concerning patent law and stem cell research.[48] The Commission recommended that IP Australia should develop Stem Cell Examination Guidelines along the lines of the Practice Note issued by the United Kingdom Patent Office. It observed: 'In developing the proposed Stem Cell Examination Guidelines, the distinctions drawn by the UK Patent Office between totipotent and pluripotent cells may provide a helpful way to approach the application of section 18(2) of the *Patents Act* to inventions involving embryonic stem cell technologies.'[49]

In light of the decisions of the Deputy Commissioner of Patents and the report of the Commission, IP Australia has revised its examination guidelines regarding stem cells. The IP Australia Manual provides this advice to its examiners:

> The approach to be taken deals with whether the cells have an inherent capability to produce a human being. Human totipotent cells . . . possess the ability to form the entire range of cell types present in a human being, in the placenta and in other supporting tissues necessary for development of an embryo *in utero*. As a consequence . . . totipotent cells are excluded under s.18(2). Although . . . similar to stem cells in that they have the ability to give rise to a wide range of cells types . . . they are not generally accepted as such. Human stem cells are generally pluripotent or multipotent and unlike totipotent cells do not have the inherent capability to produce a human being. Human embryonic stem cells are pluripotent cells . . . they are not capable of producing cells for the placenta or other supporting tissues necessary for foetal development. Therefore isolated pluripotent human embryonic stem cells *per se* are not excluded under s.18(2).[50]

This approach to patent law and stem cell, with its demarcation between pluripotent, totipotent and multipotent stem cells, seems to be closely modelled on that of the United Kingdom Patent Office.

II AN 800-POUND GORILLA: THE WISCONSIN ALUMNI RESEARCH FOUNDATION AND GERON CORPORATION

The United States Patent and Trademark Office (USPTO) has long had a policy of banning human being and human embryo patents on the grounds that they would violate the 13th Amendment prohibiting slavery.[51]

In July 2003, Congressman Dave Weldon, MD, a Republican Doctor from Florida, attached a rider to the *Consolidated Appropriations Act* 2004 (US) that would prohibit patents on human organisms. The wording of the bill stated: 'None of the funds appropriated or otherwise made available under this Act may be used to issue patents on claims directed to or encompassing a human organism.' Congressman Weldon commented on the intent of this bill:

> Technology proceeds at a rapid rate, bringing great benefits to humankind, from treatments of disease to greater wealth and greater knowledge of our world. However, sometimes technology can be used to undermine what is meant to be human, including the exploitation of human nature for the purpose of financial gain. I recognize that there are many institutions . . . that have extensive patents on human genes [and] human stem cells. This would not affect any of those current, existing patents.[52]

The *Consolidated Appropriations Act* 2004 (US) was passed by the United States Congress, and approved by President George W. Bush. In 2006, the President vetoed a Congressional Bill, the *Stem Cell Research Enhancement Act* 2005 (US), which would have allowed for public funding in respect of embryonic stem cell research. He commented:

> Like all Americans, I believe our Nation must vigorously pursue the tremendous possibilities that science offers to cure disease and improve the lives of millions. Yet, as science brings us ever closer to unlocking the secrets of human biology, it also offers temptations to manipulate human life and violate human dignity. Our conscience and history as a Nation demand that we resist this temptation. With the right scientific techniques and the right policies, we can achieve scientific progress while living up to our ethical responsibilities.[53]

As a result of this veto, private biotechnology companies retained their vanguard position in respect of embryonic stem cell research in the United States.

A Wisconsin Alumni Research Foundation

The USPTO has adopted the position that purified and isolated stem cells are patentable subject matter.[54] The organization has been criticized for granting broad patents in respect of stem cell research. Seth Shulman observes: 'Perhaps the biggest lesson of all, though, surrounds the chronic myopia of the U.S. Patent and Trademark Office in awarding such needlessly all-encompassing patents as it has in this field.'[55]

The University of Wisconsin was the main beneficiary of President George W. Bush's decision to limit federal funding of stem cell research to existing stem cell lines, because it held extensive patents in respect of primate and human embryonic stem cell research. In November 1998, Dr James Thomson of the University of Wisconsin first isolated and cultivated pluripotent human embryonic stem cells. His team established five unmodified human embryonic stem cell lines. Through the Wisconsin Alumni Research Foundation (WARF), he filed for a patent on 26 June 1998. After overcoming initial doubts from the patent examiners,[56] Thomson was issued on 13 March 2001 with US Patent No: 6 200 806, with the title 'Primate Embryonic Stem Cells'.[57] The patent broadly covers both the method of isolating human embryonic stem cells and the five unmodified stem cell lines themselves. Thomson also received patents in respect of related technologies: US Patent No: 5 843 780 and US Patent No: 7 029 913.[58] The technology transfer unit, WARF, was proud of its substantial intellectual property holdings.[59]

Prompted by a moratorium on federal funding of human embryonic stem cell research, WARF licensed the patent to the private firm Geron Corporation in return for research funding. Under a first licence agreement, WARF granted to Geron exclusive rights to develop and commercialize the unmodified stem cell lines isolated by Thomson into six specific modified stem cell lines, relating to liver, muscle, nerve, pancreas, blood and bone cells. The Foundation retained the right to distribute its unmodified stem cell lines to the academic research community.

On 13 August 2001, WARF filed a lawsuit against Geron, contesting the company's rights to additional human embryonic stem cell types.[60] This legal action was prompted, in part, by government pressure and media scrutiny. As Rai and Eisenberg observe: 'Exclusive licenses on research tools with potentially broad applications threaten to throttle scientific progress by limiting the number of players in a developing field.'[61]

On 26 July 2001, Geron exercised an option contained in the first licence to claim 12 additional stem cell types.[62] WARF argued that the option had expired a week earlier and that the use of the option could be denied at WARF discretion. The dispute was settled out of court.[63] On 9 January 2002, WARF and Geron signed a new licence that gives Geron

(a) Exclusive rights to develop therapeutic and diagnostic products from three types of human embryonic stem cells (nerve, cardiac muscle and pancreas cells);
(b) Non-exclusive rights to develop therapeutic and diagnostic products from three further human embryonic stem cell types (blood, cartilage and bone cells); and
(c) Non-exclusive rights to develop research products in six human embryonic stem cell types.

Furthermore, WARF and Geron agreed to grant research rights to existing human embryonic stem cells patents and patent filings to academic and governmental researchers without royalties or fees.

David Earp, Geron's vice president of intellectual property, discusses the intellectual property portfolio licensed to Geron by the WARF: 'Our patent portfolio includes issued U.S. patents for primate and human embryonic stem (ES) cells and human embryonic germ (EG) cells, as well as over 50 patent applications pending around the world covering many aspects of human embryonic stem cell culture, production, differentiation and uses in cellular reprogramming.'[64]

However, the managing director of WARF, Carl Gulbrandsen, has sought to allay such fears:

> I don't want people to see us as an 800-pound gorilla. We will work very hard with the government to make sure that there is access to this technology and that our patents are not an impediment to researchers.[65]

On 5 September 2001, the National Institutes of Health signed an agreement with the WiCell Research Institute.[66] The memorandum of understanding covered both access to intellectual property and tangible property held by the WARF. The Parties agreed that Wisconsin Patent Rights are to be made available without cost for use in the biomedical research programme subject to a number of conditions. First, the Patent Rights only may be used in certain programmes in compliance with the law. Additionally, NIH researchers have to send a yearly notification saying that they are using the cells in accordance with the law. Second, WiCell agreed to allow the Patent Rights to be used in research programmes involving other materials. Finally, the parties agreed that the Wisconsin Patent rights may be used in public research to make patentable inventions, which themselves may eventually be the basis of commercial products that benefit human health.

B The Foundation for Taxpayer and Consumer Rights and the Public Patent Foundation

In 2004, 59.1 per cent of the California electorate endorsed Proposition 71, also known as the *Stem Cell Research and Cures Act* 2004 (California).[67] This measure established a new state medical research institute, the Californian Institute for Regenerative Medicine, and authorized the issuance of $3 billion in state general obligation bonds to provide funding for stem cell research and research facilities in California.[68] Bioethics expert Glenn McGee observed that Californian voters were upset that such funding of stem cell research could be hampered by patents:

The taxpayers of California are none too pleased either. The $3 billion, they are beginning to gripe, was to go to stem cell research, and they're furious that someone who filed a patent on looking at the human embryo could collect royalties every time an embryonic stem cell is made, or used in a discovery, regardless of whether that cell came from Wisconsin.[69]

WARF general counsel Elizabeth Donley said that WARF would demand licence fees and payments on all embryonic stem cell research in California, funded under Proposition 71.[70]

In 2006, the Los Angeles-based Foundation for Taxpayer and Consumer Rights and the New York-based Public Patent Foundation (the Foundations) requested the USPTO to engage in a re-examination of the claims of three of the Thomson patents: US Patent Nos: 5 843 780, 6 200 806 and 7 029 913.[71] The two organizations were concerned that the patents were causing significant public harm, because they had proved to be an impediment to stem cell research:

> By demanding significant financial consideration before allowing research to be performed, the owner of the '780, '806 and '913 patents is impeding, and in some cases literally stopping, domestic human ES cell research at its infancy. This not only harms scientific advance here in the United States, it also has a harmful economic impact on Americans by diverting taxpayer dollars meant for research to pay for licensing fees. In the words of one industry insider, this aggressive patent assertion is 'stifling industrial research and investment.'[72]

The Foundations argued that the patents 5 843 780 and 6 200 806 were invalid because they had been anticipated, and, in any case, were obvious. The organizations submitted that patent 7 029 913 was obvious in light of the prior art. The Foundations maintain that four pieces of prior art – one prior patent and three scientific publications – called into question the validity of the Thomson patents, because they demonstrated how to derive embryonic stem cells in various animals, including mice, pigs and sheep.

The Foundations referred to a patent granted to an Australian inventor, Robert Lindsay Williams, who was working for AMRAD, on 24 November 1992.[73] The submission notes: 'Williams taught a method for isolating ES cells of various animals, including specifically humans.'[74] The Foundations argued that the Williams patent anticipates both the Thomson patents 5 843 780 and 6 200 806, which claimed the earliest priority date of 20 January 1995. The submission provides tabular comparisons of the patent claims of Williams and Thomson.

The Foundations also cited as prior art two book chapters, written in 1983 and 1987, by Elizabeth Robertson, now at Oxford University in England; and a research paper published in 1990 by Jorge Piedrahita, now at North Carolina State University in Raleigh.[75] The submission noted that

Robertson had conducted significant research on isolating mice embryonic stem cells. 'More than a decade before the initial application leading to the '780 patent was filed, Robertson 1983 taught a step-by-step process for isolating pluripotential mammalian ES cells.'[76] It added: 'A few years later, Robertson 1987 again taught the step-by-step process for isolating pluripotential mammalian ES cells, this time giving even further detail regarding each specific step.'[77] Similarly, Piedrahita taught a method of isolating murine (rodent), porcine (pig) and ovine (sheep) embryonic stem cells.

Dr Jeanne Loring, a stem cell scientist at the Burnham Institute for Medical Research, filed a statement supporting the groups' challenges.[78] She observed: 'The real discovery of embryonic stem cells was by Martin Evans, Matt Kaufman and Gail Martin in 1981, and none of these scientists considered patenting them.'[79] Loring concluded: 'It is outrageous that WARF claimed credit for this landmark discovery nearly 15 years after it was made.'[80]

James Thomson has defended the validity of the patents. He has emphasized that no scientists were able to grow the cells from humans before he did:

> Although sometimes things seem obvious in retrospect, it is curious that no one accomplished the derivation of human (embryonic stem) cells between 1981 . . . and 1998. Some very good, simple ideas only seem obvious afterwards.[81]

He maintained that the opponents of WARF's patents wanted the California Institute for Regenerative Medicine to profit from future patents on discoveries made through its stem-cell research initiative: 'The only real difference is that the CIRM-funded patents will largely benefit California, and the WARF patents largely benefit Wisconsin.'[82]

C United States Patent and Trademark Office Ruling

On 30 March 2007, the USPTO upheld the challenges by the consumer advocates to the three patents on human embryonic stem cells held by WARF, finding that the inventions were obvious in light of the prior art and the declaration by Dr Jeanne Loring. Its key decision in respect of US Patent No: 5 843 780 said:

> It would have been obvious to one skilled in the art at the time the invention was filed to the method of isolating ES cells from primates and maintaining the isolated ES cells on feeder cells for periods longer than one year. A person skilled in the art would have been motivated to isolate primate (human) ES cells, and maintained in undifferentiated state for prolonged periods, since ES cells are pluripotential and can be used in gene therapy.[83]

Similar findings were made in respect of US Patent Nos: 6,200,806 and 7,029,913.[84]

The Foundation for Taxpayers and Consumer Rights was triumphant at the decision. John M. Simpson observed: 'This is a great day for scientific research.'[85] Dr Jeanne Loring was also pleased by the outcome:

> The real discovery of embryonic stem cells was by Martin Evans, Matt Kaufman, and Gail Martin in 1981, and none of these scientists considered patenting them. It is outrageous that WARF claimed credit for this landmark discovery nearly 15 years after it was made.[86]

The director of the Public Patent Foundation, Dan Ravicher, was hopeful that WARF would drop all its claims in light of the ruling of the USPTO: 'This rejection is substantial, and it will cause a significant deterioration in the impact the patents will have in the marketplace.'[87] WARF intends to challenge the rulings of the USPTO. The managing director, Carl Gulbrandsen, commented:

> WARF has absolute confidence in the appropriateness and legitimacy of these patents. It is inconceivable to us that Dr. Thomson's discovery, which *Science* magazine heralded as one of the greatest scientific discoveries in history, would be found to not be worthy of a patent.[88]

The patents will remain in force until the matter is resolved. If the decision of the USPTO stands, both United States and international researchers will have the freedom to conduct stem cell research, without fear of infringing these key WARF patents.

III A SECULAR CLOISTER: THE EUROPEAN PATENT OFFICE

One-time Swiss Patent examiner, Albert Einstein, observed in his correspondence that the patent office was a 'secular cloister'.[89] This metaphor is an apt description. It helps evoke the insular attitude of patent administrations to matters of public policy.

The dominant sentiment within the courts and the patent offices is that ethical considerations are necessarily extrinsic to patent law. As Benjamin Enerson has observed: 'The patent system should not become a theater for judging the morality of controversial inventions.'[90] Furthermore, Bently and Sherman observe:

> One of the defining features of patent law, at least up until its encounter with biotechnology, was that it was treated as if it was hermetically sealed, closed off

from external considerations. Modern patent law is characterized not only by its highly technical and specialized nature but also by its startling and marked isolation from matters cultural, political and ethical.[91]

IP Australia has baulked at taking into account ethical considerations in the examination of patent applications. The Australian Patent Office Manual of Practice and Procedure of November 1999 asserts that it is inappropriate for the patent office to deal with matters of ethics and social policy.[92] Similarly, the United States Patent and Trademark Office has been reluctant to take into account questions of public order and morality. Famously, it avoided dealing with such questions when Stuart Newman and Jeremy Rifkin put forward patent applications in respect of chimeras.[93]

By contrast, the European Union makes explicit provision for opposition on the grounds of public order and morality under the *European Union Directive on the Legal Protection of Biotechnological Inventions* 1998 (EU). Article 6(1) provides: 'Inventions shall be considered unpatentable where their commercial exploitation would be contrary to ordre public or morality; however, exploitation shall not be deemed to be contrary merely because it is prohibited by law or regulation.' Article 6(2) stipulates that certain particular inventions shall be considered to be unpatentable, including:

(a) processes for cloning human beings;
(b) processes for modifying the germ line genetic identity of human beings;
(c) uses of human embryos for industrial or commercial purposes;
(d) processes for modifying the genetic identity of animals which are likely to cause them suffering without any substantial medical benefit to man or animal, and also animals resulting from such processes.

Such ethical considerations were evident in patent applications by Edinburgh university, the WARF and CalTech.

In 2002, the European Group on Ethics in Science and New Technologies released its report on 'The Ethical Aspects of Patenting involving Human Stem Cells'.[94] The majority of the Group was of the opinion that isolated stem cells which have not been modified do not, as product, fulfil the legal requirements to be seen as patentable, especially with regard to industrial applications.[95] In addition, such isolated cells are so close to the human body, to the foetus or to the embryo they have been isolated from that their patenting may be considered as a form of commercialization of the human body. When unmodified stem cell lines are established, they can hardly be considered as a patentable product. Such unmodified stem cell lines, indeed, do not have a specific use but a very large range of potential undescribed uses. Therefore, to patent such unmodified stem cell lines would also result in broadly framed patents. The Group maintained that only stem cell lines which have been modified by in vitro treatments, or genetically modified so that they have acquired characteristics for specific industrial application, fulfil the legal requirements for patentability.[96] As to the patentability of processes involving human stem cells, whatever their source, there is no specific ethical obstacle, in so far as they fulfil the requirements of patentability: the criteria of novelty, inventive step and industrial application.

The recommendations of the European Group on Ethics in Science and New Technologies have been met with some hostility from members of the patent profession. R. Stephen Crespi is particularly critical that the distinction drawn between modified and unmodified stem cells has no basis in patent law.[97] The patent attorney laments that the 'report has, in my view, side-stepped the most difficult questions and has settled for devising its own patentability criteria, of which the key opinion is that only stem cells that have been modified by in vitro treatment or genetic modification can be considered fit subject-matter for patents'.[98]

In 2005, the president of the European Patent Office, Alain Pompidou, commented that the European Patent Office will not patent any embryonic stem-cell technology for the time being, because 'there are too many ethical aspects that have not been resolved at the political level'.[99] He observed that there was a lack of consensus about the topic of stem cell patents amongst both the members of the European Patent Office and the European Union: 'The European Patent Office needs to take note of the European Union's political climate.'[100] Pompidou noted that the European Patent Office had not approved any of the three patent applications in respect of stem cell research. He expressed concern that any patent granted in respect of stem cell research would trigger protests similar to those that surrounded patents for genetically modified organisms. The European Patent Office has been worried that it would not receive political backing from the European Commission to defend itself against such action in the wake of a human embryonic stem-cell patent.

A The Edinburgh Patent

The 'Edinburgh' patent is European patent No. EP 0695351, with the title 'Isolation, selection and propagation of animal transgenic stem cells'.[101] The patent relates to an invention in the field of developmental biology. It describes a method of using genetic engineering to isolate stem cells (including embryonic stem cells) from more differentiated cells in a cell culture in order to obtain pure stem cell cultures. Holders of the patent are Austin Smith at the University of Edinburgh and Peter Mountford, chief scientific officer at Stem Cell Sciences.[102]

The granting by the European Patent Office of this patent to the University of Edinburgh in December 1999 led to fierce protests and triggered a major public debate on the patenting of stem cell technology. Greenpeace staged dramatic protests against this particular patent at the European Patent Office in Munich, Germany.[103] The spokesperson for the group, Christopher Then, declared: 'The EPO is selling out the right to use animals, plants and humans to the genetic engineering industry.'[104] Ninety activists managed to shut down the EPO by bricking up the main entrance and the basement garage. Climbers hoisted a banner reading, 'Stop breeding human beings. No patents on life!'[105] Greenpeace dismissed the apologies of the EPO: 'Apologies, promises, and cosmetic corrections are not enough.'[106] Christopher Then emphasized: 'The patenting of human beings is not the result of the EPO's carelessness – it's a cold-blooded policy.'[107]

The patent was opposed by 14 parties, demanding that it be revoked. Ten of the opponents took part in the hearing. The governments of Germany, Italy and the Netherlands, and the German branch of Greenpeace, were among the parties that have lodged oppositions to the patent. The opponents alleged, among other things, that the 'Edinburgh' patent contravened Article 53(a) of the *European Patent Convention* 1973 (EU) which precluded the patenting of inventions whose exploitation would be contrary to 'ordre public' or morality.

The disputed part of the Edinburgh patent was claim 48, which related to a 'method of preparing a transgenic animal'. Although the practical examples mention mice, the scope of the patent had the potential to encompass humans. The description of the patent mentions that 'in the context of this invention, the term "animal cell" is intended to embrace all animal cells, especially of mammalian species, including human cells'.

The University of Edinburgh, the patent proprietor, has made it clear that it never intended the patent to cover creation of genetically altered humans. It therefore requested that the patent be limited. In a statement, Dr Peter Mountford denied any intention to patent or develop technologies for human genetic engineering:

> The techniques described in this patent represent a significant advance in the culture of stem cells, making available populations of specific types of stem cells for numerous research and clinical applications. While we at Stem Cell Sciences are delighted to be at the forefront of this exciting technology, we are however concerned that our techniques may have been misunderstood.[108]

After a three-day public hearing at the European Patent Office, the Opposition Division decided that the 'Edinburgh' patent should be maintained in an amended form as introduced by the patent proprietor during the oral proceedings.[109] It no longer includes human or animal embryonic stem cells, but still covers modified human and animal stem cells other than embryonic stem cells.

In 2002, the Opposition Division ruled that the patent application suffered from an insufficiency of disclosure under Article 83(a) of the *European Patent Convention* 1973 (EU).[110] The Opposition Division took heed of the objections of the opponents:

> According to the analysis of the Opponents the present patent specifies that a source of animal stem cells is maintained under culture conditions as a homogenous population in undifferentiated state. The cultured stem cells are thus amenable to in vitro genetic manipulation. The source of animal stem cells is very broadly defined in the description of the contested patent, covering cells from all animal species, including mammals such as humans as well as non-mammalian animals but also covering sources for all sorts of stem cells, including embryos, embryonic tissues as well as later stage materials comprising stem cells for particular tissue lineages . . . In their view the results obtained for mouse ES cells cannot be extrapolated across species.[111]

The Opposition Division 'agrees with the opponents in that there are serious doubts as to whether the results obtained for mouse ES cells as exemplified in the contested patent can be extrapolated across species, i.e. to human ES cells'.[112]

The Opposition Division also considered objections to the contested patent under Article 53(a) of the *European Patent Convention* 1973 (EU), which excludes from patentability 'inventions the publication or exploitation of which would be contrary to "ordre public" or morality, provided that the exploitation shall not be deemed to be so contrary merely because it is prohibited by law or regulation in some or all of the Contracting States'. Furthermore, it took notice of Rule 23d(c), which provides that uses of human embryos for industrial and commercial purposes are excluded from patentability.

The Opposition Division commented that there was a lack of consensus amongst the members of the European Union in respect of embryonic stem cell research:

> Regarding the national situation for the subject-matter under investigation no consistent law(s) or regulation(s) exist. Whereas in some European countries research on human ES cells does not pose any problems, in some other countries this kind of research was only accepted with certain restrictions on the research itself, and other countries are still in the process of arriving at a conclusion on this issue.[113]

The Opposition Division added that 'neither the evaluation of the national legislation nor the assessment of the conventionally accepted standards of conduct of European culture has revealed a uniform approach with regard to human ES cells'.[114]

In the 'Edinburgh Patent' case, the Opposition Division pointedly disregarded the recommendations of the European Group on Ethics in Science and New Technologies.[115] First, the Opposition Division complained that 'the opinion creates new patentability criteria which do not exist in the EPC and therefore cannot be taken into account' and that 'classical concepts of patent law are misinterpreted and confused'.[116] Second, the Opposition Division complained of defects and inconsistencies in the reasoning. Third, the Opposition Division cavilled: 'The conclusion of the Opinion that processes involving human stem cells should be patentable without further limitations but product claims shall be patentable only under certain conditions (modified to an extent that a specific industrial application is visible) cannot be followed logically.'[117] The Opposition Division concluded: 'Due to its many inconsistencies, logical flaws, and incompatibility with existing patent law and the EU Directive, the Opinion must be disregarded in toto.'[118]

The Opposition Division observed: 'The crucial question is whether the legislator introducing this Rule into the EPC in September 1999 has intended to ban from patenting human embryos as such or human embryos together with the cells being retrieved therefrom by destruction of the embryos, namely human ES cells.'[119] The Opposition Division was obliged to take a faithful reading of the rule: 'When weighing up all the above legal considerations, one has to come to the conclusion that only a broad interpretation of Rule 23d(c) EPC can have been intended.'[120] The Opposition Division concluded: 'In consequence, Rule 23d(c) EPC in order to have a purpose exceeding the one of Rule 23e(1) EPC has to be interpreted broadly to encompass not only the industrial or commercial use of human embryos but also the human ES cells retrieved therefrom by destruction of human embryos.'[121]

In December 2003, the University of Edinburgh lodged an appeal against the decision of the Opposition Division. The University of Edinburgh claimed that the decision concerning morality was wrong: 'The Opposition Division has exceeded the mandate of the EPO by independently assuming the mantle of moral censor, a responsibility that has been explicitly delegated to the European Group on Ethics in Science and New Technologies (EGE) and which is in addition a matter for the governments of individual contracting states.'[122] Furthermore, the University of Edinburgh complained that the decision in respect of sufficiency was also wrong, because the Opposition Division made the ruling 'with an eye on the moral issues'.[123] The Technical Board of Appeals is yet to hear this appeal, which will no doubt be a complicated matter, given the large number of opponents.

Furthermore, the 'Edinburgh patent' prompted the European Parliament to pass a resolution in 1999 condemning the cloning of human beings.[124] First, it stressed that it was 'deeply shocked at the granting of a patent to the University of Edinburgh, which includes a technique for the genetic modification of the germ line of human embryos and of the embryos themselves, a patent on isolation, selection, and propagation of animal and transgenic stem cells, which could be used for the cloning of human beings'. Second, it undertook 'to file without delay an objection to patent number EP 695 351 if legally possible, and calls on the other institutions of the European Union and Member State governments to do likewise'. Third, it 'demands a review of the operations of the EPO to ensure that it becomes publicly accountable in the exercise of its functions, and to amend its operating rules to provide for it revoking a patent on its own initiative'.

B Wisconsin Alumni Research Foundation

WARF's patents in respect of embryonic stem cell research have attracted much controversy in the European Union. In 1996, the WARF filed a patent application in respect of 'primate embryonic stem cells'.[125] Claim 1 concerns 'A cell culture comprising primate embryonic stem cells which (i) are capable of proliferation in vitro culture for over one year, (ii) maintain a karyotype in which all chromosomes normally characteristic of the primate species are present and are not noticeably altered through culture for over one year, (iii) maintain the potential to differentiate to derivatives of endoderm, mesoderm, and ectoderm tissues throughout the culture, and ... (iv) are prevented from differentiating when cultured on a fibroblast feeder layer.' The patent application describes the removal of embryonic stem cells from the inner cell mass of monkey blastocysts and the culturing of those stem cells on feeder cells. The application discloses that the same procedures and growth conditions will allow the isolation and growth of human embryonic stem cells and such cells have been deposited with the NIH. The cells are described as being pluripotent.

In June 2004, the Examining Division rejected the patent application filed by the WARF under Article 97(1) of the *European Patent Convention* 1973 (EU).[126] The Division held that claims 1 to 7, 9 and 10 did not comply with the requirements of Article 53(a) in conjunction with Rule 23d(c) of the *European Patent Convention* 1973 (EU) because the use of human embryos as starting material was described in the application as originally filed as being indispensable. The Examining Division held that the use of a human embryo as starting material for the generation of a product of industrial application was prohibited under Rule 23d(c) and Article 53(a) of the *European Patent Convention* 1973 (EU). The description provided only one source of starting cells, namely a pre-implantation embryo. Accordingly, the Examining Division held that it was irrelevant that the claimed subject-matter related to cell cultures and not to a method of production of said cultures. The exception to the exclusion from patentability with regard to the use of human embryos derivable from Recital 42 of the *European Union Directive on the Legal Protection of Biotechnological Inventions* 1998 (EU) did not apply to the present case because the generated cell cultures did not serve any therapeutic or diagnostic purpose useful to the embryo that gave rise to the said cultures.

In November 2005, the Technical Board of Appeal of the European Patent Office handed down its interlocutory decision.[127] The Technical Board of Appeal considered in its judgment 'whether Rule 23d(c) EPC should be construed *narrowly* (thereby excluding from patentability only applications whose claims were directed to the use of human embryos) or *broadly* (thereby extending the exclusion to products whose isolation necessitated the direct and unavoidable use of human embryos)'.[128] It canvassed the various arguments in favour of a narrow and a broad construction. The Technical Board of Appeal recognized that the question of the patentability of stem cells was a controversial one in the European Union:

> In relation to inventions directed to human embryonic stem cells, one should consider that there was no consensus amongst Contracting States as to the ethical acceptability of using human embryonic stem cells. The development of human embryonic stem cells from supernumerary embryos was permitted in several EPC states and there was an ongoing debate as to the ethics of using human embryonic stem cells, it being clear that moral attitudes were changing as was e.g. shown by the fact that in November 2003 the European Parliament voted to permit public funding for human embryonic stem cell research. Where matters relevant to morality arose, any decision based on them should be based on facts as substantiated at the date of the decision. Inventions relating to human embryonic stem cells were clearly not of the type that was so abhorrent that the grant of patent rights would be inconceivable.[129]

In the end, the Technical Board of Appeal abstained from making a conclusive ruling on this matter. It referred a number of outstanding questions to the Enlarged Board of Appeal for their deliberation. Most importantly, the Enlarged Board of Appeal has been asked to ruled whether 'Rule 23d(c) [of the *European Patent Convention* 1973 (EU)] forbids the patenting of claims directed to products (here: human embryonic stem cell cultures) which – as described in the application – at the filing date could be prepared exclusively by a method which necessarily involved the destruction of the human embryos from which the said products are derived, if the said method is not part of the claims'.

The patent attorney, R. Stephen Crespi, has noted the significance of the determination of these issues by the Enlarged Board of Appeal:

> The present discussion seems to have reached a point from which it cannot proceed further without reference to fundamental belief systems that compete for acceptance by society at large. Whether the Enlarged Board of Appeal will enter this moral minefield remains to be seen, but the notion that patent law can accommodate such profoundly difficult questions when society remains divided and confused over them has only to be stated to see how impossible it must be in the fast-developing technologies that are a feature of bioscience today.[130]

Nonetheless, Crespi notes that certain jurisdictions, such as the United Kingdom, may take their own approach, rather than conform to whatever the Enlarged Board of Appeal rules.

C California Institute of Technology Patent Application

In 1993, David J. Anderson of the California Institute of Technology (Caltech) and Derek Stemple of the Cardiovascular Research Center Massachusetts General Hospital East filed a patent application for a method to isolate neural stem cells from embryonic tissue.[131] Claim 1 concerned 'a method of proliferating in vitro a clonal population of neural crest stem cells to produce a population of neural crest stem cells and differentiated progeny thereof, wherein the neural crest stem cells are characterised as being capable of self-renewal in the culture medium and capable of differentiation to progeny cells that are peripheral nervous system neuronal or glial cells'.[132]

In a decision on 17 October 2003, the Examining Division refused the patent application by Caltech on a number of grounds.[133] The Examining Division held that the patent application offended Article 53(a) and Rule 23d(c) of the *European Patent Convention* 1973 (EU). The Examining Division rejected the attempts of Caltech to distinguish its application from the 'Edinburgh patent' case:

> According to the Applicant, Rule 23d(c) EPC is not relevant since the neural crest cells are not embryonic stem (ES) cells like in the 'Edinburgh' case, as they do not have any capacity to develop into an embryo or mammal. Furthermore, even if the invention requires the use of an embryo as source material, such use would be a positive one allowing for the furthering of medical research for the benefit of mankind. Such a use would be ethically acceptable, especially if the embryos were created for IVF and would otherwise be destroyed.[134]

The Examining Division, though, observed that it was not relevant whether or not the neural crest cells were pluripotent, not totipotent, stem cells: 'The relevant question is whether the claimed cells comprise *human embryonic* cells since then the invention involves the use of a human embryo.'[135] Accordingly, the Examining Division ruled that, 'since neural crest cells are embryonic cells, the isolation of which involves the use of an embryo (see, e.g., example 1) and since the scope of claim 1 includes human neural crest cells, the subject matter of claim 1 is to be excluded from patentability under Rule 23d(c) EPC'.[136]

The Examining Division also ruled that the patent application lacked novelty under Article 54 of the *European Patent Convention* 1973 (EU): 'Apart from the main focus of D2, which relates to the preparation of a clonal immortalized neural crest cell line (NCM-1), D2 also contains a disclosure of non-immortalized neural crest cells which affects the novelty of claim 1.'[137]

In 2004, Caltech appealed against the decision of the Examining Division, claiming that it had erred in its interpretation of Rule 23d(c) of the *European Patent Convention* 1973 (EU) and its finding that the claimed subject-matter was contrary to Article 53(a) of the *European Patent Convention* 1973 (EU). Caltech protested that there was no step or feature in the claim reciting use of an embryo or even removal of cells from an embryo. Caltech maintained that the claims were directed to a method of proliferating a clonal population of cells in vitro, and argued that the Examining Division's reference to the Opposition Division in respect of the 'Edinburgh patent' was misjudged. Caltech submitted that the exclusions under the *European Patent Convention* 1973 (EU) should be interpreted narrowly. This appeal is still on foot.

CONCLUSION

There is a need for national governments to reform patent law to better address dynamic developments in the field of embryonic stem cell research. Matthew Herder from Dalhousie University has commented:

> Governments must begin to work through and devise strategies to respond to these proliferating patent problems, calculating the benefits and risks associated with each approach, in order to preserve the primary public interest in hESC research – the development of affordable, widely beneficial health care technologies. Diverse strategies are available to address issues of access, patent (over)breadth, and return on public investment ... It is not obvious what approach will work best to ensure that hESC research delivers on its promise. Different responses may be needed for different times.[138]

The academic concludes that 'society's expectations for the field – expectations generated in large measure by those who seek, grant, own and enforce intellectual property rights over hESC inventions – should be tempered not only while scientific challenges remain but also while the challenges posed by patenting are left unaddressed'.[139]

Surveying the positions of Australia, the United States and the European Union, it is unclear to what extent stem cell research is patentable subject matter. In Australia, section 18(2) of the *Patents Act* 1990 (Cth) is uncertain and indeterminate, because it provides little guidance as to whether stem cells fall foul of the prohibition against patenting 'human beings and the biological processes for their generation'. IP Australia has sought to alleviate this uncertainty by following the approach of the United Kingdom Patent Office. In the United States, an appropriations bill provides the indirect guidance: 'None of the funds appropriated or otherwise made available under this Act may be used to issue patents on claims directed to or encompassing a human organism.' However, such a clause is of little help in determining whether stem cells are patentable and, if so, to what extent. The challenge against WARF's patents by the Foundation for Taxpayer and Consumer Rights and the Public Patent Foundation may provide a better indication as to the approach of the USPTO. The European Patent Office has placed a moratorium on stem cell patents, expressing concerns that they contravene the *European Patent Convention* 1973 (EU) and the *European Union Directive on the Legal Protection of Biotechnological Inventions* 1998 (EU). The European Group on Ethics in Science and New Technologies has recommended that a distinction be made between modified stem cells (which would be patentable) and unmodified stem cells (which would not). The United Kingdom Patent Office has instead drawn a distinction between pluripotent stem cells (which are patentable) and totipotent stem cells (which are not).

Furthermore, there is a need to ensure that the granting of patents in respect of stem cell research will not impair research and development in the field, or prevent equitable access to therapies and drugs derived from this work. The scope of patent protection for stem cell research could be limited by the strict application of patent criteria. Kenneth Taymor, Christopher Thomas Scott and Henry Greely wonder whether it will be possible for researchers to 'invent around' patented technology:

> Scientific innovation nonetheless is creating new research venues beyond those patents' scope. As this research unfolds we will learn whether it has been a creative advance or an unfortunate resource-diverting attempt to 'invent around' patented technology.[140]

National governments would be well-advised to create a research exemption to give third parties access to stem cell products and research tools. They should also modernize the compulsory licensing provisions to enable parties to obtain access to stem cell research without need for authorization from the patent owners. It would also be worthwhile addressing fears that the field of stem cell research will be monopolized by a small number of commercial biotechnology companies. It must open intellectual property law to oversight under competition law. National governments must also consider patent law and stem cell research within the prism of the debate over the ethics of patenting life forms. It should seek to include public policy considerations (such as ethical considerations) in an assessment of patent applications.

NOTES

1. Piccinini, P., http://www.patriciapiccinini.net/.
2. Piccinini, P. (2002), 'Still life with stem cells', Biennale of Sydney, http://patriciapiccinini.net/essay.php?id=24.
3. Biotechnology Industry Organization (1999), 'Statement for hearing regarding commercial development of pluripotent stem cells', Subcommitttee on Labor, Health and Human Services, Education of the Senate Appropriations Committee, http://www.bio.org/bioethics/background/stemcell_testimony.asp, 12 January.
4. Ibid.
5. Ibid.
6. Ibid.
7. Rai, A. (2002), 'Stem cell research: an NPR Special Report, Á "Virtual roundtable" On federal funding', http://www.npr.org/programs/specials/stemcells/viewpoints.rai.html.
8. United Kingdom Stem Cell Bank, http://www.ukstemcellbank.org.uk/; and also see McLennan, Alison (2007), 'Which bank? A guardian model for regulation of embryonic stem cell research in Australia', *Journal of Law and Medicine*, **15**, 45–76.
9. United States Department of Health and Human Services (2001), 'Memorandum of understanding between the WiCell Research Institute and the Public Health Service', http://www.nih.gov/news/stemcell/WicellMOU.pdf, 5 September.
10. Neville, W. (2003), 'Submission to the Australian Law Reform Commission issues paper on gene patenting and human health', 29 October.
11. *Fertilitescentrum AB and Luminis Pty Ltd* [2004] APO 19; and *Woo-Suk Hwang* [2004] APO 24.
12. Foundation for Taxpayer and Consumer Rights and the Public Patent Foundation (2006), 'Request for re-examination in respect of US Patent No: 5,843,780, http://www.pubpat.org/assets/files/warfstemcell/780Request.pdf; Foundation for Taxpayer and Consumer Rights and the Public Patent Foundation (2006), 'Request for Re-examination in Respect of US Patent No: 6,200,806, http://www.pubpat.org/assets/files/warfstemcell/90008139granted.pdf; and Foundation for Taxpayer and Consumer Rights and the Public Patent Foundation (2006), 'Request for Re-examination in Respect of US Patent No: 7,029,913', http://www.pubpat.org/assets/files/warfstemcell/913Request.pdf.
13. *Greenpeace Deutschland e.V. v The University of Edinburgh*, Opposition Division, European Patent Office, (24 July 2002); Wisconsin Alumni Research Foundation patent application in respect of 'Primate embryonic stem cells', T 1374/04–3.3.08. Interlocutory decision of the Technical Board of Appeal 3.3.08 of 18 November 2005; and Caltech Patent Application, Examining Division, European Patent Office, T522/04–338 (17 October 2003).
14. The European Group on Ethics in Science and New Technologies to the European Commission (2002), *Opinion on the Ethical Aspects of Patenting Inventions Involving Human Stem Cells*, Opinion Number 16, 7 May.

15. Harradine, B. (1990), 'Patents Bill 1990 (Cth), in committee', Senate Hansard, 21 September, p. 2564.
16. Coulter, J. (1990), 'Patents Bill 1990 (Cth), in committee', Senate Hansard, 20 September, p. 2653.
17. IP Australia (2001), 'House of Representatives Standing Committee On Legal And Constitutional Affairs Inquiry Into The Scientific, Ethical And Regulatory Aspects Of Human Cloning', Submission 274, http://www.aph.gov.au/house/committee/laca/humancloning/sub274.pdf.
18. Ibid, p. 4.
19. *Fertilitescentrum AB and Luminis Pty Ltd* [2004] APO 19; and *Woo-Suk Hwang* [2004] APO 24.
20. *Fertilitescentrum AB and Luminis Pty Ltd* [2004] APO 19; (2004) 62 IPR 420.
21. Robertson, S., M. Wikland and C. Sjoblom (1999), 'Method and medium for in vitro culture of human embryos', Australian Patent Application No: 44916/99, PCT No: WO/1999/067364.
22. Robertson, S., M. Wikland and C. Sjoblom (1999), 'Method and medium for in vitro culture of human embryos', Australian Patent Application No: 44916/99, PCT No: WO/1999/067364.
23. *Fertilitescentrum AB and Luminis Pty Ltd* [2004] APO 19; (2004) 62 IPR 420.
24. *Fertilitescentrum AB and Luminis Pty Ltd* [2004] APO 19; (2004) 62 IPR 420.
25. *Fertilitescentrum AB and Luminis Pty Ltd* [2004] APO 19; (2004) 62 IPR 420.
26. Hwang, W.-S. et al. (2004) 'Evidence of a pluripotent human embryonic stem cell line derived from a cloned blastocyst', *Science*, **303**, 1669–70.
27. Hwang, W.-S. et al. (2005), 'Patient-specific embryonic stem cells derived from human SCNT blastocysts', *Science*, **308**, 1777.
28. Hwang, W.S. (2003), 'A method of inter-species nuclear transplantation', Patent Application No: 2003204164.
29. Hwang, W.S. (2003), 'A method of inter-species nuclear transplantation', Patent Application No: 2003204164.
30. *Woo-Suk Hwang* [2004] APO 24.
31. *Woo-Suk Hwang* [2004] APO 24.
32. *Woo-Suk Hwang* [2004] APO 24; see also *Dow Chemical Co v Ishihara Sangyo Kaisha Ltd*, 5 IPR 415.
33. IP Australia (2003), Chapter 8 of *Patent Manual of Practice and Procedure: Patentable Subject Matter*, http://www.ipaustralia.gov.au/pdfs/patents/manual/Part208.PDF, December.
34. *Woo-Suk Hwang* [2004] APO 24.
35. *Woo-Suk Hwang* [2004] APO 24.
36. S.51(1)(a) of the *Patents Act* 1990 (Cth); and *Dow Chemical Co v Ishihara Sangyo Kaisha Ltd.* 5 IPR 415.
37. *Bristol-Myers Squibb v FH Faulding* (2000) 46 IPR 553; and *Welcome Real-Time SA v Catuity Inc* (2001) 51 IPR 327.
38. Imrie, Katherine (2006), 'The Hwang scandal: scientific misconduct and research integrity', honours thesis, ANU College of Law.
39. Cronin, D. (2006), 'Stem cell experts warn of alarmists', *The Canberra Times*, 23 September, 13.
40. Ebert, L. (2006) 'Lessons to be learned from the Hwang matter', *Journal of the Patent and Trademark Office Society*, **88**, 239–55 at 255.
41. Schatten, G., C. Simerly and C. Navara (2004), 'Methods for correcting mitotic spindle defects associated with somatic cell nuclear transfer in animals', US Patent Application 20040268422; and Check, E. (2006), 'Schatten in the spotlight', *Nature News*, http://www.nature.com/news/2006/060109/full/060109-7.html, 11 January.
42. Rimmer, M. (2003), 'The attack of the clones: patent law and stem cell research', *Journal Of Law And Medicine*, **10**(4), 488–505.
43. Casell, J. (2001), 'Lengthening the stem: allowing Federal funded researchers to derive human pluripotent stem cells from embryos', *University of Michigan Journal of Law*

Reform, **34**(3), 547–72; Bruce, A. (2002), 'The search for truth and freedom: ethical issues surrounding human cloning and stem cell research', *Journal of Law and Medicine*, **9**(3), 323–35; Nicol, D., D. Chalmers and B. Gogarty (2002), 'Regulating biomedical advances: embryonic stem cell research', *Macquarie Law Journal*, **2**, 31–60; and Gogarty, B. and D. Nicol (2002), 'The UK's cloning laws: a view from the antipodes'. *Murdoch University Electronic Journal of Law*, **9**(2), http://www.murdoch.edu.au/elaw/issues/v9n2/gogarty92.html.

44. House of Representatives Standing Committee on Legal and Constitutional Affairs (2001), *Human Cloning: Scientific, Ethical and Regulatory Aspects of Human Cloning and Stem Cell Research*, Canberra: Australian Parliament.
45. Community Affairs Legislation Committee (2002), *Provisions of the Research Involving Embryos and Prohibition of Human Cloning Bill 2002: Supplementary Report*. Canberra: Australian Parliament.
46. United Kingdom Patent Office (2003), 'Inventions involving human embryonic stem cells', http://www.patent.gov.uk/patent/notices/practice/stemcells.htm, April.
47. Ibid.
48. Australian Law Reform Commission (2004), *Genes and Ingenuity: Gene Patenting and Human Health, Report 99*, Sydney: Australian Commonwealth, http://www.austlii.edu.au/au/other/alrc/publications/reports/99/, June. For a discussion of the Commission's deliberations, see Rimmer, M. (2004), 'The last taboo: patenting human beings', *Expert Opinion on Therapeutic Patents*, **14**(7), 1061–74.
49. Australian Law Reform Commission (2004), *Gene Patenting and Human Health, Discussion Paper 68*, Sydney: Australian Commonwealth, http://www.austlii.edu.au/au/other/alrc/publications/dp/68/, February, p. 463.
50. IP Australia (2006), *Patent Manual of Practice and Procedure*, http://www.ipaustralia.gov.au/pdfs/patentsmanual/WebHelp/National/Patentable/2.9.5.1_stem_cells.htm, January.
51. Kincaid, S. (2003), 'Oh, the places you'll go: the implications of current patent law on embryonic stem cell research', *Pepperdine Law Review*, **30**, 553.
52. Wilkie, D. (2004), 'Stealth stipulation shadows stem cell research: a provision in the US Appropriations Bill bans patents on human organisms', *The Scientist*, **18**(4), 42.
53. Bush, G.W. (2006), 'Message to the House of Representatives: H.R. 810, the Stem Cell Research Enhancement Act of 2005', The White House, http://www.whitehouse.gov/news/releases/2006/07/20060719-5.html, 19 July.
54. Dickinson, T. (1999), 'Statement of the Commissioner of Patents and Trademarks before the Subcommittee on Labor, Health and Human Services, Education and Related Agencies of the Senate Appropriations Committee', 12 January.
55. Shulman, S. (2001), 'The morphing patent problem', Owning the Future, *Technology Review*, November.
56. Regalado, A. and M. Louis (2001), 'Ethical concerns block patents of useful embryonic advances', *The Wall Street Journal*, 20 August.
57. Thomson, J. (1998), 'Primate embryonic stem cells', US Patent No: 6,200,806.
58. Thomson, J. (1996), 'Primate embryonic stem cells', US Patent No: 5,843,780 and Thomson, J. (2001), 'Primate embryonic stem cells', US Patent No: 7,029,913.
59. Penn, M. (2002), 'Agency's aggressive patent management protects public, professors', *On Wisconsin*, The University of Wisconsin–Madison, 15 May.
60. *Wisconsin Alumni Research Foundation v Geron Corporation* (2002) Case No. 01-C-0459-C.
61. Rai, A. and R. Eisenberg (2003), '*Bayh–Dole* reform and the progress of biomedicine', *Law and Contemporary Problems*, **66**, 289–314.
62. For a full account of this dispute, see Gratton, B. (2002), 'The Wisconsin Alumni Research Foundation and Geron Corporation', in The European Group On Ethics in Science and New Technologies to the European Commission, *Opinion on the Ethical Aspects of Patenting Inventions Involving Human Stem Cells*, Opinion Number 16, 7 May, p. 61.
63. WARF and Geron Corporation (2002), 'WARF and Geron Resolve Lawsuit and Sign New License Agreement', Press Release, 9 January.

64. Geron Corporation (2001), 'Geron Reports Issuance Of U.S. Patent For Human Embryonic Stem Cells', Menlo Park, California, 13 March. For a current list of Geron Corporation's patents, see http://www.geron.com/showpage.asp?code=prodpa.
65. Stolberg, S. (2001), 'Patent laws may determine shape of stem cell research', *The New York Times*, Washington Report, 16 August.
66. United States Department of Health and Human Services (2001), 'Memorandum of understanding between the WiCell Research Institute and the Public Health Service', http://www.nih.gov/news/stemcell/WicellMOU.pdf, 5 September.
67. California Proposition 71, http://www.smartvoter.org/2004/11/02/ca/state/prop/71/.
68. The Californian First District Court of Appeal has ruled that Proposition 71 is indeed constitutional: *Californian Family Bioethics Council v California Institute for Regenerative Medicine* (26 February 2007 A114195, Alameda County, Super. Ct. No. HG05 206766), http://www.courtinfo.ca.gov/opinions/documents/A114195.DOC.
69. McGee, G. (2006) 'Working with stem cells? Pay up', *The Scientist Magazine*, 1 November.
70. Vanden, J. (2006), 'Request to re-examine WARF stem cell patents escalates war of words', *Wisconsin Technology Network*, 19 July.
71. Foundation for Taxpayer and Consumer Rights and the Public Patent Foundation (2006), 'Request for re-examination in respect of US Patent No: 5,843,780, http://www.pubpat.org/assets/files/warfstemcell/780Request.pdf; Foundation for Taxpayer and Consumer Rights and the Public Patent Foundation (2006), 'Request for re-examination in respect of US Patent No: 6,200,806, http://www.pubpat.org/assets/files/warfstemcell/90008139granted.pdf; and Foundation for Taxpayer and Consumer Rights and the Public Patent Foundation (2006), 'Request for re-examination in respect of US Patent No: 7,029,913', http://www.pubpat.org/assets/files/warfstemcell/913Request.pdf.
72. Foundation for Taxpayer and Consumer Rights and the Public Patent Foundation (2006), 'Request for re-examination in respect of US Patent No: 5,843,780, http://www.pubpat.org/assets/files/warfstemcell/780Request.pdf.
73. Williams, R., N. Gough and D. Hilton (1989) 'In vitro propagation of embryonic stem cells', US Patent No: 5,166.065.
74. Foundation for Taxpayer and Consumer Rights and the Public Patent Foundation (2006), 'Request for re-examination in respect of US Patent No: 5,843,780, http://www.pubpat.org/assets/files/warfstemcell/780Request.pdf.
75. Robertson, E., M. Kaufman, A. Bradley and M. Evans (1983), 'Isolation, properties and karyotype analysis of pluripotential (EK) cell lines from normal and parthenogenetic embryos', *Teratocarcinoma Stem Cells*, Cold Spring Harbor Laboratory, Cold Spring Harbor, **10**, 647663; Robertson, E. (1987), 'Embryo derived stem cell lines'. *Teratocarcinomas and Embryonic Stem Cells; A Practical Approach*, Oxford: 1RL Press. ch. 4,71112 and Piedrahita, J., G. Anderson and R. Bondurant (1990), 'On the isolation of embryonic stem cells: comparative behavior of murine, porcine and ovine embryos', *Theriogenology*, **34**(5), 879901.
76. Foundation for Taxpayer and Consumer Rights and the Public Patent Foundation (2006), 'Request for re-examination in respect of US Patent No: 7,029,913', http://www.pubpat.org/assets/files/warfstemcell/913Request.pdf.
77. Foundation for Taxpayer and Consumer Rights and the Public Patent Foundation (2006), 'Request for re-examination in Respect of US Patent No: 7,029,913', http://www.pubpat.org/assets/files/warfstemcell/913Request.pdf.
78. Loring, J. (2006), 'Declaration on US Patent No: 5,843,780', http://www.pubpat.org/assets/files/warfstemcell/LoringDeclarations.pdf.
79. Ibid.
80. Ibid.
81. Ibid.
82. Ibid.
83. *Ex Parte Re-examination of US Patent No. 5,843,780* (No. 90/008102, 30 March 2007). http://www.pubpat.org/assets/files/warfstemcell/780rejected.pdf, p. 20.

84. *Ex Parte Re-examination of US Patent No. 6,200,806* (No. 90/008319, 30 March 2007), http://www.pubpat.org/assets/files/warfstemcell/806rejected.pdf; and *Ex Parte Re-examination of US Patent No. 7,029,913* (No. 95/000, 154, 30 March 2007), http://www.pubpat.org/assets/files/warfstemcell/913rejected.pdf.
85. Public Patent Foundation (2007), 'PTO rejects human stem cell patents at behest of consumer groups: re-examination was initiated by Foundation for Taxpayer and Consumer Rights and Public Patent Foundation', http://pubpat.org/warfstemcellpatentsrejected.htm, 2 April.
86. Ibid.
87. Pollack, A. (2007), '3 patents on stem cells are revoked in initial review', *The New York Times*, 3 April.
88. Ibid.
89. Albert Einstein, Swiss Patent Office technical expert third class, fondly spoke of the patent office as 'that secular cloister where I hatched my most beautiful ideas': Einstein, Albert (1972), 'Letter to Michele Besso: 12 December 1919' in Pierre Speziali (ed.), 'Michele Besso, Correspondence 1903–1955', Paris: Hermann, pp. 147–9.
90. Enerson, B. (2004), 'Protecting society from patently offensive inventions: the risk of reviving the moral utility doctrine', *Cornell Law Review*, **89**, 685–720, at 720.
91. Sherman, B. and L. Bently (1998), 'The question of patenting life', in S. Maniatis (ed.) *Intellectual Property and Ethics*, London: Sweet & Maxwell, p. 111.
92. IP Australia (1999), *Manual of Practice and Procedure*, November, Part 8.1.
93. Dickson, D. (1998), 'Legal fight looms over patent bid on human/animal chimaeras', *Nature*, **392**, 423; and Chick, E. (2003), 'Biotech critic tries to sew up research on chimaeras', *Nature*, **421**, 4.
94. The European Group on Ethics in Science and New Technologies to the European Commission (2002), *Opinion on the Ethical Aspects of Patenting Inventions Involving Human Stem Cells*, Opinion Number 16, 7 May.
95. Ibid., p. 16.
96. Ibid.
97. Crespi, S. (2001/2002), 'Patenting and ethics: a dubious connection', *Bio-science Law Review*, **5**, 271, http://pharmalicensing.com/features/disp/1046280396_3e5cf8cc0fb40.
98. Ibid.
99. Schubert, S. (2005), 'Europe halts decisions on stem cell patents', *Nature* **435**, 720–21, 9 June.
100. Ibid.
101. Smith, A. and P. Mountford (1996), 'Isolation, selection and propagation of animal transgenic stem cells', European Patent No: 0695351.
102. There is already an Australian patent (No. AU678233) which appears to be similar to the one in Europe.
103. AFP (2000), 'Greenpeace paralyses Patent Office in "human clone" protest', Munich, Germany, 22 February.
104. Ibid.
105. Ibid.
106. Ibid.
107. Ibid.
108. Salleh, A. (2000), 'A patent mistake', ABC Science Online, http://www.abc.net.au/science/news/stories/s102681.htm, 23 February.
109. European Patent Office Press Release (2002), '"Edinburgh" patent limited after European Patent Office opposition hearing', Munich, 24 July.
110. *Greenpeace Deutschland e.V. v The University of Edinburgh*, Opposition Division, European Patent Office (24 July 2002).
111. Ibid., 11–12.
112. Ibid., 14–15.
113. Ibid., 20–21.
114. Ibid., 21.
115. Ibid., 23.

116. Ibid., 25.
117. Ibid.
118. Ibid., 26.
119. Ibid., 22.
120. Ibid.
121. Ibid.
122. University of Edinburgh (2003), 'Statement of grounds of appeal under Article 108 EPC', European Patent Office, 8 December, 11–12.
123. University of Edinburgh (2003), 'Statement of grounds of appeal under Article 108 EPC', European Patent Office, 8 December, 11.
124. European Parliament (2000) 'Resolution on the decision by the European Patent Office with regard to Patent No EP 695351 granted on 8 December 1999' (Document BS-0288, issued on 30 March 2000, Official Journal EC-L- 29 December 2000, 378/95).
125. Thomson, J. (1996), 'Primate embryonic stem cells', European Patent No: EP0770125.
126. Wisconsin Alumni Research Foundation patent application in respect of 'primate embryonic stem cells', Examining Division, June 2004.
127. Wisconsin Alumni Research Foundation patent application in respect of 'primate embryonic stem cells', T 1374/04–3.3.08 Interlocutory decision of the Technical Board of Appeal 3.3.08 of 18 November 2005.
128. Wisconsin Alumni Research Foundation Patent Application in Respect of 'Primate embryonic stem cells', T 1374/04-3.3.08 Interlocutory decision of the Technical Board of Appeal 3.3.08 of 18 November 2005, p. 5.
129. Wisconsin Alumni Research Foundation Patent Application in Respect of 'Primate embryonic stem cells', T 1374/04–3.3.08 Interlocutory decision of the Technical Board of Appeal 3.3.08 of 18 November 2005, pp. 11–12.
130. Crespi, S. (2006), 'The human embryo and patent law: a major challenge ahead', *European Intellectual Property Review*, **28**(11), 569–75.
131. Anderson, D. and D. Stemple (1993), 'Mammalian multipotent neural stem cells', European Patent Application No: EP0658194.
132. Anderson, D. and D. Stemple (1993), 'Mammalian multipotent neural stem cells', European Patent Application No: EP0658194.
133. Caltech Patent Application, Examining Division, European Patent Office, T522/04–338 (17 October 2003).
134. Caltech Patent Application, Examining Division, European Patent Office, T522/04–338 (17 October 2003), 5.
135. Caltech Patent Application, Examining Division, European Patent Office, T522/04–338 (17 October 2003), 6.
136. Caltech Patent Application, Examining Division, European Patent Office, T522/04–338 (17 October 2003), 5.
137. Caltech Patent Application, Examining Division, European Patent Office, T522/04–338 (17 October 2003), 4.
138. Herder, M. (2006), 'Proliferating patent problems with human embryonic stem cell research', *Bioethical Inquiry*, **3**, 69–79 at 77.
139. Ibid.
140. Taymor, K., C.T. Scott and H. Greely (2006), 'The paths around stem cell intellectual property', *Nature Biotechnology*, **24**, 411–13.

Conclusion. Blue sky research: patent law and frontier technologies

In its report on *Gene Patenting and Human Health*, the Australian Law Reform Commission predicted that the patent system would face new challenges from emerging fields, such as bioinformatics, proteomics, pharmacogenomics and nanotechnology:

> The patent system is over 400 years old. It has accommodated the arrival of many new technologies including inventions associated with mechanics in the industrial revolution; electricity and electronics; industrial and chemical materials; food production and agriculture; scientific instruments and devices; transportation and energy; warfare; medical devices and pharmaceutical products; computing and information technology; and business methods. In the past 20 years, inventions in the field of biotechnology have become a new focus of the patent system, particularly in relation to genetic materials and technologies . . . Once patent examiners have become familiar with genetics, they will, no doubt, be met with a new range of challenges from emerging disciplines such as bioinformatics, pharmacogenomics, proteomics and nanotechnology.[1]

There has been much debate as to the extent to which the existing regimes of intellectual property need to be revised in order to accommodate new, frontier technologies. Robert Hahn has commented: 'New technologies with commercial potential inevitably raise unique questions about the appropriate degree of intellectual property protection and the appropriate methods for getting from here to there.'[2] He maintains that 'investing now in designing the right legal structure for today's technologies will almost certainly pay dividends in the form of a faster, more-sophisticated resolution of IP problems raised by tomorrow's innovations.[3]

The political debate over patent law and frontier technologies has mirrored earlier struggles over patent law and pharmaceutical drugs and biotechnological inventions. Patent loyalists maintain that the patent system can accommodate new, frontier technologies through the application of a broad and flexible approach to the interpretation of patentable eligibility. They insist that the patent system is technology-neutral, and applies protection equally to all forms of inventions. The rubric in *Diamond v Chakrabarty* that patent protection extends to the 'anything under the sun that is made by man' has been applied with full force to a new generation of biological inventions.

Under this logic, the patent system has accepted such hybrid fields as bioinformatics, proteomics, pharmacogenomics and nanotechnology as patentable subject matter. A number of policy makers argue that the patent system can accommodate new technologies within its framework, but only through the flexible use of patent doctrines and administrative guidelines. Dan Burk and Mark Lemley maintain that the patent system is technologically specific in the way that it deals with new technologies.[4] They suggest that Patent Offices and courts should make use of existing policy levers within patent law to address and respond to new technologies: 'The great flexibility in the patent statute presents an opportunity for courts to take account of the needs and characteristics of different industries.'[5] By contrast, law reform bodies have advocated substantive changes to patent law, policy and practice. A number of commentators favour sui generis regimes of intellectual property to accommodate new technologies and scientific developments.[6] Such critics suggest that it would be better to create new regimes to deal with bioinformatics, proteomics, pharmacogenomics and nanotechnology. Patent abolitionists have reiterated their concerns about patent protection being extended to frontier technologies. Notably, the ETC Group has been vocal in their opposition to the privatization of new scientific fields.[7]

This conclusion considers how patent policy, law and practice will deal with the emergence of new scientific fields and disciplines in the life sciences. Section one considers the burgeoning field of bioinformatics, which cuts across both information technology and biotechnology. Section two explores patents in the field of proteomics – the study of proteins and their functions. Section three examines the potential of pharmacogenomics – which involves the use of genetic predisposition testing to develop personalized medicines. Section four investigates the cluster of patents in respect of nanotechnology – which embraces devices, assemblies and systems which are larger than 1 nanometer. The epilogue concludes that important lessons can be learned about regulating frontier technologies from the policy debate over intellectual property and biotechnology.

I BEYOND BLUE GENE: INTELLECTUAL PROPERTY AND BIOINFORMATICS

Bioinformatics is the art and science of using computer systems to store, manage and analyse biological information.[8] It brings together the diverse disciplines of mathematics, statistics, engineering and computer science to map and model genes and proteins. Bioinformatics played a critical role in mapping the human genome in both the large public and commercial projects. Robert Cook-Deegan notes:

Databases, computers, and mathematical algorithms proved as important as DNA sequencing, cloning, and other more obviously biological techniques. As geneticists produced a deluge of data during the 1990s and beyond, those who understood hardware and software would play an increasingly important role.[9]

The public consortium relied upon cloning methods to map the location of genes, dividing the genome into small blocks. The private efforts lead by Celera Genomics engaged in whole genome shotgun sequencing, fracturing the DNA of an organism into small fragments and then using powerful computer sequencing machines to identify the base pairs at the end of each fragment. The sequencing, storage and retrieval of genetic information have generated new possibilities for understanding the function and structure of genes and proteins.

In the wake of the Human Genome Project, there was great controversy about the patenting of genes and gene sequences.[10] However, it is worth recognizing that biotechnology companies have not been exclusively interested in the patenting of genes and genetic sequences. Biotechnology firms applied for patents over databases of genetic information, and other proprietary informatics systems for storing and analysing genomic variation data. They have sought patents for computer software and computer hardware related to the life sciences, as well as underlying instrumentation such as microarrays and gene chips. Such entities have also applied for patents over novel business methods that utilize technologies for providing genomic services to the pharmaceutical and biotechnology industry. In addition to patent protection of bioinformatics software and hardware, biotechnology companies have also deployed a range of other forms of intellectual property protection in respect of scientific publications and genetic databases, such as copyright law and the protection of confidential information.[11]

Professor Rebecca Eisenberg from the University of Michigan has questioned whether bioinformatics software and hardware is an appropriate subject matter for a patent claim: 'I believe that patent claims to DNA sequences stored in computer-readable medium represent a fundamental departure from the traditional patent bargain of exclusionary rights to tangible inventions in exchange for free disclosure of information, and should not be allowed.'[12] She added: 'The claim to the sequence in computer-readable medium, in effect, gives the patent holder the right to restrict the ability of others to use the information in a computer-readable medium and thus precludes others from perceiving and analyzing the sequence information itself.'[13] Eisenberg draws analogies between the patenting of genes and genetic sequences and the recent controversies over business methods. She observes: 'Recent decisions concerning the patentability of computer-implemented inventions may provide more guidance than prior decisions in the life sciences in predicting whether DNA sequence information stored in computer-readable medium may be patented.'[14]

The United States courts have also recognized that business methods are patentable subject matter. In *State Street Bank And Trust Co v Signature Financial Group Inc*, the Court of Appeals for the Federal Circuit held that the use of a mathematical algorithm, formula or calculation to produce numbers will be patentable, as long as the result is 'useful, concrete and tangible'.[15] Further, the Court of Appeals for the Federal Circuit expressly rejected the existence of a business method exception to patentability. In *AT & T Corp v Excel Communications Inc*, the Court of Appeals for the Federal Circuit reaffirmed the decision that Internet business methods were patentable.[16] In *Amazon.com Inc v Barnes And Noble. Com Inc*, the court applied this reasoning in relation to electronic commerce.[17] One might add to this list the case of *Welcome Real-Time v Catuity Inc*,[18] in which the Federal Court of Australia affirmed the decision in *State Street Bank v Signature Financial Group* as persuasive in this jurisdiction.[19] However, in his dissenting judgment in *Laboratory Corp. of America Holdings v Metabolite Laboratories, Inc.*, Breyer J questioned past Federal Circuit authorities, which had asserted patents could be granted in respect of business methods, as long as there was a 'useful, concrete, and tangible result'.[20] His Honour cited with approval an article by Malla Pollack on 'The multiple unconstitutionality of business method patents', suggesting that the judge might have underlying constitutional objections to the grant of business method patents.[21]

Patent attorneys and lawyers have hailed the decision in *State Street Bank* as opening the way forward for the patenting of bioinformatic inventions.[22] Ernest Buff is perhaps representative in his enthusiasm: '*State Street* and its progeny will likely change the way in which biotechnology and bioinformatics industries do business'.[23] However, as Stephen Lesavich comments, such patents were well available before the *State Street Bank And Trust Co v Signature Financial Group Inc*.[24] Most bioinformatic inventions, such as those related to software methods, software systems, data structures, the Internet and other software features, were capable of receiving patent protection with software patents under United States patent law long before the *State Street Bank And Trust Co v Signature Financial Group Inc* and *AT & T Corp v Excel Communications Inc* cases were decided.

A number of preliminary studies have considered to what extent patents have been filed in respect of bioinformatics. A study conducted by London-based consulting firm Silico Research in 2001 found that only 50 software-related patents had been issued by the USPTO between 1996 and 2001 to companies operating in the pharmaceutical, biotechnology and genomics research.[25] Another study, published by *Nature Biotechnology* in 2000, provides a sharper image of the changing marketplace of bioinformatics.[26] Paolo Saviotti and his collaborators found that the number of bioinformatics-related patents has been increasing steadily from 1979 to 1997, after which there was a notable boom in patent applications, with a peak of 60 patents in 1998. The USPTO reported in 2001 that the actual number of pending bioinformatics patents was relatively small, with 11 examiners in the bioinformatics unit processing a total of around 200 patents.[27]

In a 2005 study, Iain Cockburn reflected that the number of patents in respect of bioinformatics inventions was still relatively meagre:

> Using these fairly generous definitions, approximately 48,000 'molecular biology' patents and approximately 220,000 'software' patents were issued in the United States between 1990 and 2003 but only 305 fall in both sets. Of these, about half claim the results of using computational techniques to identify specific genes or therapeutic compounds. Only 148 of these 305 patents appear to be 'pure' bioinformatics inventions in the sense of claiming general purpose algorithms or methods.[28]

Similarly, he noted that the litigation activity in bioinformatics had been quite limited to date. There had been legal disputes in relation to Affymetrix's ownership of microarray technologies, but such matters were subsequently settled with no legal findings as to validity and enforceability.[29]

Iain Cockburn comments that the field of bioinformatics has tested the flexibility and plasticity of the patent regime:

> Bioinformatics illustrates some general difficulties in establishing appropriate IP rules for new, rapidly evolving technologies. Any new technology presents the patent system with transitional difficulties in determining standards of patentability, establishing procedural requirements, and developing legal doctrine to address idiosyncratic aspects. These may be satisfactorily resolved with the passage of time and sufficient accumulation of experience, but while courts and administrative entities wrestle with these difficult questions, large numbers of applications can build up in the Patent Office and eventually start issuing. Under the statutory presumption of validity, these patents can strongly affect the nature of ongoing research, the development of the technology, and the cast of players. In this case, the situation is further complicated by the new area inheriting unresolved patent issues from its parent disciplines, compounded by the boundary-spanning nature of bioinformatics.[30]

Cockburn expressed concerns that bioinformatics patents will experience similar controversies to those that have affected software patents and business method patents: 'These include difficulties in searching prior art, poor quality control and uneven standards of review at the Patent Office, and rejection by the relevant community of practice of the standards applied by the Patent Office and the courts to evaluate non-obviousness and novelty.'[31] The developments in relation to business methods have been a spur for a reform of patent administration. Most notably, Merges was prompted to consider a number of initiatives to improve the operation of the patent office.[32] However, bioinformatics has unique and important implications in respect of patent administration, and the examination of prior art. Therefore there will be a need for patent offices to develop a cadre of examiners who can deal with the combination of information technology and biotechnology.[33]

Alternatively, there should be greater collaboration between examiners from the sections dealing with information technology and biotechnology.

Recently, a number of mainstream information technology companies, such as IBM, Microsoft and Google, have invested in bioinformatics. Commentators wonder what effect the entrance of these new players might have on the market for bioinformatics:

> The movement of these IT-based entrants into the market is important because they are very large and powerful firms capable of shaking up the industrial structure of bioinformatics. It is tempting to speculate that the expertise of these companies in other industries might be rapidly translated to software solutions that provide the kind of standardization, integration, and analysis of the data so sorely needed.[34]

Furthermore, the information technology firms will have an important impact upon the field of bioinformatics. Such companies have shown great talent in fully exploiting both copyright law and patent law in managing the protection of computer software and hardware.[35] They may be able to translate such tactics and strategies in the management of intellectual property to the field of bioinformatics. Similarly, public researchers and even private organizations have much to learn from open source software and peer to peer technology. Such strategies may provide the means to resist the privatization of genetic information. Furthermore, there should be a consideration of the implications of intellectual property and bioinformatics for related fields of information science.[36]

The firm, IBM, has been involved in the life sciences, both as a participant in large-scale biology projects and as a consultant to biotechnology companies and pharmaceutical drug-makers. A project called Blue Gene has come to symbolize both the promise and the hype of bioinformatics.[37] IBM has devoted $US100 million to build a super-computer which will seek to analyse protein folding. It boasts that Blue Gene is 1000 times more powerful than Deep Blue, the machine that defeated world chess champion Garry Kasparov, and can map the human genome. IBM has also been a partner in the Genographic Project in collaboration with National Geographic. IBM has also created a consulting division for its biotechnology and pharmaceutical customers. It has also set up a new organization, called Blueprint Worldwide, which will generate a public database of bioinformatics and biomedical data.[38] Such grand ambitions herald the marriage of life sciences and information technology. They also highlight the importance of intellectual property to the field of bioinformatics.

The Chairman of Microsoft, Bill Gates, has had a long-standing interest in biotechnology. He observed back in 1996:

> Like software, biotechnology will change the world. I expect to see breathtaking advances in medicine over the next two decades, and biotechnology researchers and companies will be at the center of that progress. The biggest breakthroughs in medicine will result from the mapping and understanding of the human genome.[39]

Bill Gates provided funding for the Department of Molecular Biotechnology at the University of Washington and helped recruit such stellar geneticists as BRCA1 and BRCA2 geneticist, Mary-Claire King, and systems biologist, Leroy Hood;[40] he has invested in a range of start-up biotechnology companies and he has also established the Bill & Melinda Gates Foundation to help promote health-care in developing countries.[41] In 2006, Microsoft announced the development of the BioIT Alliance.[42]

The network is designed to facilitate partnerships between the information technology company and various organizations in the life sciences, including pharmaceutical companies and biotechnology firms. The company has observed of the venture: 'The BioIT Alliance is designed to enable collaboration among organizations in the Life Sciences field in order to shorten the time between discovery of new biological data and the application of that knowledge to human health.'[43]

The search engine, Google, has long-term plans to expand into the fields of biology and genomics. Using the map of the Human Genome Project, Google plans to develop a genetic database, analyse it, and find meaningful correlations for individuals and populations. Co-founders Larry Page and Sergey Brin believe that the search engine can play a role in enhancing 'the ability for cellular biologists and other kinds of medical researchers to be able to use data clusters like we have at Google, and certainly like the ones we're going to have in a decade or two decades' time, and be able to do completely new things that we weren't able to dream of before'.[44] The Mountain View Company has entered into a collaborative venture with J. Craig Venter, formerly of Celera Genomics. Venter has observed:

> We need to use the largest computers in the world. Larry and Sergey have been excited about our work and about giving us access to their computers and their algorithm guys and scientists to improve the process of analysing data. It shows the broadness of their thinking. Genetic information is going to change the

world. Working with Google, we are trying to generate a gene catalogue to characterize all the genes on the planet and understand their evolutionary development. Geneticists have wanted to do this for generations.[45]

Predictably, such plans have alarmed anti-biotechnology groups. The ETC Group awarded a Captain Hook prize for biopiracy to Google for daring to collaborate with Venter: 'Google took the prize for the "Biggest Threat to Genetic Privacy" for its collaboration with J. Craig Venter – another of today's winners in a solo effort as "Greediest Biopirate" – to create a searchable online database of all the genes on the planet.'[46] Such concerns seem mere speculation, at present, given the lack of public detail about the project. Nonetheless, there have been concerns raised about Google, and information privacy, in other contexts.

II THE PROTEIN ATLAS OF THE HUMAN GENOME: INTELLECTUAL PROPERTY AND PROTEOMICS

In 1994, Australian scientist Marc Wilkins coined the term 'proteome' to refer to the protein complement encoded by a genome.[47] The derivative phrase, 'proteomics' refers to the 'the large-scale study of proteins in a cell or organism'.[48] In 1999, the scientific magazine, *Nature*, predicted that the field of proteomics would be a burgeoning area of interest:

> Analysing the entire set of proteins of an organism is a far bigger challenge than anything in genomics. The technological obstacles and biological complexities require, for now, a steady approach to that necessary goal. The inside of a cell is a crowded and dynamic place, where proteins are perpetually being created and discarded. Understanding the structures, interactions and functions of all of a cell's or organism's proteins is one of the grand goals of the post-genomic era, and has been given a disciplinary title of its own: proteomics. There are even some who want to develop a human proteome project.[49]

The editorial in the magazine cautioned: 'Researchers and funding agencies need to beware of hype, but should be conscious of the great potential in this research, and keep themselves abreast of the key techniques and technologies.'[50] The Harvard Business School observes that 'Proteomics had become the new buzzword on Wall Street in the post-genomic era.'[51]

The field of proteomics has posed particular issues and problems in the field of patent law. As Robert Bohrer observed: 'Just as the era of gene sequencing produced innumerable questions and challenges for the legal system, so will the era of proteomics.'[52] Richard Warburg commented: 'The patenting of inventions in the fields of proteomics and genomics has been prolific in the last few years and will continue in that vein for some more years.'⁵³ An analysis of the USPTO Patent register as at 2007 mentions 721 patents which mention 'proteomics' in their specifications.

There has been discussion about whether there will be litigation between the holders of gene patents and their competitors in the field of proteomics. Stanford Professor John Barton expressed concerns about patents in various emerging areas of post-genomic research such as proteomics and pharmacogenomics. He observed: 'These quasi-genomic patents go quite far, and clearly much further than the patents on genes that have been the source of much concern.'⁵⁴ Leslie Misrock observed that 'the question of litigation in genomics will pale with respect to litigation in proteomics'.⁵⁵ He predicted that there would be conflict between the holders of gene patents and holders of protein patents.⁵⁶ The patent attorney warned industry leaders about the impact of litigation in this field: 'If there's a series of patents that issues, and [the proteomics companies] aggressively want to enforce those patents to get commercial advantage, this will act to divert assets that should be put into the business and the science, to be spent on lawyers.'⁵⁷

As an illustration of the issues raised by patents in the field of proteomics, it is worth focusing upon Oxford Glycosciences (OGS), as a case study. OGS was established in 1988 as a spin-off company from Oxford University to identify and analyze glycoproteins.⁵⁸ The company was transformed into a biotechnology company applying proteomics technologies and glycobiology to the discovery, development and commercialization of novel therapeutic products. OGS raised £33 million to finance a 'landgrab' to patent proteins.⁵⁹ It developed an ambitious plan to file 5000 to 10 000 patent applications in relation to disease-specific proteins. In *Scientific American*, journalist Carol Ezzell commented upon this bold development:

> Oxford GlycoSciences in England is betting that it can tie up the patent rights to a significant portion of the human genome and proteome using proteomics data. Last December the company filed patent applications for 4,000 human proteins, a move that could shake up how intellectual property is defined in biotechnology. In the past, companies sought to patent DNA sequences and the single protein that they predicted would be encoded by them. But because the same gene can make a range of proteins, claims based on the proteins themselves could be more valuable and offer a way to get around patents on the DNA sequences held by competitors. If so, the courts could be one more arena where genes will have to move in favor of proteins.⁶⁰

The plans of OGS were widely reported, featuring in such respected publications as *Nature*, the *Guardian* and *The Times*.⁶¹ The company's stock price was significantly boosted by this public announcement about its plans for patent prosecution. OGS assembled a considerable patent portfolio dealing with proteomics technology, therapeutic proteins and drug discovery. First of all, OGS has sought to protect its 'Core Proteomics Technology'. Its website boasted:

> Our proteomics technology is protected by a series of issued patents and pending patent applications. The most significant of these, U.S. 6,064,754, covers the use of our industrialized proteomics technology. A European counterpart of this patent is pending. Supporting this key patent is a series of issued and pending patent applications covering aspects of proteomics analytical technologies, including fluorescent dyes for labeling proteins and carbohydrate components of glycoproteins, as well as the analysis of these carbohydrate components. In addition, we have filed a patent application relating to methods for the high throughput automated use of mass spectrometry for proteomics analysis.[62]

OGS was awarded its key patent for 'Computer-assisted methods and apparatus for identification and characterization of biomolecules in a biological sample' in 2000.[63] The patent received accolades from various luminaries. Randal Scott of Incyte Genomics observed: 'This patent confirms OGS as the pioneer in high-throughput proteomics technology and promises a bright future for our joint development of the best proteomics databases in the world.'[64] Similarly, Leroy Hood noted: 'The patent confirms OGS's leading position in industrial proteomics and their contribution to the science of proteomics which is the next step of importance with genome closure at hand.'[65] Leslie Misrock, an eminent patent attorney, commented: 'Biotechnology innovators receive broad patents when they deserve them and recipients of such broad patents typically use such assets to great advantage.'[66]

Second, OGS sought to develop a large portfolio of protein patents, filing applications for over 4000 disease-associated proteins, genes and protein isoforms. Raj Parekh, Chief Scientific Officer, said: 'By concentrating principally on proteins in relevant cells, tissues and fluids, we avoid so much of the ambiguity inherent in a genomics-only approach and can discover the protein molecules that are altered in clinical samples from patients with specific diseases.'[67] OGS proclaimed:

> By comparing the amino acid sequences of each disease-associated protein that we identify with our proteomics technology to the many freely available, public domain databases of gene and protein information, we can determine whether the protein from which these amino acids are derived has been observed and publicly disclosed before.
> If no match is found for the protein in any accessible database of gene or protein information, then it is likely that we have identified a previously unknown, or novel, protein or isoform of the protein. This offers us the opportunity to file a patent application relating to that protein or isoform and, if the encoding gene has also not been publicly disclosed before, on the gene as well.

These patent applications, if granted, could be expected to cover both the composition of matter of the protein and its gene and their uses.[68]

Although OGS may well have filed hundreds of such preliminary protein patent applications, the company in the end only obtained a small portfolio of 11 granted patents.[69] The company may well have lacked the financial resources to follow through in its plans because of financial difficulties.

Third, OGS was vigilant in its opposition to trade mark registrations of the term 'proteomics'. In particular, OGS was involved in litigation with MDS Proteomics over its efforts to trade mark the term 'Proteomics'.[70]

Finally, OGS developed a database of proteins called the 'Protein Atlas of the Human Genome™'. It hoped to become an information provider along the lines of Lexis/Nexis. OGS articulated its ambitions in a press release: 'As the next logical step beyond the Human Genome Project (HUGO), the Protein Atlas will provide researchers with a powerful tool to interpret genomic data, especially in the areas of target identification, validation and protein diagnostics.'[71] OGS announced a joint venture with Marconi, a global provider of innovative communications solutions, to be known as Confirmant Limited.[72] It combined OGS' proteome databases, in particular the Protein Atlas of the Human Genome and its data analysis software tools, with Marconi's broadband data transmission and hosting capabilities. Commentators were sceptical whether the joint venture of Confirmant Limited would be a viable business model.[73] Such doubts seem well-founded. In March 2004, Confirmant folded, after efforts to sell it failed.

After substantial losses from 2000 to 2003, OGS was the subject of a number of rival bids.[74] In February 2003, the English company Celltech Group Plc launched a hostile takeover of OGS. It announced a cash offer for the entire issued share capital of OGS and in the April the offer was announced unconditional in all respects. Following the integration of OGS, Celltech identified a number of high-quality programmes and personnel, particularly in the oncology area, which were absorbed into its own operations. Several assets, identified as non-core to Celltech's operations, were divested.[75]

This takeover is reflective of a wider rationalization in respect of biotechnology companies specializing in proteomics. Patent attorney Leslie Misrock commented that mergers, acquisitions and insolvencies will affect the industry:

> The number of proteomics companies is going to devolve to about two or three, [including] certainly Celera. Many companies are bulking up in this field by acquiring other companies, or hoping for a series of mergers where there is a complementary technology as opposed to supplementary technologies.[76]

OGS has not been the only proteomics company to struggle, with a downturn in technology stocks. The United States proteomics company, the Large Scale Biology Corporation, has acquired a significant portfolio of 32 issued United States patents, many in the field of proteomics; however, the company has found it difficult to be profitable.[77] The Canadian proteomics company, MDS Proteomics, was restructured and reorganized in 2004 into the bio-marker company, Protana Inc.[78] That entity went into receivership the following year (2005), and was sold off to Transition Therapeutics Inc.[79] The Australian company, Proteome Systems, holds 12 United States patents; the company was listed on the Australian Stock Exchange in 2004; it has been forced to refocus its operations, with significant operating losses.[80]

Some law reform bodies have recommended that there should be reforms in respect of patent law to better accommodate proteomics. The Nuffield Council on Bioethics maintained that patents on therapeutic proteins should be narrowly defined: 'By this we mean that the rights to the DNA sequence should extend only to the protein described.'[81] Going further, J. Jason Williams has argued that the patent system is unable to deal properly with the challenges posed by proteomics. He expresses concerns that the current patent law encourages patent races in the proteomics field, leaving those with the most resources to control the industrial field. Williams contends that a sui generis scheme would be preferable:

> In order to accommodate this revolutionary system, as well as to anticipate the further merger of biology and the technological arts, something beyond tinkering with the existing system may be necessary. Sui generis protection represents the best scenario for encompassing this constantly expanding science and for providing a check to the potential volumes of litigation this area will surely engender.[82]

He concludes: 'Without special considerations for proteomics (and probably genomics as well) that fine-tune the particular problems presented, the future direction of scientific research and the allocation of scientific resources may rest in the hands of a select few.'[83]

III PHARMACOGENOMICS

The Californian lawyer Qin Shi discusses how intellectual property laws have been challenged by the emerging field of pharmacogenomics:

> Pharmacogenomics concerns the application of genomics discoveries in the development of pharmaceuticals. The field was born on the heels of the completion of

the human genome project. That project delivered the sequence of the entire human genome. Pharmacogenomics is defined by the International Society of Pharmacogenomics (ISP) as 'the influence of the human genome on response to medication.' It covers drug response markers that link individual genomic variations (DNA polymorphisms) to drug target, drug metabolism, clinical responses, and side effects. Pharmacogenomics thus intrinsically relates to the promise of personalized medicine.[84]

Shi comments: 'Insofar as the principal asset derived from pharmacogenomics research is information on the relationships between individual genomic variations and individual responses to a particular drug compound, the field poses interesting questions on the desirability and strategy of protecting such information.'[85] She suggests: 'Protection for valuable pharmacogenomics information may also be sought in connection with other patentable subjects, including, for example, methods of statistical analysis of genomic or expression data, methods for identifying biomarkers for drug responses, biomarkers and biomarker kits, and methods of use related to biomarkers.'[86] A search of the specifications and description of patents in the USPTO database reveals that there are 772 patents dealing with 'pharmacogenomics' as at March 2007.

The Royal Society of the United Kingdom cautions that the take-up of pharmacogenomics within the clinical setting will be a slow and gradual process:

> It is unlikely therefore that there will be an immediate change in clinical practice based on pharmacogenetics. Rather, there is likely to be a gradual increase in its clinical applications; its true potential may not become apparent for 15–20 years. during which time a great deal more information may become available about the practicalities of applying information derived from complex multifactorial systems in the clinic.[87]

The Royal Society commented: 'The increasing availability of high quality collections of DNA samples with associated phenotypic data will continue to support the trend of industry using population-based genetic association studies, rather than susceptibility gene hunting approaches, to help validate the disease association of novel drug targets in the early discovery process.'[88] The group also notes: 'The future impact of pharmacogenetics will be linked to the development of [reliable and rapid genetic predisposition] tests that can rapidly deliver useful diagnostic data to healthcare professionals on much larger numbers of tests.'[89]

Duke University academic Arti Rai wonders whether the developments in the field of pharmacogenomics might provide an opportunity for the revision of patent law, especially with regard to the special incentives that exist in relation to pharmaceutical drugs:

> The increasing integration of genomics into drug development has the potential to change the economic structure of drug development, primarily by reducing the time and money it takes to discover drugs. Equally important, it is likely to change the structure of health care delivery, by increasing the number of diseases that can be addressed by drugs.[90]

Rai suggests: 'To the extent that the biotechnology and pharmaceutical industries can internalize fully the efficiency benefits of digital technology, reforms that align the structure of intellectual property for pharmaceuticals more closely with that of other innovation might be considered.'[91] She submits: 'Equally important will be regulation that imposes cost-effectiveness requirements on pharmaceutical innovation.'[92]

IV THE AMBRI BIOSENSOR: PATENT LAW AND NANOTECHNOLOGY

Nanotechnology is the field of science and technology which is focused on the hundred nanometer scale downwards.[93] It refers to devices whose parts can be measured in nanometres, or billionths of a metre. Lawyer Veronica Mullally and physicist David Winn define the field of nanotechnology in these broad terms:

> Nanotechnology embraces objects, mechanisms, assemblies, and systems based on size scales smaller than the micrometer/micron and larger than 1 nanometer (nm) or about 10 atomic diameters. Some points of reference for this scale are a human hair, which is about 80,000nm in diameter, and a red blood cell, which is about 1000nm. The prototype, and perhaps the ultimate, nanotechnology system is found in nature: a virus or a living cell. A cell uses energy and forces, senses its environment, modifies its environment, communicates by chemical or even light messengers, moves about, reproduces, and manufactures (antibodies, hormones etc.) all by sub-cellular structures, mechanisms and macro chemicals the scale of 1nm to 100nm. Living cells do all of this in an assembly of several microns in size. The dream of nanotechnologists is to create mechanisms and processes on the scale of a single cell.[94]

Nanotechnology is a multi-disciplinary field of research. This area of science and technology cuts across the traditional disciplinary fields of biology, chemistry, physics and engineering. Nanotechnology often involves bio-mimicry.[95] Scientists in this field often study nature's models and then imitate or take inspiration from these designs and processes to solve human problems.[96]

As an illustrative example, a team of scientists working at CSIRO obtained a family of patents in respect of the 'Ambri Biosensor'.[97] The device is designed to detect substances with extreme sensitivity. Its central component is a tiny electrical switch, an ion-channel, 1.5 billionths of a metre in size. One of the inventors, Dr Bruce Cornell, commented that the biosensor's sensitivity was equivalent to detecting the increase of the sugar content of Sydney Harbour after throwing a sugar cube from a ferry. He explained the nature of the invention:

> This biosensor is a unique blend of the ability of biology to identify individual types of molecule in complex mixtures, with the speed, convenience and low cost of microelectronics. It consists of a synthetic membrane that we make ourselves, chemically tethered to a thin metal film coated onto a piece of plastic. This membrane behaves like the outer skin of the cells of the human body in its ability to sense other molecules. As we evolved from the sea, it is not surprising that ions (single atoms) in sea water such as sodium and potassium play a role in human cell signalling and sensory systems. These depend on ion currents that flow across certain cell membranes. When the membrane detects its target molecule, it turns these currents on or off by opening or closing molecular channels that pass through the otherwise insulating membrane. We have made a synthetic version of this mechanism, that is stable, inexpensive and convenient to use as a molecular detector.[98]

The research team hoped that biosensors have a huge range of potential uses, especially in medicine, for detecting drugs, hormones, viruses and pesticides and to identify gene sequences for diagnosing genetic disorders. In the pharmaceuticals industry the device may also be used to identify new drugs and medically-active compounds. The technology transfer manager, Keith Daniel, commented: 'Because of their low cost, sensitivity and ease of use, they will probably also find particular application in on-site measurements, such as ensuring food safety and quality, in environmental monitoring and drug detection in athletes.'[99]

The research team hoped that the patents in respect of the nanotechnology device would help develop a new field of industrial manufacturing. One of the inventors, Vijoleta Braach-Maksvytis, explained the ambitions behind the research: 'Literally, from the very early discussions about the concept for the biosensor, the purpose of the project was not just to create new scientific knowledge, but to also produce a product and a new manufacturing base for Australia.'[100] Dr Bruce Cornell stressed, 'We now have a chance to be in at the start of a new generation of technologies, such as our biosensor – devices which operate on the molecular or nanometre scale.'[101] He opined: 'We need to take full scientific and commercial advantage of this early lead.'[102] A spin-off company, Ambri, has been established to develop products and devices related to the Ambri Biosensor.[103] However, it has proved difficult to bring commercial products in the field of medicine to the stage of marketing.

There has been great public and private support for research into nanotechnology. In January 2002, the Australian Government announced that the Australian Research Council would devote one-third of its budget to four priority areas, one of which was nanotechnology.[104] In the United States, President George W. Bush signed the *21st Century Nanotechnology Research and Development Act* 2003 (US) which authorized $3.7 billion in funding for federal nanotechnology research and development from 2005 to 2009. There has been much government and industry investment in nanotechnology in the members of the European Union,[105] and Japan.

There has been an exponential increase in the number of patent applications filed by both public and private entities in respect of nanotechnology inventions. A USPTO patent examiner, Vivek Koppikar, gives a sense of the early history of this field:

> In many cases, technology which was first conceived in the 1970s and early 1980s, often as academic curiosities, have now become a major area of commercial development under the nanotechnology rubric. Several examples are noted below reflecting different areas of nanotechnology. The atomic force microscope (AFM) is a powerful, fundamental nanotechnology tool and was first patented in 1988 by Bennig and IBM. By 1994, over 100 patents issued per year and, by 2003, over 500 patents were issuing per year referring to this tool. Quantum dots and dendrimers, similarly, are examples of nanomaterials first patented in the mid-1980s. By 1994, over ten patents issued per year and, by 2003, over 100 patents were issuing per year referring to each of these materials.[106]

A recent search of the USPTO database in respect of patents in March 2007 reveals that there are 4550 patents granted within the strict classification field of nanotechnology.[107]

Research and industry groups have expressed concerns about the examination of nanotechnology patent applications. Mark Modzelewski of the NanoBusiness Alliance observed:

> Our big issues are making sure that the United States Patent and Trademark Office understands nanotechnology, so when people come with their patents, examiners understand what are reasonable boundaries. We would not like to see, within nanotechnology patents, some of the things we've seen in recent technology waves, where there have been concept patents awarded, which allow people to lock up huge areas.[108]

In response to such concerns, the USPTO has created a specialist unit to examine nanotechnology patents. As part of its efforts to improve the ability to search and examine nanotechnology-related patents, USPTO established a new cross-reference digest for nanotechnology designated Class 977/Dig.1, entitled 'Nanotechnology'.[109] The USPTO defines the term 'nanostructure' to mean 'an atomic, molecular, or macromolecular structure that (a) has at least one physical dimension of approximately 1–100 nanometers; and

(b) possesses a special property, provides a special function, or produces a special effect that is uniquely attributable to the structure's nanoscale physical size'.[110] The Nanotechnology art collection provides for disclosures related to 'nanostructure and chemical compositions of nanostructure', 'devices that include at least one nanostructure', 'mathematical algorithms, e.g., computer software, etc., specifically adapted for modeling configurations or properties of nanostructure', 'Methods or apparatus for making, detecting, analyzing, or treating nanostructure'; and 'specified particular uses of nanostructure.'[111]

In extra-judicial comments, Gajarsa J, of the Court of Appeals for the Federal Circuit, and the author of the decision in *Madey v Duke University*, has expressed confidence that the patent system will accommodate nanotechnological inventions:

> What does nanotechnology mean for technological development? It will provide us with the ability to develop computers the size of a dime. It will combine biology with electronics. It will give us the means to meet tomorrow's massive computing challenges.[112]

The judge is of the firm view that the intellectual property system will promote economic progress and the dissemination of scientific information: 'Our intellectual property system . . . will continue to add dynamism to our new information-based economy because the dialectic balances of providing patent protection for these new technologies do not limit economic progress.'[113]

In a thoughtful piece for the *Stanford Law Review*, Mark Lemley identifies a number of features of nanotechnology which will pose particular issues for patent law.[114] First, he notes that, unlike other fields, the basic blocks of nanotechnology were patented very early on: 'It is too early to tell for sure how significant nanotech building-block patents will turn out to be or how they will be enforced, but it is quite possible that more of the fundamental building blocks of nanotechnology will be patented than in any of the [other] industries discussed.'[115] Second, he notes that nanotechnology is a cross-disciplinary field: 'Nanotech is not confined to a single field of endeavor, but exploits the peculiar properties of matter at the nanoscale across many different fields of modern engineering.'[116] Third, Lemley notes that 'nanotechnology patents . . . are held in surprisingly large proportions by universities'.[117] He concludes that the field of nanotechnology is worth watching, because of the combination of these three factors:

> Nanotechnology patents bear watching. They have characteristics that may well turn out to be fundamentally different than patents in any other industry in the last eighty years. How the market responds to these characteristics will determine

whether and how the law must step in and tailor the rules of patent law to the needs of this nascent industry. It will also give us broader insight into the role of patents in enabling technologies. Nanotechnology is a natural experiment that can teach us whether we have learned anything since the days of the Wright Brothers about how to license and enforce patents without restricting innovation, or whether the absence of patent protection for the enabling technologies of the last century was a series of fortunate events.[118]

Lemley suggests that there are a number of possible solutions if the field of nanotechnology becomes cluttered with strong and broad patents. He notes that the utility requirement could be applied stringently in respect of nanotechnology inventions: 'If there is a significant risk that nanotechnology innovation will be retarded by broad upstream patents, we can replicate by law the result we got by accident in the biotech, software, hardware, and Internet industries – freedom to use basic tools and processes with patents only on downstream implementations.'[119] Moreover, he suggests that there could be scope for government funding agencies to encourage non-exclusive licensing in respect of nanotechnology inventions.

Lemley maintains, though, that it is premature to intervene in the field, until the science and industry has matured: 'Restricting nanotech patents is also premature because we have not yet had an opportunity to see how significant the patents will turn out to be, how they will be licensed, and how industry participants will react.'[120] Other academic commentators have raised a number of options to deal with the patent thicket in nanotechnology, such as patent pooling,[121] experimental use, compulsory licensing and Crown use.

By contrast, Siva Vaidhyanathan of New York University is somewhat more sceptical as to whether the patent regime is appropriately adapted for the field of nanotechnology: 'The dream of nanotechnology – engineering substances at the scale of one nanometer – reveals many of the dangers of an overprotective patent system.'[122] He comments: 'Paradoxically, an overprotective patent system threatens the potential benefits of a fully realized nanotechnology industry.'[123] Vaidhyanathan wonders whether a sui generis system of protection would be more appropriate for nanotechnology:

> Regardless, it's worth considering whether a special set of rules should apply to nanotechnology. Theoretically, the patent system is supposed to be nondiscriminatory. It should operate the same way under the same principles regardless of the type of technology at hand. But in practice, different fields do work differently in the patent system. Perhaps nanotechnology would grow more equitably, efficiently, and predictably if its patents worked for a shorter time, perhaps 10 years instead of 20. Perhaps there should be a global nanotechnology patent database run through the United Nations.[124]

Vaidhyanathan expressed concerns that a 'tragedy of the anti-commons' would afflict the nanotechnology field, much as Heller and Eisenberg predicted of biomedical research. He submitted that there is a need to encourage greater openness in this field of science: 'The proprietary and competitive nature of the current nanotechnology community does not bode well for transparency and equity.'[125]

The Canadian-based ETC Group has expressed ethical and political objections to the patenting of nanotechnology: 'Nano-scale manipulation in all its forms offers unprecedented potential for sweeping monopoly control of elements and processes that are fundamental to biological function and material resources.'[126] The ETC Group fears that 'Atomtech will eclipse genetic engineering because it involves *all* matter – both living and non-living.'[127] The ETC Group expounds on the topic:

> The prohibition on product of nature patents was rendered vacuous by the 1980 Supreme Court decision. Today, with the world's largest corporations gearing up to work down at the nano-level, it is only a matter of time before industry convinces patent examiners that the genetically-engineered microbe of twenty-two years ago is no different from the atomically-engineered elements of today. Between nuclear colliders, atomic force microscopes, and cameras that can photograph light as it meanders through a retina, the nanotech industry will be in a political position to argue that any tinkering with the elemental products of nature is patent-worthy.[128]

The ETC Group objected to patents being granted in respect of nanotechnology, suggesting that it was akin to patenting elements of the Periodic Table: 'Patenting at the nano-scale can mean monopolizing the basic elements that make life possible.'[129] The civil society organization maintained that nanotechnological devices should not be conceived of as patentable subject matter because they are products of nature and scientific discoveries. The ETC group believes that a moratorium should be placed on research involving molecular self-assembly and self-replication.

EPILOGUE

This book has argued that courts, law makers, and policy makers can learn some important lessons from the policy debates over intellectual property and biotechnology. Under the nostrum that 'anything under the sun made by man' is patentable, eligible subject matter has been defined in a broad and flexible fashion to accommodate a range of new technologies, including biotechnological research. Calling for an end to the default position established in *Diamond v Chakrabarty*, Helen Berman and Rochelle Cooper

Dreyfuss maintain that there is a need for a systematic re-evaluation of the way patent law applies to genes, proteins and related inventions:

> *Chakrabarty* essentially changed the default position on protecting life-sciences materials. Its broad holding – that the subject matter of patent law extends to 'anything under the sun made by man' – means that developments in these fields are now presumed to be patentable. As a result, Congress has had little occasion to intervene, or even consider, patent law issues regarding genomics or proteomics. And even Supreme Court involvement has become minimal, for at around the same time that biotechnological research began to flourish, Congress established the Court of Appeals for the Federal Circuit.[130]

This book has suggested that patent law should be technology-specific, especially when dealing with the demands of particular fields of biotechnology.[131] Thus it is worth tailoring the rules and principles of patent law to cater for the peculiarities of plants, animals and micro-organisms, as well as medical, gene, stem cell and drug patents, plus access to genetic resources and traditional knowledge. Greater efforts should be made to preserve and conserve what Breyer J of the Supreme Court of the United States called the 'Storehouse of Knowledge' – the public domain and the intellectual commons. There is a need to preserve certain traditional exclusions from patentable subject matter, especially in respect of scientific discoveries, abstract ideas, products of nature and business methods. As the judge has noted: 'Patent law seeks to avoid the dangers of overprotection just as surely as it seeks to avoid the diminished incentive to invent that underprotection can threaten.'[132]

This book has reinforced the allegation made by economists Adam Jaffe and Josh Lerner that the patent system is 'broken' and needs to be reformed.[133] There should be further impetus to the attempts in the United States Congress and international fora to reform the patent system.[134] There is a need to improve the capacity of the patent office in dealing with applications for new technologies, especially in biotechnology and adjacent fields. 'Patent trolls' should not be allowed to flourish and to hold public and private investors in research and development to ransom. There needs to be greater scope for challenging patent applications by civil society and public interest groups, such as the Public Patent Foundation – 'patent-busting' if you will.[135] The criteria for patentability should be applied strictly in respect of new technologies. Not only should the requirement for utility be strengthened, but patent offices and courts should apply the tests for novelty and inventive step in a stringent fashion. There should be greater creativity and problem-solving abilities attributed to a 'person skilled in the art' to ensure that the patent system rewards more than merely nominal improvements to the scientific knowledge and art in the public domain.[136] Helen Berman and Rochelle Cooper Dreyfuss contend: 'At the very least, the courts must develop a method for dealing with "moving target" issues, such as keeping track of the actual level of skill in the art, the degree to which research functions have been automated, the availability of fundamental data, and the agreements that scientists have made about how such data should be represented.'[137]

Given the expansion of the scope of patentable subject matter, and the concomitant threat of the creation of an 'anti-commons' in respect of scientific research, there is a need in turn to broaden the range of exceptions to patent infringement. Graeme Dinwoodie and Rochelle Cooper Dreyfuss observe that members of the World Trade Organization should take advantage of the flexibilities within the *TRIPS Agreement* 1994 so as to protect the scientific commons:

> The public domain of science is likely shrinking, but more through the efforts of technological change than through legal efforts to privatize culture. International law heavily circumscribes the capacity to redraw the public/private boundaries in ways that ensure an optimal public domain. Scholars might thus view international law as an obstacle around which national patent policymakers must navigate. But the function of international intellectual property law should be conceptualized more broadly. Informed by the value of a strong domain of accessible knowledge, international law could help member states resist scientific and technological commodification.[138]

A broad interpretation should be provided in respect of the defence of experimental use. The defence should be defined along the lines of the European Union model. In addition, there should be a defence in respect of personal, non-commercial use of patented inventions. Countries should provide for a broad safe harbour for research into pharmaceutical drugs, along the lines articulated by the Supreme Court of the United States in *Merck KGaA v Integra Lifesciences I, Ltd*. There is also a need for special defences for particular industries, such as a limited liability for medical practitioners in respect of patent infringement, a defence in respect of farm-saved seed, and an innocent bystander presumption. Such measures are necessary to deal with problems associated with biological inventions. Market-based solutions such as cross-licensing and patent pooling should be encouraged. There is also a demand to reform the compulsory licensing and Crown use, so that it allows for competition and access to essential medicines. There is a role for competition regulators to supervise and monitor the impact of patents in respect of pioneering technologies. It is recommended that the bioethical principles of informed consent and benefit sharing be incorporated into patent law in a number of fields.

Such recommendations will not only help resolve the existing disputes over intellectual property and biotechnology, but they will better prepare patent offices, courts and legislatures in the regulation of the next generation of frontier inventions and pioneer technologies.

Particularly noteworthy is the work of the J. Craig Venter Institute and its new spin-off company, Synthetic Genetics, in the field of synthetic biology.[139] In November 2002, J. Craig Venter and Nobel Laureate Hamilton Smith received a US$3 million grant from the US Energy Department to create a new, 'minimalist' life form in the laboratory – a single-celled, partially human-made organism. On 12 October 2006, the J. Craig Venter Institute Inc. filed a patent application in respect of a 'minimal bacterial genome'.[140] The abstract of the patent application observes:

> The present invention relates, eg., to a minimal set of protein-coding genes which provides the information required for replication of a free-living organism in a rich bacterial culture medium, wherein (1) the gene set does not comprise the 101 genes listed in Table 2; and/or wherein (2) the gene set comprises the 381 protein-coding genes listed in Table 3 and, optionally, one of more of: a set of three genes encoding ABC transporters for phosphate import (genes MG410, MG411 and MG412; or genes MG289, MG290 and MG291); the lipoprotein-encoding gene MG185 or MG260; and/or the glycerophosphoryl diester phosphodiesterase gene MG293 or MG385.[141]

This patent application has already caused consternation amongst anti-biotechnology groups. The ETC Group observed that 'Venter's work poses ethical and environmental concerns about the use of biodiversity to build new life forms from scratch.'[142] The civil society organisation fears that synthetic biology will have attendant environmental risks: 'The extraordinary appeal of solving the world's energy problems by harnessing new, engineered life forms, tends to eclipse the very real concerns about potential negative consequences.'[143] The organisation suggested: 'Intellectual property claims on human-made life also pose concerns about *ordre public*.'[144] The policy debate over intellectual property and biotechnology will no doubt inform future discussions over the patenting of artificial and synthetic life.

NOTES

1. Australian Law Reform Commission (2004), *Genes and Ingenuity: Gene Patenting and Human Health, Report 99*, Sydney: Australian Commonwealth, http://www.austlii.edu.au/au/other/alrc/publications/reports/99/, June, p. 64.
2. Hahn, Robert (2005), 'Introduction' in Robert Hahn (ed.), *Intellectual Property Rights in Frontier Industries: Software and Biotechnology*, AEI-Brookings Joint Center for Regulatory Studies, p. 7.

3. Ibid.
4. Burk, D. and M. Lemley (2003), 'Policy levers in patent law', *Virginia Law Review*, **89**, 1575–1696 at 1577.
5. Ibid., 1641.
6. Palombi, Luigi (2004), 'The patenting of biological materials in the context of the Agreement on Trade Related Aspects of Intellectual Property', PhD thesis, University of New South Wales, Sydney, http://cgkd.anu.edu.au/menus/PDFs/PhDThesisFinal.pdf.
7. ETC Group (2002), 'Patenting elements of nature', http://www.etcgroup.org/upload/publication/220/01/nanopatentsgeno.rtf.pdf, 25 March.
8. Brown, N., A. Nelis, B. Rappert and A. Webster (1999), 'Bioinformatics: a technology assessment of recent developments in bioinformatics and related areas of research and development including high-throughput screening and combinational chemistry', Final Report for the Science and Technological Options Assessment Unit, European Parliament.
9. Cook-Deegan, Robert (1994), *The Gene Wars: Science, Politics, and the Human Genome*, New York and London: W.W. Norton, p. 288.
10. Eisenberg, R. (1987), 'Proprietary rights and the norms of science', *Yale Law Journal*, **97**, 177–223; and Eisenberg, R. (2000), 'Re-examining the role of patents in appropriating the value of DNA sequences', *Emory Law Journal*, **49**, 783–800.
11. For extensive discussion of copyright law and genetic databases, see Rimmer, M. (2003), 'Beyond blue gene: intellectual property and bioinformatics', *International Review of Industrial Property and Copyright Law*, **34**(1), 31–49; Rimmer, M. (2005), 'Japonica rice: intellectual property, scientific publishing, and data-sharing', *Prometheus*, **23**(3), 325–47; and Rimmer, Matthew (2007), 'Chapter 8, remix culture: the creative commons and its discontents', *Digital Copyright and the Consumer Revolution*, Cheltenham, UK and Northampton, MA, USA: Edward Elgar.
12. Eisenberg, R. (2002), 'Molecules vs. information: should patents protect both?', *Boston University Journal of Science and Technology Law*, **8**, 190–217 at 200.
13. Ibid.
14. Eisenberg, R. (2000), 'Re-examining the role of patents in appropriating the value of DNA sequences', *Emory Law Journal*, **49**, 783–800 at 791.
15. *State Street Bank And Trust Co v Signature Financial Group Inc* 149 F 3d 1368 (1998).
16. *AT & T Corp v Excel Communications Inc* 172 F 3d 1352 (1999).
17. *Amazon.com Inc. v Barnes and Noble.com, Inc.*, 239 F.3d 1343 (Fed.Cir.2001).
18. *Welcome Real-Time v Catuity Inc* [2001] FCA 445.
19. *State Street Bank And Trust Co v Signature Financial Group Inc* 149 F 3d 1368 (1998).
20. *Laboratory Corp. of America Holdings v Metabolite Laboratories, Inc.* 126 S.Ct. 2921 at 2928 (2006).
21. Pollack, M. (2002), 'The multiple unconstitutionality of business method patents: common sense, congressional consideration, and constitutional history', *Rutgers Computer and Technology Law Journal*, **28**, 61–120.
22. Sung, L. and D. Pelto (1998), 'Bioinformatics may get boost from "State Street": software that can manipulate vast libraries of genetic data may receive patent protection', *The National Law Journal*, **21**(8), 19 October.
23. Buff, E. and L. Restaino (1999), 'State street alters landscape of biotechnology and bioinformatics', *New Jersey Law Journal*, **157**(4), http://www.riker.com/articles/index.php?id=3282.
24. Lesavich, S. (2000), 'Bioinformatic tools', *The National Law Journal*, 16 October, B10.
25. Power, E. (2001), 'Pharmaceutical companies are "failing to patent new technologies"', Silico Research, London, 2 May.
26. Saviotti, P., M-A. de Looze, S. Michelland and D. Catherine (2000), 'The changing marketplace of bioinformatics', *Nature Biotechnology*, **18**(12), 1247–9.
27. Toner, B. (2001), 'Bioinformatics patents remain a rarity in IP-heavy biopharmaceutical industry', *Genomeweb.com*, 4 July.
28. Cockburn, Iain (2005), 'State street meets the Human Genome Project: intellectual property and bioinformatics' in Robert Hahn (ed.), *Intellectual Property Rights in*

Frontier Industries: Software and Biotechnology, AEI-Brookings Joint Center for Regulatory Studies, pp. 109–30 at 111.
29. There have been battles over the ownership and commercialization of gene chip or microarray technology. Oxford Gene Technology brought a legal action against Affymetrix Inc. At first, the English High Court held that Affymetrix, Inc. was not licensed to use Oxford Gene Technology's DNA microarray patents in Europe and the United States, numbers EP 373 203 and US 5,700,637. However, the English Court of Appeal reversed the decision of the High Court. It decided that Affymetrix had successfully purchased Beckman's business in DNA microarrays and was therefore licensed from the date of purchase. Finally, the US District Court for the District of Delaware gave judgment in a patent infringement action brought by Oxford Gene Technology against Affymetrix, Inc. A jury of eight decided that Affymetrix's process for making and using its GeneChip arrays infringed Oxford Gene Technology's DNA microarray patent, number US 5,700,637. They were also opposition proceedings against Affymetrix's European patent No. EP 0 619 321. In the end, Oxford Gene Technology, Ltd. (OGT) and Affymetrix, Inc. announced the settlement of all existing litigation between the two companies.

Furthermore, Affymetrix argued that Incyte Pharmaceuticals had infringed several of its microarray-related patents (U.S. Patent Nos: 5,445,934, 5,744,305, 5,800,992, 5,871,928, 6,040,193), and challenged Incyte's RNA amplification patents (U.S. Patent Nos: 5,716,785, 5,891,636).

The companies reached a settlement, agreeing to cross-licences under their respective intellectual property portfolios. Financial terms were not disclosed. For a summary, see Affymetrix (2001), 'Quarterly report: legal proceedings', http://sec.edgar-online.com/2001/11/13/0000912057-01-539101/Section11.asp, 23 March.
30. Cockburn, Iain (2005), 'State street meets the Human Genome Project: intellectual property and bioinformatics' in Robert Hahn (ed.), *Intellectual Property Rights in Frontier Industries: Software and Biotechnology*, AEI-Brookings Joint Center for Regulatory Studies, pp. 109–30 at 111.
31. Ibid., pp. 114–15.
32. Merges, R. (1999), 'As many as six impossible patents before breakfast: property rights for business concepts and patent system reform', *Berkeley Technology Law Journal* Spring, **14**(2), 577–615, http://www.law.berkeley.edu/institutes/bclt/pubs/merges/.
33. Vondran, C. and R. Florence (2002), 'Bioinformatics: patenting the bridge between information technology and the life sciences', *The Journal of Law and Technology*, **42**, 93–131.
34. Saviotti, P., M-A. de Looze, S. Michelland and D. Catherine (2000), 'The changing marketplace of bioinformatics', *Nature Biotechnology*, **18**(12), 1247–9.
35. Likhovski, M., M. Spence and M. Molineaux (2000), 'The first mover monopoly: a study on patenting business methods in Europe', *Oxford Electronic Journal of Intellectual Property Rights*, November.
36. Reichman, J.H. (2001), 'A contractually reconstructed research commons for scientific data in a highly protectionist intellectual property environment', *Law & Contemporary Problems*, **66**, 315–440.
37. IBM (1999), 'IBM announces $100 million research initiative to build world's fastest supercomputer: blue gene to tackle protein folding grand challenge', Yorktown Heights, New York, 6 December; and IBM, 'Blue Gene', http://domino.research.ibm.com/comm/research_projects.nsf/pages/bluegene.index.html.
38. Blueprint Worldwide (2001), 'Science and technology leaders launch blueprintworldwide Inc., giving researchers unprecedented access to vital biomolecular interaction data; new company founded by IBM and MDS Proteomics will help accelerate drug discovery with the world's most comprehensive collection of bioMolecular data', http://www-03.ibm.com/press/us/en/pressrelease/1240.wss, 30 May.
39. Washington Biohistory (2007), 'The Microsoft effect', http://www.wabio.com/biohistory_microsoft.htm?pp=1.
40. Department of Molecular Biotechnology, the University of Washington, http://www.gs.washington.edu/.

41. Bill and Melinda Gates Foundation, http://www.gatesfoundation.org/default.htm.
42. BioIT Alliance, http://bioitalliance.org/default.aspx.
43. BioIT Alliance, http://bioitalliance.org/FAQ.aspx.
44. Vise, David (2005), *The Google Story*, New York: Random House, p. 286.
45. Ibid., 285.
46. ETC Group and the coalition against biopiracy (2006), 'Captain Hook and Cog Awards announced in Curitiba', http://www.etcgroup.org/en/materials/publications.html?id=20, 24 March.
47. Formanek, Jr., R. (2005), 'Proteomics: moving beyond the human genome', *FDA Consumer*, **39**(6), 22–5, http://www.fda.gov/fdac/features/2005/605_proteomics.html.
48. West, J. (2001), 'Proteome systems', Harvard Business School, 11 October.
49. Editorial (1999), 'The promise of proteomics', *Nature*, **402**, 703.
50. Ibid.
51. Ibid.
52. Bohrer, R. (2003), 'Proteomics: the next phase in the biotechnology revolution and the next challenge for biotechnology law', *Biotechnology Law Report*, **22**(3), 263.
53. Warburg, R., A. Wellman, T. Buck and A. Ligler Schoenhard (2003), 'Patentability and maximum protection of intellectual property in proteomics', *Biotechnology Law Report*, **22**(3), 264–72 at 264.
54. Barton, J. (2002), 'United States law of genomic and post-genomic patents', *International Review of Industrial Property and Copyright Law*, **33**(7), 779–89.
55. Misrock, L. (2001), 'Genomeweb Q&A: biotech patent guru discusses possible proteomics patent wars', 13 March.
56. Ibid.
57. Ibid.
58. Oxford Glycosciences, www.ogs.com, website archived on the Wayback Machine, http://www.archive.org/web/web.php; see also Oxford University, 'Oxford Glycosciences', http://www.isis-innovation.com/spinout/oxglycosciences.html.
59. Durman, P. (2000), 'Oxford glycosciences protein patent', *The Times*, 26 February, p. 26.
60. Ezzell, C. (2002), 'Proteins rule', *Scientific American*, April, 41–7.
61. Durman, P. (2000), 'Oxford glycosciences protein patent', *The Times*, 26 February, p. 26; Clark, A. (2001), 'Glycosciences makes "land grab" for proteins', *The Guardian*, 16 March; and Gavaghan, H. (2000), 'Companies of all sizes are prospecting for proteins', *Nature*, **404**, 684–5.
62. Oxford Glycosciences, 'Intellectual property', http://www.ogs.com/Proteomics/ProteomicsIntellectualProperty/, website archived on the Wayback Machine, http://www.archive.org/web/web.php.
63. Parekh, R., R. Amess, J. Bruce, S. Prime, A. Platt and R. Stoney (1997), 'Computer-assisted methods and apparatus for identification and characterization of biomolecules in a biological sample', US Patent No: 6,064,754.
64. Oxford Glycosciences (2000), 'OGS announces issuance of major US patent central to identification of disease-associated proteins', 16 May.
65. Ibid.
66. Ibid.
67. Oxford Glyosciences (2001), 'OGS announces achievement of major protein patenting milestone: completes discovery and patent filings for over 4,000 disease-associated proteins, genes and protein isoforms', 7 December.
68. Oxford Glycosciences, 'Intellectual property', http://www.ogs.com/Proteomics/ProteomicsIntellectualProperty/, website archived on the Wayback Machine, http://www.archive.org/web/web.php.
69. Courtney, S. (1997), 'Therapeutic compounds with pyrimidine base', US Patent No: 5,945,406; Parekh, R., R. Amess, J. Bruce, S. Prime, A. Platt and R. Stoney (1997), 'Computer-assisted methods and apparatus for identification and characterization of biomolecules in a biological sample', US Patent No: 6,064,754; Parekh, R., R. Amess, J. Bruce, S. Prime, A. Platt and R. Stoney (2000), 'Computer-assisted isolation and characterization of proteins', US Patent No: 6,278,794; Pennington, M., D. Scopes and

Conclusion   305

M. Orchard (1999), 'Quinolinium- and pyridinium-based fluorescent dye compounds', US Patent No: 6,335,446; Parekh, R., R. Amess, J. Bruce, S. Prime, A. Platt and R. Stoney (2000), 'Methods for computer-assisted isolation of proteins', US Patent No: 6,459,994; Parekh, R., R. Amess, J. Bruce, S. Prime, A. Platt and R. Stoney (2000), 'Robotic device for removing selected portions of a polyacrylamide gel', US Patent No: 6,480,618; Parekh, R., P. Goulding and D. Pfost (1999), 'Assaying and storing labelled analytes', US Patent No: 6,534,321; Townsend, R., R. Parekh, S. Prime and N. Wedd (1997), 'Method for de novo peptide sequence determination', US Patent No: 6,582,965; Orchard, M. and J. Neuss (2003), 'Thiazolidine derivatives and its use as antifungal agent', US Patent No: 6,740,670; Townsend, R. and A. Robinson (2001), 'Automated identification of peptides', US Patent No: 6,963,807; Orchard, M. (2003), 'Benzylidene thiazolidinediones and their use as antimycotic agents', US Patent No: 7,105,554.
70. *Oxford Glycosciences v MDS Inc.*, Case R 397/2000-1 (2000), Office for Harmonization in the Internal Market (Trade Marks and Designs), http://oami.europa.eu/legaldocs/boa/2000/en/R0397_2000-1.pdf.
71. Marconi and Oxford Glycosciences (2001), 'Marconi and OGS form joint venture to provide proteomics data and hosting solutions: major initiative in pharmaceutical bioinformatics as proteomics and communications converge', 15 June.
72. Ibid.
73. Philipkoski, K. (2002), 'An "Atlas" to Count the Genes', *Wired Magazine*, 27 February; and Ferranti, M. (2002), 'Confirmant set to release protein atlas', *Bio-IT Worlds News*, http://www.bio-itworld.com/news/040702_report184.html, 4 July.
74. Black, D. (2003), 'New bidders enter fray for Oxford GlycoSciences', *The Guardian*, 27 March.
75. Celltech Group (2003), 'PLC Annual General Meeting Statement', 22 May.
76. Misrock, L. (2001), 'Genomeweb Q&A: biotech patent guru discusses possible proteomics patent wars', 13 March.
77. Large Scale Biology Corporation, 'Intellectual property', http://www.lsbc.com/ip.html.
78. MDS Inc. (2004), 'MDS Inc. announces completion of proteomics reorganization: MDS proteomics renamed Protana Inc.', 29 July.
79. Transition Therapeutics, http://www.transitiontherapeutics.com/.
80. Proteome Systems, http://www.proteomesystems.com/.
81. Tarcza, J. (2001), 'Protein claiming in the age of proteomics', Human Proteome Project, http://www.xensei.com/users/chi/2001/hpr/, April.
82. Williams, J.J. (2005), 'Protecting the frontiers of biotechnology beyond the genome: the limits of patent law in the face of the proteomics revolution', *Vanderbilt Law Review*, **58**, 955–93 at 993.
83. Ibid., 989.
84. Shi, Q. (2005), 'Patent system meets new sciences: is the law responsive to changing technologies and industries', *New York University Annual Survey of American Law*, **61**, 316–46 at 341.
85. Ibid., 342.
86. Ibid., 342.
87. Royal Society of the United Kingdom (2005), *Personalised Medicines: Hopes and Realities*, London: Royal Society, http://www.royalsoc.ac.uk/displaypagedoc.asp?id=17049, p. 41.
88. Ibid.
89. Ibid.
90. Rai, A. (2001), 'The information revolution reaches pharmaceuticals: balancing innovation incentives, cost, and access in the post-genomics era', *University of Illinois Law Review*, 173–210 at 189.
91. Ibid., 209.
92. Ibid., 209.
93. For a discussion of the historical development of nanotechnology, see Peterson, C. (2004), 'Nanotechnology: from Feynman to the grand challenge of molecular

manufacturing', *IEEE Technology and Society Magazin*, **23**(4), 9–15; Kulinowski, K. (2004) 'Nanotechnology: from "Wow" to "Yuck"?', *Bulletin of Science, Technology & Society*, **24**(1), 13–20; and Drexler, E. (2004), 'Nanotechnology: from Feynman to funding', *Bulletin of Science, Technology & Society*, **24**(1), 21–7.

94. Mullally, M.V. and D. Winn (2004), 'Patenting nanotechnology: a unique challenge to the IP bar', *The New York Law Journal*, http://www.orrick.com/fileupload/324.pdf, 6 July.
95. Benyus, Janine (1997), *Biomimicry: Innovation Inspired by Nature*, New York: Harper Perennial.
96. 'Sydney technologist Dr Vijoleta Braach-Maksvytis gets her inspiration from nature. She marvels at the tiny hairs on the feet of geckos that allow them to walk upside down across a smooth glass ceiling. She is intrigued by the microscopic structures on the wings of some butterflies that make them iridescent . . . Braach-Maksvytis says the inspirations for the device were the membranes around the hardy bacteria that can survive in boiling temperatures near volcanos, and the methods our own bodies use to detect invading microbes'; Smith, D. (2000), 'Adventures in Lilliput', *The Sydney Morning Herald*, 11 December.
97. The key patent is Raguse, B., B. Cornell, V. Braach-Maksvytis and R. Pace (1997), 'Biosensor membranes', US Patent No: 5,798,030.
98. CSIRO (1997), 'Australian scientists create a nanomachine', Press Release, 5 June.
99. Ibid.
100. Braach-Maksvytis, V. (2003), 'Interview with author', 6 November.
101. CSIRO (1997), 'Australian scientists create a nanomachine', Press Release, 5 June.
102. Ibid.
103. Ambri, http://www.ambri.com.au/.
104. Australian Government (2005), *Australian Nanotechnology: Capability and Commercial Potential*, 2nd edn, Invest Australia, http://www.investaustralia.gov.au/media/IR_Nano_NanotechReport.pdf.
105. European Commission (2007), 'Nanotechnology, http://cordis.europa.eu/nanotechnology/'.
106. Koppikar, V., S. Maebius and J.S. Rutt (2004), 'Current trends in nanotech patents: a view from inside the patent office', *Nanotechnology Law and Business*, **1**, 24–30 at 25.
107. Search conducted in March 2007, using the designated Class 977/Dig.1, entitled 'Nanotechnology'.
108. NanoBusiness Alliance (2002), 'US patent examiners may not know enough about nanotech', *Small Times*, http://www.smalltimes.com/articles/stm_print_screen.cfm?ARTICLE_ID=267691.
109. United States Patent and Trademark Office (2004), 'New cross-reference digest for nanotechnology', http://www.uspto.gov/web/patents/biochempharm/crossref.htm.
110. United States Patent and Trademark Office (2004), 'New cross-reference digest for nanotechnology', http://www.uspto.gov/web/patents/biochempharm/crossref.htm.
111. United States Patent and Trademark Office (2004), 'New cross-reference digest for nanotechnology', http://www.uspto.gov/web/patents/biochempharm/crossref.htm.
112. Gajarsa, A. (2002), 'Quo vadis? The fifth annual Honorable Helen Wilson Nies memorial lecture in intellectual property law', *Marquette Intellectual Property Law Review*, **6**, 1 at 8–9.
113. Ibid.
114. Lemley, M. (2005), 'Patenting nanotechnology', *Stanford Law Review*, **58**, 601–30.
115. Ibid., 613.
116. Ibid., 614.
117. Ibid., 615.
118. Ibid., 630.
119. Ibid., 628.
120. Ibid., 629.
121. Lee, A. (2006), 'Examining the viability of patent pools for the growing nanotechnology patent thicket', *Nanotechnology Law and Business*, **3**, 317–27.

122. Vaidhyanathan, Siva (2006), 'Nanotechnology and the law of patents: a collision course', Hunt, Geoffrey and Michael Mehta (eds), *Nanotechnology: Risk, Ethics, Law*, London: Earthscan Publications, http://ssrn.com/abstract=740550.
123. Ibid.
124. Ibid.
125. Ibid.
126. ETC Group (2003), *The Big Down: From Genomes to Atoms; Atomtech: Technologies Converging at the Nano-Scale*, http://www.etcgroup.org/documents/TheBigDown.pdf, 30 January.
127. Ibid.
128. Ibid.
129. ETC Group (2002), 'Patenting elements of nature', http://www.etcgroup.org/upload/publication/220/01/nanopatentsgeno.rtf.pdf, 25 March.
130. Berman, H. and R. Cooper Dreyfuss (2006), 'Reflections on the science and law of structural biology, genomics, and drug development', *UCLA Law Review*, 53, 871–908 at 873.
131. Burk, D. and M. Lemley (2002), 'Is patent law technology-specific?', *Berkeley Technology Law Journal*, 17, 1155–1202.
132. *Laboratory Corp. of America Holdings v Metabolite Laboratories, Inc.* 126 S.Ct. 2921 at 2922 (2006).
133. Jaffe, Adam and Josh Lerner (2004), *Innovation and its Discontents: How our Broken Patent System is Endangering Innovation and Progress, and What to do About It*, Princeton: Princeton University Press.
134. United States House of Representative Subcommittee on Courts, the Internet, and Intellectual Property (2007), 'American innovation at risk: "The Case for Patent Reform"', Oversight Hearing, http://judiciary.house.gov/oversight.aspx?ID=27115 February.
135. Public Patent Foundation, http://www.pubpat.org/About.htm.
136. Gold, E.R. and K. Durell (2005), 'Innovating the skilled reader: tailoring patents to new technologies', *Intellectual Property Journal*, **19**(1), 189–226.
137. Berman, H. and R. Cooper Dreyfuss (2006), 'Reflections on the science and law of structural biology, genomics, and drug development', *UCLA Law Review*, 53, 871–908 at 908.
138. Dinwoodie, Graeme and Rochelle Cooper Dreyfuss (2006), 'Patenting science: protecting the domain of accessible knowledge', in Lucie Guibault and P. Bernt Hugenholtz (eds) (2006), *The Future of the Public Domain: Identifying the Commons in Information Law*, Netherlands: Kluwer Law International, 191–21 at 221.
139. Rosenwald, M. (2006), 'J. Craig Venter's next little thing: the man who mapped the human genome has a new focus – using microbes to create alternative fuels', *The Washington Post*, 27 February.
140. Glass, J., H. Smith, C. Hutchison, N. Alperovich and N. Assad-Garcia (2006), 'Minimal bacterial genome', US Patent No: 20070122826.
141. Ibid.
142. ETC Group (2004), 'Playing God in the Galapagos: J. Craig Venter, master and commander of genomics, on global expedition to collect microbial diversity for engineering life', *Communique*, **84**, March/April.
143. Ibid.
144. Ibid.

Bibliography

REFERENCES

Abbott, A. (2004), 'Clinicians win fight to overturn patent for breast-cancer gene', *Nature*, **429**, 329, 27 May.

Abbott, A. (2005), 'Genetic patent singles out Jewish women', *Nature*, **426**, 12, http://www.nature.com/nature/journal/v436/n7047/full/436012a.html, 7 July.

Advisory Council on Intellectual Property (2002), *Innovation Patent – Exclusion Of Plant And Animal Subject Matter*, Canberra: Department of Industry, http://www.acip.gov.au/library/Innovation%20Patent%20Issues%20Paper.PDF.

Advisory Council on Intellectual Property (2004), *Patents and Experimental Use: Options Paper*, Canberra: Commonwealth Government, http://www.acip.gov.au/library/Experimental%20Use%20Options%20Paper%20A.pdf.

Advisory Council on Intellectual Property (2005), *Patents and Experimental Use: Final Report*, Canberra: Commonwealth Government, http://www.acip.gov.au/library/ACIP%20Patents%20&%20Experimental%20Use%20final%20report%20FINAL.pdf.

Advisory Council on Intellectual Property (2005), *Review of Crown Use Provisions for Patents and Designs*, Canberra: Commonwealth Government, http://www.acip.gov.au/library/review_of_Crown_Use_provisions.pdf.

Affymetrix (2001), 'Quarterly report: legal proceedings', http://sec.edgar-online.com/2001/11/13/0000912057-01-539101/Section11.asp, 23 March.

AFP (2000), 'Greenpeace paralyses Patent Office in "Human Clone" Protest', Munich, Germany, 22 February.

Amanor-Boadu, V., M. Freeman and L. Martin (1995), 'The potential impacts of patenting biotechnology on the animal and agri-food sector', Intellectual Property Policy Directorate, Industry Canada, http://strategis.ic.gc.ca/pics/ip/martinef.pdf.

Ambri, http://www.ambri.com.au/.

American Anti-Vivisection Society, 'Stop animal patents!', http://www.stopanimalpatents.org/.

Andrews, E. (1990), 'Kastenmeier's loss will have a major impact on inventors', *The New York Times*, 18 November, H3.
Andrews, L. (2002), 'The gene patent dilemma: balancing commercial incentives with health needs', *Houston Journal of Health Law & Policy*, **2**, 65–106.
Andrews, L. (2005), 'Harnessing the benefits of biobanks', *Journal of Law, Medicine and Ethics*, **33**, 22–8.
Andrews, L., J. Paradise, T. Holbrook and D. Bochneak (2006), 'When patents threaten science', *Science*, **314**, 1395–6.
Aoki, K. (2003), 'Weeds, seeds and deeds: recent skirmishes in the seed wars', *Cardozo Journal of International and Comparative Law*, **11**, 247–311.
Armstrong, D. (2001), 'The arguments of law, policy and practice against Swiss-type patent claims', *Victoria University Of Wellington Law Review*, **32**(1), 201–54.
The Association of American Medical Colleges (2006), 'Re: request for comments on interim guidelines for examination of patent applications for patent subject matter eligibility', http://www.uspto.gov/web/offices/pac/dapp/opla/comments/ab98/aamc.pdf, 31 July.
Atkinson, R. (2005), 'Mixed messages: Canada's stance on patentable subject matter in biotechnology', *Intellectual Property Journal*, **19**, 1–27.
Attorney General's Department (2002), 'Inquiry into human genetic property issues,' Press Release, http://www.ag.gov.au/agd/WWW/attorneygeneralHome.nsf/Page/Media_Releases_2002_December_2002_Inquiry_into_human_genetic_property_issues_(17_December_2002), 17 December.
Australian Broadcasting Corporation (2001), 'Fears genetic testing may be confined to the rich', *7:30 Report* (transcript), 14 March.
Australian Broadcasting Corporation (2003), 'Genius of junk (DNA)', *Catalyst*, http://www.abc.net.au/catalyst/stories/s898887.htm, 10 July.
Australian Constitutional Commission (1988), *Final Report of the Constitutional Commission*, Canberra: Australian Government Publishing Service, **2**.
Australian Government (2005), *Australian Nanotechnology: Capability and Commercial Potential*, 2nd edn, Invest Australia, http://www.investaustralia.gov.au/media/IR_Nano_NanotechReport.pdf.
Australian Law Reform Commission (2003), *Essentially Yours: The Protection of Human Genetic Information*, Sydney: Australian Commonwealth, http://www.austlii.edu.au/au/other/alrc/publications/reports/96/.

Australian Law Reform Commission (2003), *Gene Patenting and Human Health, Issue Paper 27*, Sydney: Australian Commonwealth, http://www.austlii.edu.au/au/other/alrc/publications/issues/27/, July.

Australian Law Reform Commission (2004), *Gene Patenting and Human Health, Discussion Paper 68*, Sydney: Australian Commonwealth, http://www.austlii.edu.au/au/other/alrc/publications/dp/68/, February.

Australian Law Reform Commission (2004), *Genes and Ingenuity: Gene Patenting and Human Health, Report 99*, Sydney: Australian Commonwealth, http://www.austlii.edu.au/au/other/alrc/publications/reports/99/, June.

Bagley, M. (2003), 'Patent first, ask questions later: morality and biotechnology in patent law', *William and Mary Law Review*, **45**, 469–547.

Bagley, M. (2004–05), 'Stem cells, cloning and patents: what's morality got to do with it?', *New England Law Review*, **39**, 501–9.

Bancroft Library, http://bancroft.berkeley.edu/Exhibits/Biotech/25.html.

Bancroft Library, 'Bioscience and biotechnology: resources for historical research, 1999–2002', http://sunsite.berkeley.edu:2020/dynaweb/teiproj/oh/science/.

Barton, J. (2002), 'United States law of genomic and post-genomic patents', *International Review of Industrial Property and Copyright Law*, **33**(7), 779–89.

Becerra, X. and D. Weldon (2007), 'Representatives Becerra and Weldon introduce bill to ban the practice of gene patenting', United States Congress, http://weldon.house.gov/News/DocumentSingle.aspx?DocumentID=57930, 9 February.

Benson, J. (2002), 'Resuscitating the patent utility requirement, again: a return to *Brenner v Manson*', *University of California Davis Law Review*, **36**, 267–95.

Bently, Lionel and Brad Sherman (2004), *Intellectual Property*, 2nd edn, Oxford: Oxford University Press.

Benyus, Janine (1997), *Biomimicry: Innovation Inspired by Nature*, New York: Harper Perennial.

Berman, H. and R. Cooper Dreyfuss (2006), 'Reflections on the science and law of structural biology, genomics, and drug development', *UCLA Law Review*, **53**, 871–908.

Beyleveld, D. and R. Brownsword (2002), 'Is patent law part of the EC legal order? A critical commentary on the interpretation of Article 6(1) of Directive 98/44/EC in Case C-377/98', *Intellectual Property Quarterly*, **1**, 97–110.

Bhattacharjee, Y. (2007), 'In the courts', *Science*, **315**, 581b.

Biagioli, Mario (ed.) (1999), *The Science Studies Reader*, New York: Routledge.

Biagioli, Mario (2003), 'Rights or rewards? Changing frameworks of scientific authorship', in Mario Biagioli and Peter Galison (eds), *Scientific Authorship: Credit and Intellectual Property in Science*, New York: Routledge.

Biagioli, Mario and Peter Galison (eds) (2003), *Scientific Authorship: Credit and Intellectual Property in Science*, New York: Routledge.

Bill and Melinda Gates Foundation, http://www.gatesfoundation.org/default.htm.

Binns, R. and B. Driscoll (1999), 'Are the generic companies winning the battle?', *Managing Intellectual Property*, **86**, 36.

BioIT Alliance, http://bioitalliance.org/default.aspx.

BIOTECanada (2002), 'BIOTECanada responds to Supreme Court decision on Harvard mouse case', Ottawa, 5 December.

Biotechnology Industry Organization, 'The importance of intellectual property', http://www.bio.org/ip/.

Biotechnology Industry Organization (1999), 'Statement for hearing regarding commercial development of pluripotent stem cells', Subcommitttee on Labor, Health and Human Services, Education of the Senate Appropriations Committee, http://www.bio.org/bioethics/background/stemcell_testimony.asp, 12 January.

Biotechnology Newswatch (1989), 'Author of animal patents bill hears dire warnings pro, con', *Biotechnology Newswatch*, 2 October, **9**(19), 4.

Black, D. (2003), 'New bidders enter fray for Oxford GlycoSciences', *The Guardian*, 27 March.

Blanton, K. (2002), 'Corporate takeover', *Boston Globe*, 24 February.

Blueprint Worldwide (2001), 'Science and technology leaders launch blueprintworldwide Inc., giving researchers unprecedented access to vital biomolecular interaction data, new company founded by IBM and MDS proteomics will help accelerate drug discovery with the world's most comprehensive collection of bioMolecular data', http://www-03.ibm.com/press/us/en/pressrelease/1240.wss, 30 May.

Boehm, Klaus with Aubury Silberston (1967), *The British Patent System*, Cambridge: Cambridge University Press.

Bohrer, R. (2003), 'Proteomics: the next phase in the biotechnology revolution and the next challenge for biotechnology law', *Biotechnology Law Report*, **22**(3), 263.

Botstein, D., R. White, M. Skolnick and R. Davis (1980), 'Construction of a genetic linkage map in man using restriction fragment length polymorphisms', *American Journal of Human Genetics*, **32**(3), 314.

Braach-Maksvytis, V. (2003), 'Interview with author', 6 November.

Brandt-Rauf, S., V. Raveis, N. Drummond, J. Conte and S. Rothman (2006), 'Ashkenazi Jews and breast cancer: the consequences of linking ethnic identity to genetic disease', *American Journal of Public Health*, **96**(11), 1979.

Breen Smith, M. (2002), 'An end to gene patents? The Human Genome Project versus the United States Patent and Trademark Office's 1999 Utility Guidelines', *University of Colorado Law Review*, **73**, 747–85.

Breyer, S. (2000), 'Genetic advances and legal institutions', *Journal of Law, Medicine & Ethics*, **28**, 23–8.

Brickley, P. (2003), 'Patent rights wrangle puts law in question', *The Scientist*, **17**(42), http://www.the-scientist.com/yr 2003/mar/prof2_030310.html, 10 March.

Brignati, M. (2006), 'Access to the safe harbor: bioterrorism, influenza, and the Supreme Court's interpretation of the research exemption from patent infringement', *Journal of Intellectual Property Law*, **13**, 375–404.

Brogan, J. (2005), 'Federal Court rules gene fragments not patentable', Cooley Alert, http://www.cooley.com/files/tbl_s 24News%5CPDFUpload152%5C1646%5CALERT_GeneFragPatents.pdf.

Brookes, Martin (2002), *Fly: The Unsung Hero of 20th Century Science*, New York: Ecco.

Brown, N., A. Nelis, B. Rappert and A. Webster (1999), 'Bioinformatics: a technology assessment of recent developments in bioinformatics and related areas of research and development including high-throughput screening and combinational chemistry', Final Report for the Science and Technological Options Assessment Unit, European Parliament.

Bruce, A. (2002), 'The search for truth and freedom: ethical issues surrounding human cloning and stem cell research', *Journal Of Law And Medicine*, **9**(3), 323–35.

Buff, E. and L. Restaino (1999), 'State street alters landscape of biotechnology and bioinformatics', *New Jersey Law Journal*, **157**(4), http://www.riker.com/articles/index.php?id=3282.

Burk, D. and M. Lemley (2002), 'Is patent law technology-specific?', *Berkeley Technology Law Journal*, **17**, 1155–1202.

Burk, D. and M. Lemley (2003), 'Policy levers in patent law', *Virginia Law Review*, **89**, 1575–1696.

Burrell, R. and S. Hubicki (2005), 'Patent law and genetic drift: *Schmeiser v Monsanto Canada Inc*', *Environmental Law Review*, **7**(3), 278–98.

Burton, K. and J. Varghese (2004), 'The PBS and the Australia–US Free Trade Agreement', Parliament of Australia Research Note no. 3 2004–05, 22 July.

Bush, G.W. (2006), 'Message to the House of Representatives: H.R. 810, the Stem Cell Research Enhancement Act of 2005', The White House, http://www.whitehouse.gov/news/releases/2006/07/20060719-5.html, 19 July.

Cambrosio, Alberto and Peter Keating (1998), 'Monoclonal antibodies: from local to extended networks', in Arnold Thackray (ed.), *Private Science: Biotechnology and the Rise of the Molecular Sciences*, Philadelphia: University of Pennsylvania Press, pp. 165–81.

Campbell, E., B. Clarridge, M. Gokhale, L. Birenbaum, S. Hilgartner, N. Holtzman and D. Blumenthal (2002), 'Data withholding in academic genetics: evidence from a national survey', *Journal of the American Medical Association*, **287**(4), 473–80.

Canadian Biotechnology Advisory Committee (2002), *Patenting Of Higher Life Forms*, Ottawa: Canadian Biotechnology Advisory Committee, June.

Canadian Biotechnology Advisory Committee (2003), *Advisory Memorandum: Higher Life Forms And The Patent Act*, Ottawa: Canadian Biotechnology Advisory Committee, 24 February.

Canadian Biotechnology Advisory Committee (2006), *Human Genetic Materials, Intellectual Property and the Health Sector*, Ottawa: Canadian Biotechnology Advisory Committee, http://cbac-cccb.ca/epic/internet/incbac-cccb.nsf/en/ah00578e.html.

Canadian Broadcasting Corporation (2000), 'The impact of gene patents on health care and medical research: the case of breast cancer genetic screening', 21 March.

Canadian Canola Growers Association (2003), 'Canadian Canola Growers Association standing up for the interests of Canadian canola growers', *Seed Quest*, http://www.seedquest.com/News/releases/2003/december/7224.htm, 8 December.

Casell, J. (2001), 'Lengthening the stem: allowing federal funded researchers to derive human pluripotent stem cells from embryos', *University of Michigan Journal of Law Reform*, **34**(3), 547–72.

Caulfield, T. (2000), 'Underwhelmed: hyperbole, regulatory policy, and the genetic revolution', *McGill Law Journal*, **45**, 437–60.

Caulfield, T. (2005), 'Policy conflicts: gene patents and health care in Canada', *Community Genetics*, **8**, 223–7.

Caulfield, Tim and Bryn Williams-Jones (eds) (1999), *The Commercialization of Genetic Research: Ethical, Legal, and Policy Issues*, New York: Kluwer Academic.

Caulfield, T., R.C. Cook-Deegan, F.C. Kieff and J. Walsh (2006), 'Evidence and anecdotes: an analysis of human gene patenting controversies', *Nature Biotechnology*, **24**(9), 1091–4 at 1093.

Celera Genomics, http://www.celera.com/.

Celltech Group (2003), 'PLC Annual General Meeting statement', 22 May.

Centre for Intellectual Property Policy (2005), *Genetic Patents and Health Care in Canada: An International Comparison of Patent Regimes of Canada and its Major Trading Partners*, Montreal: McGill Centre for

Intellectual Property Policy, http://www.cipp.mcgill.ca/data/publications/00000015.pdf.

Chakrabarty, Ananda (2002), 'Patenting of life forms: from a concept to reality', in David Magnus, Arthur Caplan and Glenn McGee (eds), *Who Owns Life?*, Amherst, NY: Prometheus Books, pp. 17–24.

Chakrabarty, Ananda (2003), 'Patenting life forms: yesterday, today, and tomorrow', in F. Kieff Scott (ed.), *Perspectives on Properties of the Human Genome Project*, Amsterdam: Elsevier, pp. 3–11.

Chalmers, R. (2006), 'Evergreen or deciduous? Australian trends in relation to the "evergreening" of patents', *Melbourne University Law Review*, **30**, 29–61.

Chan, K. and D. Fernandez (2003), 'Patents in proteomics: possibilities and precautions', *Biotechnology Law Report*, **22**(3), 273–8.

Check, E. (2006), 'Schatten in the spotlight', *Nature News*, http://www.nature.com/news/2006/060109/full/060109-7.html, 11 January.

Chick, E. (2003), 'Biotech critic tries to sew up research on chimaeras', *Nature*, **421**, 4.

Cho, M., S. Illangasekare, M. Weaver, D. Leonard and J. Merz (2003), 'Effects of patents and licenses on the provision of clinical genetic testing services', *Journal of Molecular Diagnostics*, **5**(1), 3–8 at 3–4.

Christensen, I. (2003), 'A letter to David Sage, Chief Medical Officer of the Auckland District Health Board', available at Australian Broadcasting Corporation, http://www.abc.net.au/4corners/content/2003/20030811_patent/documents/page1.htm, 23 May, p. 1.

Christie, A. and N. Peace (1996), 'Intellectual property protection for the products of animal breeding', *European Intellectual Property Review*, **18**(4), 213–33.

Clark, A. (2001), 'Glycosciences makes "land grab" for proteins', *The Guardian*, 16 March.

Clinton, W. and T. Blair (2000), 'Joint statement by President William Clinton and Prime Minister Tony Blair of the United Kingdom', http://ipmall.info/hosted_resources/ippresdocs/ippd_44.htm, 14 March.

Cockburn, Iain (2005), 'State street meets the Human Genome Project: intellectual property and bioinformatics', in Robert Hahn (ed.), *Intellectual Property Rights in Frontier Industries: Software and Biotechnology*, AEI–Brookings Joint Center for Regulatory Studies, pp. 109–30.

Collins, F. (2000), 'Mapping the genome', Online Newshour with Jim Lehrer, http://www.pbs.org/newshour/bb/health/jan-june00/extended_collins.html.

Collins, F. (2001), 'Remarks at the press conference announcing sequencing and analysis of the human genome', http://www.genome.gov/10001379, 12 February.

Collins, F. (2003), 'A patent's place', *Bio-IT World*, http://www.bio-itworld.com/archive/081303/horizons_aussie_sidebar_1.html, 13 August.

Community Affairs Legislation Committee (2002), *Provisions of the Research Involving Embryos and Prohibition of Human Cloning Bill 2002: Supplementary Report*, Canberra: Australian Parliament.

Consumer Project on Technology, http://www.cptech.org/.

Cook, T. (2006), 'Responding to concerns about the scope of the defence from patent infringement for acts done for experimental purposes relating to the subject matter of the invention', *Intellectual Property Quarterly*, 3, 193–222.

Cook-Deegan, Robert (1994), *The Gene Wars: Science, Politics and the Human Genome*, New York and London: WW Norton & Company.

Cook-Deegan, R. (2001), 'Hype and hope', *American Scientist*, **89**, 62.

Cooper Dreyfuss, Rochelle (2003), 'Varying the course in patenting genetic material: a counter-proposal to Richard Epstein's steady course', in F. Scott Kieff (ed.), *Perspectives on Properties of the Human Genome Project*, Amsterdam: Academic Press, Elsevier, pp. 195–208.

Cooper Dreyfuss, Rochelle, Harry First and Dianne Zimmerman (eds) (2001), *Expanding the Bounds of Intellectual Property: Innovation Policy for the Knowledge Society*, Oxford: Oxford University Press.

Cornish, W. (1998), 'Experimental use of patented inventions in European states', *International Review of Industrial Property and Copyright Law*, **29**, 735–53.

Cornish, William and David Llewelyn (2003), *Intellectual Property: Patents, Copyright, Trade Marks and Allied Rights*, 5th edn, London: Sweet and Maxwell.

Cornish, William, Margaret Llewelyn and Mike Adcock (2003), *Intellectual Property Rights and Genetics: A Study into the Impact and Management of Intellectual Property Rights within the Healthcare Sector*, Cambridge: Public Health Genetics Unit, http://www.phgu.org.uk/pages/work/IP.htm.

Coulter, J. (1990), 'Patents Bill 1990 (Cth), in committee', Senate Hansard, 20 September, p. 2653.

Crespi, S. (2001/2002), 'Patenting and ethics – a dubious connection', *Bio-Science Law Review*, **5**(3), 71–8.

Crespi, S. (2006), 'The human embryo and patent law: a major challenge ahead', *European Intellectual Property Review*, **28**(11), 569–75.

Crichton, M. (2007), 'Patenting life', *The New York Times*, 13 February.

Cronin, D. (2006), 'Stem cell experts warn of alarmists', *The Canberra Times*, 23 September, 13.

CSIRO (1997), 'Australian scientists create a nanomachine', Press Release, 5 June.

Curtis, George (1873), *A Treatise on the Law of Patents for Useful Inventions* §124, 4th edn.
Dam, K. (1994), 'The economic underpinnings of patent law', *Journal of Legal Studies*, **23**, 247–270.
Davies, Kevin (2001), *The Sequence: Inside The Race For The Human Genome*, London: Weidenfeld & Nicolson.
Davies, Kevin and Michael White (1995), *Breakthrough: The Quest to Isolate the Gene for Hereditary Breast Cancer*, London: Macmillan Press.
Davis, P., J. Kelley, S. Caltrider and S. Heinig (2005), 'ESTs stumble at the utility threshold', *Nature Biotechnology*, **23**, 1227–9.
deBeer, J. (2005), 'Reconciling property rights in plants', *Journal of World Intellectual Property*, **8**(1), 5–31.
deBeer, J. (2007), 'The rights *and* responsibilities of ag-biotech patent owners', *The University of British Columbia Law Review*, **40**(1), 343–73.
Department of Molecular Biotechnology, the University of Washington, http://www.gs.washington.edu/.
Dickinson, T. (2000), 'Statement of Todd Dickinson before the Subcommittee on Courts and Intellectual Property of the House Committee on the Judiciary', http://www.house.gov/judiciary/dick0713.htm, 13 July.
Dickson, D. (1998), 'Legal fight looms over patent bid on human/animal chimaeras', *Nature*, **392**, 423.
Dinwoodie, Graeme and Rochelle Cooper Dreyfuss (2006), 'Patenting science: protecting the domain of accessible knowledge', in Lucie Guibault and P. Bernt Hugenholtz (eds), *The Future of the Public Domain: Identifying the Commons in Information Law*, Netherlands: Kluwer Law International, pp. 191–221.
Donley, B. (2003), 'Using patented materials in your research: what you should know', Wisconsin Alumni Research Foundation, 21 May.
Drahos, P. (1999), 'Biotechnology patents, markets and morality', *European Intellectual Property Review*, **21**(9), 441–9.
Drahos, P., B. Lokuge, T. Faunce, M. Goddard and D. Henry (2004), 'Pharmaceuticals, intellectual property and free trade: the case of the US–Australia Free Trade Agreement', *Prometheus*, **22**(3), 243–57.
Drahos, Peter (1994), 'Decentring communication: the dark side of intellectual property', in Tom Campbell and Wojciech Sadurski (eds), *Freedom Of Communication*, Aldershot: Dartmouth, pp. 249–79.
Drahos, Peter and John Braithwaite (2002), *Information Feudalism*, London: Earthscan Publications.
Drexler, E. (2004), 'Nanotechnology: from Feynman to funding', *Bulletin of Science, Technology & Society*, **24**(1), 21–7.

Duke University (2003), 'Duke University statement on Supreme Court action in case involving academic research', 27 May.
Durman, P. (2000), 'Oxford Glycosciences protein patent', *The Times*, 26 February, p. 26.
Dutfield, Graham (2003), *Intellectual Property Rights and the Life Science Industries*, Aldershot: Ashgate Publishing Limited.
Dutfield, G. (2003), 'Should we terminate terminator technology?', *European Intellectual Property Review*, **25**(11), 491–5.
Ebert, L. (2006), 'Lessons to be learned from the Hwang matter', *Journal of the Patent and Trademark Office Society*, **88**, 239–55.
Edelman, Bernard (1998), 'Vers Une Approche Juridique Du Vivant', in Bernard Edelman and Marie-Angele Hermitte (eds), *L'Homme, La Nature Et Le Droit*, Paris: Christian Bourgois, pp. 28–9, trans. John Frow (1997), *Time And Commodity Culture: Essays in Cultural Theory and Postmodernity*, Oxford: Oxford University Press, pp. 195–7.
Edelman, B. (2001), 'International symposium on ethics, intellectual property, and genomics', February as reported by the rapporteur M. Kirby, 'Intellectual property and the human genome', *Australian Intellectual Property Journal*, **12**, 61–81.
Edelman, Bernard and Marie-Angele Hermitte (eds) (1988), *L'Homme, La Nature ét Le Driot*, Paris: Christian Bourgois.
Editorial (1997), 'Constraints imposed in breast cancer gene patents', *Nature*, 7 November.
Editorial (1999), 'The promise of proteomics', *Nature*, **402**, 703.
Editorial (2003), 'Patenting pieces of people', *Nature Biotechnology*, **21**, 341, 1 April.
Editorial (2005), 'Hybrid too human to patent: case highlights lacks of criterion for genetically modified organisms', *Nature Review Drug Discovery*, http://www.nature.com/news/2005/050328/full/nrd1710.html, 31 March.
Edson, Margaret (1998), *Wit: A Play*, New York: Farrar, Straus and Giroux.
Edwards, B. (2001), ' "And on his farm he had a geep": patenting transgenic animals', *Minnesota Intellectual Property Review*, **2**, 89–118.
Einstein, Albert (1972), 'Letter to Michele Besso: 12 December 1919, in Pierre Speziali (ed.), 'Michele Besso, Correspondence 1903–1955', Paris: Hermann, pp. 147–9.
Eisenberg, R. (1987), 'Proprietary rights and the norms of science in biotechnology research', *Yale Law Journal*, **97**, 177–223.
Eisenberg, R. (2000), 'Genomics in the public domain: strategy and policy', *Nature Genetics*, **1**(1), 70.
Eisenberg, R. (2000), 'Re-examining the role of patents in appropriating the value of DNA sequences', *Emory Law Journal*, **49**, 783–800.

Eisenberg, R. (2002), 'Molecules vs. information: should patents protect both?', *Boston University Journal of Science and Technology Law*, **8**, 190–217.

Eisenberg, R. (2003), 'Patent swords and shields', *Science*, **299**, 1018–19.

Eisenberg, R. (2006), 'Biotech patents: looking backward while moving forward', *Nature Biotechnology*, **24**(3), 317–19.

Eisenberg, Rebecca (2006), 'The story of *Diamond v Chakrabarty*: technological change and the subject matter boundaries of the patent system', in Jane Ginsburg and Rochelle Cooper Dreyfuss (eds), *Intellectual Property Stories*, New York: Foundation Press, pp. 327–57.

Eisenberg, R. and A. Rai (2006), 'Harnessing and sharing the benefits of state-sponsored research: intellectual property rights and data sharing in California's stem cell initiative', *Berkeley Technology Law Journal*, **21**, 1187–1212.

Elliot, M. (1995), 'NIH gets a share of BRCA1 patent', *Science*, **267**, 1086.

Ellinson, D. (1988), 'The patent system – time to reflect', *Law Institute Journal*, 292–3.

Enerson, B. (2004), 'Protecting society from patently offensive inventions: the risk of reviving the moral utility doctrine', *Cornell Law Review*, **89**, 685–720.

Epstein, R. (2002), 'If it ain't broke', *FT.Com*, 2 July http://www.law.uchicago.edu/news/epstein-genome.html.

Epstein, Richard (2003), 'Steady the course: property rights in genetic material', in F. Scott Kieff (ed.), *Perspectives on Properties of the Human Genome Project*, Amsterdam: Academic Press, Elsevier, pp. 153–94.

ETC Group, http://www.etcgroup.org/en/.

ETC Group (2002), 'Patenting elements of nature', http://www.etcgroup.org/upload/publication/220/01/nanopatentsgeno.rtf.pdf, 25 March.

ETC Group (2003), *The Big Down: From Genomes to Atoms; Atomtech: Technologies Converging at the Nano-Scale*, http://www.etcgroup.org/documents/TheBigDown.pdf, 30 January.

ETC Group (2004), 'Playing God in the Galapagos: J. Craig Venter, master and commander of genomics, on global expedition to collect microbial diversity for engineering life', *Communique*, **84**, March/April.

ETC Group and the coalition against biopiracy (2006), 'Captain Hook and Cog Awards announced in Curitiba', http://www.etcgroup.org/en/materials/publications.html?id=20, 24 March.

European Commission (2007), 'Nanotechnology', http://cordis.europa.eu/nanotechnology/.

European Group on ethics in science and new technologies to the European Commission (2002), *Opinion on the Ethical Aspects of Patenting Inventions Involving Human Stem Cells*, Opinion Number 16, 7 May.

European Parliament (2000), 'Resolution on the decision by the European Patent Office with regard to Patent No EP 695351 granted on 8 December 1999', (Document BS-0288, issued on 30 March 2000, Official Journal EC-L- 29 December 2000, 378/95).

European Parliament (2001), 'European Parliament resolution on the patenting of BRCA1 and BRCA2 ("Breast Cancer") Genes', texts adopted by Parliament, Provisional Edition: 04/10/2001, B5-0633, 0641, 0651 and 0663/2001.

European Patent Office (2002), ' "Edinburgh" patent limited after European Patent Office opposition hearing', Press Release, Munich, 24 July.

European Society of Human Genetics (2005), 'EPO upholds limited patent on BRCA2 gene: singling out an ethnic group is a "dangerous precedent" says European Society of Human Genetics', http://www.eshg.org/ESHGPressRelease01July2005.pdf, 1 July.

Ezzell, C. (2002), 'Proteins rule', *Scientific American*, April, 41–7.

Faunce, T. and P. Drahos (1998), 'Trade related aspects of intellectual property rights (TRIPS) and the threat to patients: a plea for doctors to respond internationally', *Medicine And Law*, **17**, 299–310.

Federal Trade Commission (2003), *To Promote Innovation: The Proper Balance of Competition and Patent Law and Policy*, Washington, DC: Federal Trade Commission, http://www.ftc.gov/os/2003/10/innovationrpt.pdf, October.

Federico, P.J. (1951), 'Hearings on H.R. 3760 before Subcommittee No. 3 of the House Committee on the Judiciary', 82d Congress, 1st Session, 37.

Felten, E. 'Freedom to tinker', http://www.freedom-to-tinker.com/.

Ferranti, M. (2002), 'Confirmant set to release Protein Atlas', *Bio-IT Worlds News*, http://www.bio-itworld.com/news/040702_report184.html, 4 July.

Fitzgerald, B. (2001), 'Case comment: *Grain Pool Of WA v The Commonwealth*, Australian constitutional limits of intellectual property rights', *European Intellectual Property Review*, **23**(2), 103.

Formanek, Jr., R. (2005), 'Proteomics: moving beyond the human genome', *FDA Consumer*, **39**(6), 22–5, http://www.fda.gov/fdac/features/2005/605_proteomics.html.

Fowler, C. (2000), 'The Plant Patent Act of 1930: a sociological history of its creation', *Journal of the Patent and Trademark Office Society*, **82**(9), 621–44.

Fox Keller, Evelyn (1995), *Refiguring Life: Metaphors of Twentieth-Century Biology*, New York: Columbia University Press.

Fox Keller, Evelyn (2000), *The Century of the Gene*, Cambridge, Mass.: Harvard University Press.

Fox Keller, Evelyn (2002), *Making Sense of Life: Explaining Biological Development with Models, Metaphors, and Machines*, Cambridge, Mass.: Harvard University Press.

Freschi, G. (2005), 'Navigating the research exemption's safe harbor: Supreme Court to clarify Scope implications for stem cell research in California', *Santa Clara Computer and High Technology Law Journal*, **21**, 855–99.

Frow, J. (1994), 'Timeshift: technologies of reproduction and intellectual property', *Economy and Society*, **23**, 290.

Frow, John (1997), *Time and Commodity Culture: Essays in Cultural Theory and Postmodernity*, Oxford: Oxford University Press.

Fukuyama, Francis (2002), *Our Posthuman Future: Consequences of the Biotechnology Revolution*, London: Profile Books.

Furphy, Joseph (1903), *Such is Life: Being the Diary of Certain Extracts from the Diary of Tom Collins*, Melbourne: Oxford University Press.

Gad, S., M. Scheuner, S. Pages-Berhouet, V. Caux Moncoutier, A. Bensiman, A. Aurias, M. Pinto and D. Stoppa-Lyonnet (2001), 'Identification of a large rearrangement of the BRCA1 gene using colour bar code on combed DNA in an American breast/ovarian cancer family previously studied by direct sequencing', *Journal of Medical Genetics*, **38**(6), 388.

Gaglioti, F. (2000), 'Wall Street and the commercial exploitation of the human genome', http://www.wsws.org/articles/2000/apr 2000/gene-a10. shtml, 10 April.

Gajarsa, A. (2002), 'Quo vadis? The fifth annual Honorable Helen Wilson Nies Memorial lecture in intellectual property law', *Marquette Intellectual Property Law Review*, **6**, 1–9.

Gajarsa, A. and L. Cogswell (2006), 'A review of recent decisions of the United States Court of Appeals for the Federal Circuit', *American University Law Review*, **55**, 821–44.

Galison, Peter (1992), 'The many faces of big science', in Peter Galison and Bruce Hevly (eds), *Big Science: The Growth of Large-Scale Research*, Stanford, Calif.: Stanford University Press.

Galison, Peter and Bruce Hevly (eds) (1992), *Big Science: The Growth of Large-Scale Research*, Stanford, Calif.: Stanford University Press.

Garforth, K. (2005), 'Health care and access to patented technologies', *Health Law Journal*, **13**, 77–97.

Gavaghan, H. (2000), 'Companies of all sizes are prospecting for proteins', *Nature*, **404**, 684–5.

Geisel, J., B. Weisshaar, K.Oette and W. Doerfler (1988), 'A new APA LI restriction fragment length polymorphism in the Low 28 Density Lipoprotein Receptor Gene', *Journal of Clinical Chemistry and Clinical Biochemistry*, **26**(7), 429.

Gene Technology Regulator, http://www.ogtr.gov.au/.

Genetic Technologies Limited, 'Intellectual property', http://www.gtg.com.au/index.asp?menuid=060.070.190.010.

Genetic Technologies Limited (2002), 'Genetic Technologies and Myriad Genetics announce strategic licensing agreement', 28 October.

Genetic Technologies Limited (2003), 'Corporate details', http://web.archive.org/web/20030207163028/www.gtg.com.au/CorpDetails.html, February.

Genetic Technologies Limited (2003), 'Genetic susceptibility testing–third progress report', Australian Stock Exchange Company Announcements, 22 May.

Genetic Technologies Limited (2004), 'GTG reports growth in revenue from genetic testing', http://www.gtg.com.au/index.asp?menuid=060.070.130.010&artid=155, 23 July.

Genetic Technologies Limited (2004), 'GTG reports on the current status of legal action to protect its patents', http://www.gtg.com.au/index.asp?menuid=060.070.130.010&artid=158, 3 September.

Genetic Technologies Limited (2004), 'John Dawkins joins GTG Board of Directors', http://www.gtg.com.au/index.asp?menuid=060.070.130.010&artid=174, 25 November.

Genetic Technologies Limited (2004), 'GTG supports patent protection', http://www.gtg.com.au/index.asp?menuid=060.070.200.020.010&artid=205, 19 August.

Genetic Technologies Limited (2005), *Annual Report 2005*, http://esvc001057.wic005u.server-web.com/archives/1/070.130/822/Release%20of%202005%20Annual%20Report.pdf.

Genetic Technologies Limited (2005), 'GTG celebrates NASDAQ listing with the ringing of the Stock Market Closing Bell', http://www.gtg.com.au/index.asp?menuid=060.070.130&artid=292&function=NewsArticle, 7 November.

Genetic Technologies Limited (2005), 'GTG provides further details of the settlement with Applera', http://www.gtg.com.au/index.asp?menuid=060.070.130&artid=264&function=NewsArticle, 15 December.

Genetic Technologies Limited (2005), 'GTG reports final Settlement of its patent dispute with Applera Corporation', http://www.gtg.com.au/index.asp?menuid=060.070.130&artid=262&function=NewsArticle, 12 December.

Genetic Technologies Limited (2005), 'Mediation in New Zealand results in final Settlement of legal action', 7 July.

Genetic Technologies Limited (2006), 'GTG reports final dismissal of Applera suit, and plans for the future', http://www.gtg.com.au/index.asp?menuid=060.070.130&artid=266&function=NewsArticle, 4 January.

Geron Corporation (2001), 'Geron reports issuance of U.S. patent for human embryonic stem cells', Menlo Park, California, 13 March.

Gervais, Daniel (2003), *The TRIPS Agreement: Drafting History and Analysis*, 2nd edn, London: Sweet and Maxwell.

Gervais, Daniel and Elizabeth Judge (2005), *Intellectual Property: The Law in Canada*, Toronto: Thomson Carswell.

Gibson, Johanna, 'Patenting Lives', http://www.patentinglives.org/ and http://patentinglives.blogspot.com/.

Gifford, B. (2006), 'Oh, Diehr: the CAFC's troubling patent eligibility jurisprudence as applied in Metabolite Laboratories v Labcorp', *Biotechnology Law Report*, **25**, 129–46.

Ginsburg, Jane and Rochelle Cooper Dreyfuss (eds) (2006), *Intellectual Property Stories*, New York: Foundation Press.

Gitter, D. (2001), 'International conflicts over patenting human DNA sequences in the United States and the European Union: an argument for compulsory licensing and a fair use exemption', *New York University Law Review*, **76**, 1623–91.

Gitter, D. (2004), 'Ownership of human tissue: a proposal for federal recognition of human research participants' property rights in their biological material', *Washington and Lee Law Review*, **61**(1), 257–346.

Goddard, H. (2002), 'The experimental use exception: a European perspective', Center for Advanced Studies and Research on Intellectual Property, University of Washington, Seattle, **7**, http://www.law.washington.edu/casrip/Symposium/Number 7/1-Goddar.pdf.

Gogarty, B. and D. Nicol (2002), 'The UK's cloning laws, a view from the Antipodes', *Murdoch University Electronic Journal of Law*, **9**(2), http://www.murdoch.edu.au/elaw/issues/v9n2/gogarty92.html.

Gold, E.R. (2000), 'Biomedical patents and ethics: a Canadian solution', *McGill Law Journal*, **45**, 413–35.

Gold, E.R. (2003–04), 'The reach of patent law and institutional competence', *The University of Ottawa Law and Technology Journal*, **1**, 263–84.

Gold, E. Richard (1996), *Body Parts: Property Rights and the Ownership of Human Biological Materials*, Washington, DC: Georgetown University Press.

Gold, E. Richard (1999), 'Making room: reintegrating basic research, health policy, and ethics into patent law', in Tim Caulfield and Bryn Williams-Jones (eds), *The Commercialization of Genetic Research: Ethical, Legal, and Policy Issues*, New York: Kluwer Academic, p. 63.

Gold, E.R. and T. Caulfield (2002), 'The moral tollbooth: a method that makes use of the patent system to address ethical concerns in biotechnology', *The Lancet*, **359**, 2268–70.

Gold, E.R. and K. Durell (2005), 'Innovating the skilled reader: tailoring patents to new technologies', *Intellectual Property Journal*, **19**(1), 189–226.

Gorman, P. (2004), 'Legal action against GTG', *Christchurch Press*, 19 August.

Gorman, P. (2004), 'NZ health boards in patent dispute', *Christchurch Press*, 21 August, p. 3.

Gorman, P. (2005), 'Mediation causes row', *Christchurch Press*, 26 March, p. 8.

Gratton, B. (2002), 'The Wisconsin Alumni Research Foundation and Geron Corporation', in The European Group On Ethics In Science And New Technologies To The European Commission, *Opinion On The Ethical Aspects Of Patenting Inventions Involving Human Stem Cells*, Opinion Number 16, 7 May, p. 61.

Greene, R. (1989), 'Administration opposes farmer exemption', The Associated Press, 14 September.

Greenpeace, 'Patents on life', http://www.greenpeace.org/international/campaigns/genetic-engineering/ge-agriculture-and-genetic-pol/patents-on-life.

Grisham, J. (2000), 'New rules for gene patents', *Nature Biotechnology*, **18**, 921.

Grubb, Philip (2004), *Patents for Chemicals, Pharmaceuticals and Biotechnology*, 4th edn, Oxford: Oxford University Press.

Guardiola, J., A. Maffei, S. Carrel and R. Accolla (1988), 'Molecular genotyping of the HLA-DQ [α] gene region', *Immunogenetics*, **27**(1), 12.

Guibault, Lucie and P. Bernt Hugenholtz (eds) (2006), *The Future of the Public Domain: Identifying the Commons in Information Law*, Netherlands: Kluwer Law International.

Hagelin, T. (2006), 'The experimental use exemption to patent infringement: information on ice, competition on hold', *Florida Law Review*, **58**, 483–560.

Hahn, Robert (ed.) (2005), *Intellectual Property Rights in Frontier Industries*, Baltimore, MD: AEI-Brookings Press.

Hahn, Robert (2005), 'Introduction' in Robert Hahn (ed.), *Intellectual Property Rights in Frontier Industries: Software and Biotechnology*, AEI-Brookings Joint Center for Regulatory Studies, pp. 1–10.

Hall, J.M., M. Lee, B. Newman, J. Morrow, L. Anderson, B. Huey and M.C. King (1990), 'Linkage of early-onset familial breast cancer to chromosome 17q21', *Science*, **250**, 1684–9.

Hall, Stephen (1987), *Invisible Frontiers: The Race To Synthesize A Human Gene*, London: Sidgwick and Jackson.

Hamson, C.J. (1930), *Patent Rights for Scientific Discoveries*, Indianapolis: Bobbs-Merrill.

Haplomics, http://www.haplomics.com/The%20Solution.htm.

Haraway, Donna (1997), *Modest Witness, Second Millennium: FemaleMan Meets OncoMouse. Feminism And Technoscience*, New York: Routledge.

Hareid, J. (2006), 'Testing drugs and testing limits: *Merck KGAA v Integra Lifesciences I, Ltd.* and the scope of the Hatch–Waxman safe harbor provision', *Minnesota Journal of Law, Science & Technology*, **7**, 713–56.

Harradine, B. (1990), 'Patents Bill 1990 (Cth), in committee', Senate Hansard, 21 September, p. 2564.

Harvard College (2002), 'Statement regarding Canadian Supreme Court 5–4 Decision Dec. 5, 2002 denying the patentability of "Oncomouse" in Canada', Boston, 5 December.

Hayes, L. and N. Greenaway (2003), 'Profit motive', 60 Minutes, Channel Nine, http://sixtyminutes.ninemsn.com.au/sixtyminutes/stories/2003_04_20/story_806.asp, 20 April.

Heller, M. and R. Eisenberg (1998), 'Can patents deter innovation? The anticommons in biomedical research', *Science*, **280**, 698–701.

Hellman, Hall (2001), *Great Feuds in Medicine: Ten of the Liveliest Disputes Ever*, New York: John Wiley & Sons.

Helm, K. (2006), 'Outsourcing the fire of genius: the effects of patent infringement jurisprudence on pharmaceutical drug development', *Fordham Intellectual Property, Media and Entertainment Law Journal*, **17**, 153–206.

Herder, M. (2006), 'Proliferating patent problems with human embryonic stem cell research', *Bioethical Inquiry*, **3**, 69–79.

Hindmarsh, Richard and Geoffrey Lawrence (eds) (2001), *Altered Genes II: The Future?*, Melbourne: Scribe Publications.

Holmes, J. (2003), 'Interview with Dr Mervyn Jacobson', http://www.abc.net.au/4corners/content/2003/20030811_patent/int_jacobson.htm, 11 August.

Holmes, J. (2003), 'Interview with Professor John Mattick', http://www.abc.net.au/4corners/content/2003/20030811_patent/int_mattick.htm, 10 July.

Holmes, J. (2003), 'Interview with Sir John Sulston', http://www.abc.net.au/4corners/content/2003/20030811_patent/int_sulston.htm, 9 July.

Holmes, J. (2003), 'Patently a problem', *Four Corners*, Australian Broadcasting Corporation, transcript, http://www.abc.net.au/4corners/content/2003/transcripts/s 922059.htm, 11 August.

House Committee on the Judiciary (1987), *Patents and the Constitution: Transgenic Animals*, Hearings before the Subcommittee on Courts, Civil Liberties, and the Administration of Justice, 100th Congress, 1st Session.

House of Commons (1989), *Minutes of Proceedings and Evidence of the Legislative Committee on Bill C-15, An Act Respecting Plant Breeders' Rights*, **1**, 11 October, p. 1115.

House of Representatives Standing Committee on Legal and Constitutional Affairs (2001), *Human Cloning: Scientific, Ethical and Regulatory Aspects of Human Cloning and Stem Cell Research*, Canberra: Australian Parliament.

House Subcommittee on the Judiciary, Subcommittee on Courts, Intellectual Property, and the Administration of Justice (1989), *Transgenic Animal Patent Reform Act* of 1989, HR 1556, 101st Congress, 1st Session.

Human Genetics Society of Australasia (2001), 'HGSA position paper on the patenting of genes', 3.6.

Human Genome Organisation Intellectual Property Committee (1997), *Statement on Patenting Issues Related to the Early Release of Sequence Data*, http://www.hugo-international.org/PDFs/Statement%20on%20Patenting%20Issues%20Relating%20to%20Raw%20Sequence%20Dat.pdf, May.

Human Genome Organisation Intellectual Property Committee (2000), *Statement on Patenting DNA Sequences In Particular Response to the European Biotechnology Directive*, http://www.hugo-international.org/PDFs/Statement%20on%20Patenting%20of%20DNA%20Sequences%202000.pdf, April.

Hunt, Geoffrey and Michael Mehta (eds) (2006), *Nanotechnology: Risk, Ethics, Law*, London: Earthscan Publications.

Hwang, W-S. et al. (2004), 'Evidence of a pluripotent human embryonic stem cell line derived from a cloned blastocyst', *Science*, **303**, 1669–70.

Hwang, W-S. et al. (2005), 'Patient-specific embryonic stem cells derived from human SCNT blastocysts', *Science*, **308**, 1777.

IBM, 'Blue Gene', http://domino.research.ibm.com/comm/research_projects.nsf/pages/bluegene.index.html.

IBM (1999), 'IBM announces $100 million research initiative to build world's fastest supercomputer: Blue Gene to tackle protein folding grand challenge', Yorktown Heights, New York, 6 December.

Imrie, Katherine (2006), 'The Hwang scandal: scientific misconduct and research integrity', honours thesis, ANU College of Law.

Institut Curie, http://www.curie.fr/index.cfm/lang/_gb.htm.

Institut Curie (2000), 'Opposition procedure with the European Patent Office', http://www.curie.net/actualities/myriad/declaration_e.htm, 12 September.

Institut Curie (2001), 'Opposition procedure with the European Patent Office', http://www.curie.net/actualities/myriad/declaration_e.htm, 12 September.

Institut Curie (2001), 'The Institut Curie is initiating an opposition procedure with the European Patent Office', Press Release, 12 September.

Institut Curie (2004), 'The European Patent Office has revoked the Myriad patent', Press Release, http://www.curie.fr/upload/presse/190504_gb.pdf, 21 May.

Institut Curie and others (2002), 'BRCA1 gene-linked forms of breast and/or ovarian cancer: the Institut Curie, the Assistance Publique-Hôpitaux de Paris and the Institut Gustave-Roussy file a joint opposition to a second Myriad Genetics patent', Press Release, 22 February.

Institut Curie and others (2002), 'Against Myriad Genetics' monopoly on tests for predisposition to breast and ovarian cancer associated with the BRCA1 gene: third French opposition', Press Release, 26 September.

Institute for Science, Law and Technology (2006), 'Comments on the interim guidelines for examination of patent applications for patent subject matter eligibility', http://www.uspto.gov/web/offices/pac/dapp/opla/comments/ab98/islat.pdf, 31 July.

The International Human Genome Sequencing Consortium (2001), 'Initial sequencing and analysis of the human genome', *Nature*, **409**, 860–921.

International Union for the Protection of Varieties of Plants, http://www.upov.int/.

IP Australia (1999), *Patent Manual of Practice and Procedure*, November, Part 8.1.

IP Australia (2001), 'House of Representatives Standing Committee On Legal and Constitutional Affairs Inquiry into The Scientific, Ethical And Regulatory Aspects Of Human Cloning', Submission 274, http://www.aph.gov.au/house/committee/laca/humancloning/sub274.pdf.

IP Australia (2003), Chapter 8 of *Patent Manual Of Practice and Procedure*, http://www.ipaustralia.gov.au/pdfs/patents/manual/Part208.PDF, December.

IP Australia (2006), *Patent Manual of Practice and Procedure*, http://www.ipaustralia.gov.au/pdfs/patentsmanual/WebHelp/National/Patentable/2.9.5.1_stem_cells.htm, January.

Jaffe, Adam and Josh Lerner (2004), *Innovation and its Discontents: How our Broken Patent System is Endangering Innovation and Progress, and What to do About It*, Princeton: Princeton University Press.

Janis, M. (2003), 'Experimental use and the shape of patent rights for plant innovation', Economics of Innovation and Science Policy, Department of Economics, Iowa State University, http://www.econ.iastate.edu/department/seminar/ispw/Janis-seminar-Fall-03.pdf.

Janis, M. (2006), 'Rules v standards for patent law in the plant sciences', in M. Rimmer (ed.), *Patent Law and Biological Inventions, Law in Context*, **24**(1), 54–66.

Janis, M. and J. Kesan (2002), 'Intellectual property protection for plant innovation: unresolved issues after JEM v Pioneer', *Nature Biotechnology*, **20**, 1161–65.

Janis, M. and J. Kesan (2002), 'U.S. plant variety protection: sound and fury . . .?', *Houston Law Review*, **39**, 727–78.

Jardine, Lisa (1999), *Ingenious Pursuits: Building the Scientific Revolution*, London: Abacus History.
Jenks, S. (2007), 'Debate grows over patenting of genes', *Florida Today*, 10 March.
Jensen, K. and F. Murray (2005), 'Intellectual property landscape of the human genome', *Science*, **310**, 239–40.
Joly, Y. (2006), 'Wind of change: *In re Fisher* and the evolution of American biotechnology patent law', in M. Rimmer (ed.), *Patent Law and Biological Inventions, Law in Context*, **24**(1), 67–84.
Jones, B. (2006), 'Legislating AgBiotech', ISB News Report, http://www.isb.vt.edu/news/2006/artspdf/sep 0602.pdf (September).
Kan, Y. and A. Dozy (1978), 'Antenatal diagnosis of sickle-cell anaemia by D.N.A. analysis of amniotic-fluid cells', *Lancet*, **2**, 910.
Kan, Y. and A. Dozy (1978), 'Polymorphism of DNA sequence adjacent to human beta-globin structural gene: relationship to sickle mutation', *Proceedings of the National Academy of Sciences*, **75**(11), 5631, http://www.pnas.org/cgi/reprint/75/11/5631.
Kass, L. and M. Nitabach (2002), 'A roadmap for biotechnology patents? Federal Circuit precedent and the PTO's new examination guidelines', *AIPLA Quarterly Journal*, **30**(2), 233.
Kastenmeier, R. (1987), 'Patenting life', *The New York Times*, 26 July.
Kate, Kerry Ten and Sarah Laird (1999), *The Commercial Use of Biodiversity: Access to Genetic Resources and Benefit-Sharing*, London: Earthscan.
Keim, B. (2006), 'Breast cancer research neglects non-Jewish groups, experts charge: patent monopolies have skewed research on breast cancer genetics', *Nature Medicine*, http://www.nature.com/news/2006/061127/full/nm1206-1335a.html, 29 November.
Kevles, D. (1994), 'Ananda Chakrabarty wins a patent: biotechnology, law and society, 1972–1980', *Historical Studies in the Physical and Biological Sciences*, **25**, 111–135.
Kevles, D. (2002), 'A History of Patenting Life in the United States with Comparative Attention to Canada and Europe', European Group on Ethics in Science and New Technologies to the European Commission, 12 January.
Kevles, D. (2002), 'Of mice and money: the story of the world's first animal patent', *Daedalus*, **131**(2), 78–88.
Kevles, Daniel (1998), '*Diamond v Chakrabarty* and beyond', in Arnold Thackray (ed.), *Private Science: Biotechnology and the Rise of the Molecular Sciences*, Philadelphia: University of Pennsylvania Press, pp. 65–79.
Kieff, F.S. (2001), 'Facilitating scientific research: intellectual property rights and the norms of science – a response to Rai and Eisenberg', *Northwestern University Law Review*, **95**, 691–706.

Kieff, F. Scott (ed.) (2003), *Perspectives on Properties of the Human Genome Project*, Amsterdam: Elsevier.

Kienzlen, G. (2005), 'BRCA2 patent upheld', *The Scientist*, **6**(1), http://www.the-scientist.com/news/20050701/01/.

Kincaid, S. (2003), 'Oh, the places you'll go: the implications of current patent law on embryonic stem cell research', *Pepperdine Law Review*, **30**, 553.

Kirby, M. (2000), 'The human genome and patent law', *Reform*, **79**, 10–13.

Kirby, M. (2001), 'Intellectual property and the human genome', *Australian Intellectual Property Journal*, **12**, 61–81.

Kirby, M. (2002), 'Report of the International Bioethics Committee on Ethics, Intellectual Property and Genomics', International Bioethics Committee, 10 January, http://unesdoc.unesco.org/images/0013/001306/130646e.pdf#xml=http://unesdoc.unesco.org/ulis/cgi-bin/ulis.pl?database=ged&set=3F0E5B42_2_12&hits_rec=1&hits_lng=eng.

Kloppenburg, Jack (1988), *First The Seed: The Political Economy of Plant Biotechnology 1492–2000*, Cambridge: Cambridge University Press.

Knoppers, B. (2000), 'Reflections: the challenge of biotechnology and public policy', *McGill Law Journal*, **45**, 559–65.

Koppikar, V., S. Maebius and J.S. Rutt (2004), 'Current trends in nanotech patents: a view from inside the patent office', *Nanotechnology Law and Business*, **1**, 24–30.

Kulinowski, K. (2004), 'Nanotechnology: from "wow" to "yuck"?', *Bulletin of Science, Technology & Society*, **24**(1), 13–20.

Landes, William and Richard Posner (2003), *The Economic Structure of Intellectual Property Law*, Cambridge, Mass and London: Harvard University Press.

Large Scale Biology Corporation, 'Intellectual property', http://www.lsbc.com/ip.html.

Lawrence, S. (2005), 'US court case to define EST patentability', *Nature Biotechnology*, http://www.nature.com/news/2005/050502/full/nbt0505-513.html, 3 May.

Lawson, C. (2002), 'Patenting genes and gene sequences and competition: patenting at the expense of competition', *Federal Law Review*, **30**, 97–133

Lawson, C. and C. Pickering (2004), ' "TRIPs-plus" patent privileges – an intellectual property "Cargo Cult" in Australia', *Prometheus*, **22**(4), 355–77.

Lee, A. (2006), 'Examining the viability of patent pools for the growing nanotechnology patent thicket', *Nanotechnology Law and Business*, **3**, 317–27.

Leese, Mark (1996), 'Is an American mouse a European mouse? Towards a sociology of patents', in Andrew Webster and Kathryn Packer (eds),

Innovation and the Intellectual Property System, London: Kluwer Law International Ltd, p. 171.
Leiss, W. (2002), 'Higher life forms before the law', http://www.leiss.ca/index.php?option=com_content&task=view&id=41&Itemid=43.
Lemley, M. (2005), 'Patenting nanotechnology', *Stanford Law Review*, **58**, 601–30.
Lentz, E. (2004), 'Pharmaceutical and biotechnology research after Integra and Madey', *Biotechnology Law Report*, **23**, 265–76.
Lesavich, S. (2000), 'Bioinformatic tools', *The National Law Journal*, 16 October, B10.
Lesser, W. (1993), 'Animal variety protection: a proposal for a US model law', *Journal of the Patent and Trademark Office Society*, **75**, 398.
Lessig, Lawrence (1999), *Code and Other Laws Of Cyberspace*, New York: Basic Books.
Lewontin, R. (2001), 'They got the wrong key of life', *The Sunday Times*, 8 July.
Lewontin, Richard (2000), *It Ain't Necessarily So: The Dream of the Human Genome and Other Illusions*, New York: New York Review of Books.
Likhovski, M., M. Spence and M. Molineaux (2000), 'The first mover monopoly: a study on patenting business methods in Europe', *Oxford Electronic Journal of Intellectual Property Rights*, November.
Little, P. (2003), 'GTG's inventions concerning "junk" DNA', *Bio-IT World*, http://www.bio-itworld.com/archive/091103/letters.html, 11 September.
Little, P., G. Annison, S. Darling, R. Williamson, L. Camba and B. Modell (1980), 'Model for antenatal diagnosis of beta-thalassaemia and other monogenic disorders by molecular analysis of linked DNA polymorphisms', *Nature*, **283**(5761), 144.
Llewelyn, M. (1997), 'The legal protection of biotechnological inventions: an alternative approach', *European Intellectual Property Review*, **19**(3), 115–27.
Llewelyn, M. (2006), 'European bio-protection laws: rebels with a cause', in M. Rimmer (ed.), *Patent Law and Biological Inventions, Law in Context*, **24**(1), 11–33.
Llewelyn, Margaret and Mike Adcock (2006), *European Plant Intellectual Property*, Oxford: Hart Publishing.
Lopez-Beverage, C. (2005), 'Should Congress do something about upstream clogging caused by the deficient utility of expressed sequence tag patents?', *Journal of Technology Law and Policy*, **10**, 35–92.
Loughlan, P. (1995), 'Of patents and patients: new monopolies in medical methods', *Australian Intellectual Property Journal*, **6**, 5–15.
Ludlam, C. (2000), 'Biotechnology Industry Organization comment on interim utility and written description guidelines', http://www.uspto.gov/web/offices/com/sol/comments/utilitywd/bio.pdf, 22 March.

Ludowyk, F. (ed.) (1997), 'Ozwords: furphy', http://www.anu.edu.au/andc/ozwords/November_97/, November.

Lynn, R. (2006), '*Merck KgAA v Integra Lifesciences I. Ltd.*: judicial expansion of §271(e)(1) signals a need for a broad statutory experimental use exemption in patent law', *Berkeley Technology Law Journal*, **21**, 79–100.

MacLean, D. (2000), 'I am that I am', *Harper's Magazine*, **301**(1803), 18.

Magnus, David, Arthur Caplan and Glenn McGee (eds) (2002), *Who Owns Life?*, Amherst, NY: Prometheus Books.

Mandelker, B. (1998–9), 'Commentary: Harvard College v Canada', *Intellectual Property Journal*, **13**, 87–106.

Marconi and Oxford Glycosciences (2001), 'Marconi and OGS form joint venture to provide proteomics data and hosting solutions: major initiative in pharmaceutical bio-informatics as proteomics and communications converge', 15 June.

Marshall, E. (1997), 'The battle over BRCA1 goes to court; BRCA2 may be next', *Science*, **278**, 1874.

Marshall, E. (2005), 'BRCA2 patent faces new challenge', *Science*, **308**(5730), 1851, 24 June.

Marshall, E., E. Pennisi and L. Roberts (2000), 'In the crossfire: Collins on genomes, patents, and "Rivalry"', *Science*, **287**(5462), 2396–98, 31 March.

Martin, T. (2000), 'Patentability of methods of medical treatment: a comparative study', *Journal of the Patent and Trademark Office Society*, **82**(6), 381–423.

Matthijs, G. (2006), 'The European opposition against the BRCA gene patents', *Familial Cancer*, **5**, 95–102.

McGee, G. (2006), 'Working with stem cells? Pay up', *The Scientist Magazine*, 1 November.

McKeough, Jill and Andrew Stewart (1997), *Intellectual Property in Australia*, 2nd edn, Sydney: Butterworths.

McKeough, Jill, Kathy Bowrey and Philip Griffith (2007), *Intellectual Property: Commentary and Materials*, 4th edn, Sydney: Law Book Co.

McKeough, Jill, Andrew Stewart and Philip Griffith (2004), *Intellectual Property in Australia*, 3rd edn, Sydney: LexisNexis Butterworths.

McLennan, Alison (2007), 'Which bank? A guardian model for regulation of embryonic stem cell research in Australia', *Journal of Law and Medicine*, **15**, 45–76.

McNabb, D. (2005), 'Health boards defy gene test claims', *The Independent Business Weekly*, 13 July.

McPherson, J. (2006), 'The impact of the Hatch–Waxman Act's safe harbor provision on biomedical research tools after *Merck KGAA. v*

Integra Lifesciences I. Ltd', *Michigan State University Journal of Medicine & Law*, **10**, 369–83.

MDS Inc. (2004), 'MDS Inc. announces completion of proteomics reorganization: MDS proteomics renamed Protana Inc.', 29 July.

Médecins Sans Frontières, http://www.accessmed-msf.org/.

Meek, J. (2000), 'Poet attempts the ultimate in self-invention – patenting her own genes', *The Guardian*, 29 February.

Merges, Robert (1996), 'Property rights theory and the commons: the case of scientific research', in Ellen Frankel Paul, Fred Miller and Jeffrey Paul (eds), *Scientific Innovation, Philosophy and Public Policy*, Cambridge: Cambridge University Press, http://www.law.berkeley.edu/institutes/bclt/pubs/merges/rpmart4.pdf.

Merges, R. (1999), 'As many as six impossible patents before breakfast: property rights for business concepts and patent system reform', *Berkeley Technology Law Journal Spring*, **14**(2), 577–615, http://www.law.berkeley.edu/institutes/bclt/pubs/merges/.

Merges, Robert, Peter Mennell and Mark Lemley (2006), *Intellectual Property in the New Technological Age*, Frederick, MD: Aspen Publishers.

Merton, Robert (1973), *The Sociology of Science*, Chicago: University of Chicago Press.

Merz, J. and M. Cho (2005), 'What are gene patents and why are people worried about them?', *Community Genetics*, **8**, 203–8.

Merz, J., A. Kriss, D. Leonard and M. Cho (2002), 'Diagnostic testing fails the test: the pitfalls of patenting are illustrated by the case of Haemochromatosis', *Nature*, **415**, 577–9.

Merz, J., D. Magnus, M. Cho and A. Caplan (2002), 'Protecting subjects' interests in genetics research', *American Journal of Human Genetics*, **70**, 965–71.

Mgbeoji, Ikechi (2006), *Global Biopiracy: Patents, Plants, and Indigenous Knowledge*, Vancouver: University of British Columbia Press.

Migliorini, R. (2006), 'The narrowed experimental use exception to patent infringement and its application to patented computer software', *Journal of the Patent and Trademark Office Society*, **88**, 523–46.

Misrock, L. (2001), 'Genomeweb Q&A: biotech patent guru discusses possible proteomics patent wars', 13 March.

Mohammed, E.A.C. (2004), 'Cat in the hat, a mouse in the house: comparative perspectives on Harvard Mouse', *Intellectual Property Journal*, **18**, 169–85.

Monsanto (2004), 'Supreme Court finds in favor of Monsanto in *Schmeiser v Monsanto* patent infringement case', Press Release, http://www.monsanto.com/monsanto/layout/media/04/05-21-04.asp, 21 May.

Moody, Glyn (2001), *Rebel Code: Linus Torvalds, Open Source, and the War for the Soul of Software*, London: Penguin Books.

Moody, Glyn (2004), *Digital Code of Life: How Bioinformatics is Revolutionizing Science, Medicine, and Business*, Hoboken, NJ: John Wiley & Sons.

Mooney, P.R. (2001), *The Impetus for and Potential of Alternative Mechanisms for the Protection of Biotechnological Innovations*, Ottawa: Canadian Biotechnology Advisory Committee, March.

Moore, S. (2002), 'Challenge to the biotechnology directive', *European Intellectual Property Review*, **24**(3), 149–54.

Mota, S. (2006), '*Merck v Integra Lifesciences* – the Supreme Court protects the use of patented compounds in preclinical studies', *Hamline Law Review*, **29**, 53–62.

Moukheiber, Z. (2003), 'Junkyard Dogs', *Forbes*, http://www.forbes.com/forbes/2003/0929/052_print.html, 29 September, p. 2.

Mueller, J. (2001), 'No "dilettante affair": rethinking the experimental use exception to patent infringement for biomedical research tools', *Washington Law Review*, **76**, 1–66.

Mueller, J. (2004), 'The evanescent experimental use exemption from U.S. patent infringement liability: implications for university/nonprofit research and development', *Baylor Law Review*, **56**, 918–79.

Mueller, Janice (2006), *Introduction to Patent Law*, 2nd edn, Frederick, MD: Aspen Publishers.

Mullally, M.V. and D. Winn (2004), 'Patenting nanotechnology: a unique Challenge to the IP Bar', *The New York Law Journal*, http://www.orrick.com/fileupload/324.pdf, 6 July.

Myerson, George (2000), *Donna Haraway and GM Foods*, Cambridge: Icon Books.

Myriad Genetics (1998), 'Myriad Genetics obtains Oncormed's BRCA1/BRCA2 genetic testing program in patent settlement: OncorMed and Myriad settle BRCA patent disputes', 18 May.

Myriad Genetics (2001), 'Patents issued in Europe, Canada, Australia and New Zealand encourage broad availability of predictive medicine', Salt Lake City, 15 May.

Myriad Genetics (2001), *2001 Annual Report*, Salt Lake City: Myriad Genetics.

NanoBusiness Alliance (2002), 'US patent examiners may not know enough about nanotech', *Small Times*, http://www.smalltimes.com/articles/stm_print_screen.cfm?ARTICLE_ID=267691.

National Academy of Sciences (2004), *A Patent System for the 21st Century*, Washington, DC: National Academy of Sciences, http://books.nap.edu/catalog.php?record_id=10976.

National Institutes of Health, Animal Models, Zebra Fish, http://www.nih.gov/science/models/zebrafish.

National Institutes of Health (2006), 'Comments on the USPTO's interim guidelines for examination of patent applications for patent subject matter eligibility', http://www.uspto.gov/web/offices/pac/dapp/opla/comments/ab98/nih.pdf, 31 July.

National Research Council Committee on Intellectual Property Rights in Genomic and Protein Research and Innovation (2005), *Reaping the Benefits of Genomic and Proteomic Research: Intellectual Property Rights, Innovation and Public Health*, Washington, DC: National Academies Press.

Network of Concerned Farmers, http://www.non-gm-farmers.com/.

Neville, W. (2003), 'Submission to the Australian Law Reform Commission issues paper on gene patenting and human health', 29 October.

Newman, S. (2002), 'Legal column: the human chimera patent initiative', *Lahey Clinic Medical Ethics*, **9**(1), 4, 7, http://www.lahey.org/NewsPubs/Publications/Ethics/JournalWinter2002/Journal_Winter2002_Legal.asp.

Newman, S. (2002), 'The human chimera patent initiative', New York Medical College, http://www.nymc.edu/sanewman/PDFs/Lahey_Winter_2002.pdf.

New Zealand (2000), *Royal Commission on Genetic Modification*, http://www.mfe.govt.nz/issues/organisms/law-changes/commission/.

New Zealand (2002), *Report of the Royal Commission on Genetic Modification*, http://www.mfe.govt.nz/publications/organisms/royal-commission-gm/index.html, Chapter 10, http://www.mfe.govt.nz/publications/organisms/royal-commission-gm/chapter-10.pdf.

New Zealand – Minister of Health and Associate Minister of Commerce (2003), *Implications of the Granting of Patents Over Genetic Material*, Cabinet Policy Committee, available at *Internet Archive*, <http://web.archive.org/web/*/http://www.med.govt.nz/buslt/int_prop/genetic-material/cabinet/implications/implications.pdf, November/.

New Zealand, Ministry of Health and Ministry of Commerce (2004), *Memorandum to Cabinet Policy Committee: Report Back with Recommendations and Options for Addressing Genetic Material Patents*, http://www.med.govt.nz/templates/MultipageDocumentTOC____1148.aspx, 1 May.

New Zealand Herald (2005), 'GTG's bid for health board patent fees fails', *New Zealand Herald*, 8 July.

Nicol, D. (2005), 'Balancing innovation and access to healthcare through the patent system – an Australian Perspective', *Community Genetics*, **8**, 228–34.

Nicol, D. (2005), 'On the legality of gene patents', *The University of Melbourne Law Review*, **29**(3), 809–41.

Nicol, D. and J. Nielsen (2003), 'Patents and medical biotechnology: an empirical analysis of issues facing the Australian industry', Occasional Paper No. 6, Centre for Law and Genetics, University of Tasmania, http://www.ipria.org/publications/workingpapers/BiotechReportFinal.pdf.

Nicol, D., D. Chalmers and B. Gogarty (2002), 'Regulating biomedical advances: embryonic stem cell research', *Macquarie Law Journal*, **2**, 31–60.

Nielsen, J. and D. Nicol (2002), 'Pharmaceutical patents and developing countries: the conundrum of access and incentive', *Australian Intellectual Property Journal*, **13**, 289–308.

Nolch, G. (2003), 'Wallaby genome a short hop away: Professor Jenny Graves and the Centre For Kangaroo Genomics', *Australasian Science*, **24**(7), 5.

Nottenburg, C. and J. Sharples (2004), 'Analysis of "junk DNA" Patents', *Simons Haplomics*, http://www.simonsjunkdna.com/junk%20dna%20analysis.pdf, July.

Nowak, R. (1994), 'NIH in danger of losing out on BRCA1 patent', *Science*, **266**, 209.

Nowak, R. and D. Concar (2002), 'Footing the bill: should we all have to pay for one company's bright idea?', *New Scientist*, **4**, 18 May.

Nuffield Council on Bioethics (2002), *The Ethics of Patenting DNA, A Discussion Paper*, London: Nuffield Council on Bioethics, http://www.nuffieldbioethics.org/go/ourwork/patentingdna/publication_310.html.

O'Connor, D. and T. Valoir (2006), 'The Supreme Court tilts toward drug developers' drug discovery after *Merck v Integra*', *Chicago-Kent Journal of Intellectual Property*, **5**, 124–41.

O'Neill, G. (2003), 'Prior art', *Australian Biotechnology News*, 22 August, p. 8.

O'Neill, G. (2004), 'NZ health service takes on GTG over licence fees', *Australian Biotechnology News*, http://www.biotechnews.com.au/index.php?id=813052581, 24 August.

O'Neill, G. (2005), 'GTG boss reflects on "unusual" license dispute', *Australian Biotechnology News*, http://www.biotechnews.com.au/index.php?id=1192139376, 8 July.

O'Neill, G. (2005), 'GTG celebrates win over Applera in patent battle', *Australian Biotechnology News*, http://www.biotechnews.com.au/index.php?id=548699165, 12 December, p. 2.

O'Neill, G. (2005), 'Start-up haplomics starting to muscle in on gene-testing market', *Australian Biotechnology News*, http://www.bio-itworld.com/newsitems/2005/dec2005/12-01-05-news-haplomics?page:int=-1, 1 December.

Ontario State Government (2002), *Genetics, Testing and Gene Patenting: Charting New Territory in Healthcare – Draft Report to the Provinces and*

the Territories, http://www.health.gov.on.ca/english/public/publ/ministry_reports/geneticsrepoz/genetics.html.
Organisation for Economic Co-operation and Development (2002), *Genetic Inventions, Intellectual Property Rights and Licensing Practices: Evidence and Policies*, Paris: Organisation for Economic Co-operation and Development.
Organisation for Economic Co-operation and Development (2003), *Patents and Innovation: Trends and Policy Challenges*, Paris: Organisation for Economic Co-operation and Development, http://www.oecd.org/dataoecd/48/12/24508541.pdf.
O'Rourke, M. (2000), 'Toward a doctrine of fair use in patent law', *Columbia Law Review*, **100**(5), 1177–1250.
O'Rourke, Maureen (2006), 'The story of *Diamond v Diehr*: toward patenting software', Jane Ginsburg and Rochelle Cooper Dreyfuss (eds), *Intellectual Property Stories*, New York: Foundation Press, pp. 194–219.
Oxford Glycosciences, 'Intellectual property', http://www.ogs.com/Proteomics/ProteomicsIntellectualProperty/, website archived on the Wayback Machine, http://www.archive.org/web/web.php.
Oxford Glycosciences, www.ogs.com, website archived on the Wayback Machine, http://www.archive.org/web/web.php.
Oxford Glycosciences (2000), 'OGS announces issuance of major US patent central to identification of disease-associated proteins', 16 May.
Oxford Glyosciences (2001), 'OGS announces achievement of major protein patenting milestone: completes discovery and patent filings for over 4,000 disease-associated proteins, genes and protein isoforms', 7 December.
Oxford University, 'Oxford Glycosciences', http://www.isis-innovation.com/spinout/oxglycosciences.html.
Palombi, Luigi (2004), 'The patenting of biological materials in the context of the Agreement on Trade Related Aspects of Intellectual Property', PhD thesis, University of New South Wales, Sydney, http://cgkd.anu.edu.au/menus/PDFs/PhDThesisFinal.pdf.
Parthasarathy, S. (2005), 'The patent is political: the consequences of patenting the BRCA genes in Britain', *Community Genetics*, **8**, 235–42.
Paul, Ellen Frankel, Fred Miller and Jeffrey Paul (1996), *Scientific Innovation, Philosophy and Public Policy*, Cambridge: Cambridge University Press.
Penn, M. (2002), 'Agency's aggressive patent management protects public, professors', *On Wisconsin*, The University of Wisconsin–Madison, 15 May.

Peterson, C. (2004), 'Nanotechnology: from Feynman to the grand challenge of molecular manufacturing', *IEEE Technology and Society Magazine*, **23**(4), 9–15.

Philipkoski, K. (2002), 'An "atlas" to count the genes', *Wired Magazine*, 27 February.

Phillips, J., S. Panny, H. Kazazian, C. Boehm, A. Scott and K. Smith (1980), 'Prenatal diagnosis of sickle cell anemia by restriction and endonuclease analysis: Hind III polymorphisms in gamma-globin genes extend test applicability', *Proceedings of the National Academy of Sciences*, **77**(5), 2853, http://www.pubmedcentral.nih.gov/articlerender.fcgi?artid= 349503.

Phillips, V. (2005), 'Half-human creatures, plants and indigenous peoples: musings on ramifications of Western notions of intellectual property and the attempt to patent a theoretical half-human creature', *Santa Clara Computer & High Technology Law Journal*, **21**, 383–450.

Piccinini, P., http://www.patriciapiccinini.net/.

Piccinini, P. (2002), 'Still life with stem cells', Biennale of Sydney, http://patriciapiccinini.net/essay.php?id=24.

Piedrahita, J., G. Anderson and R. Bondurant (1990), 'On the isolation of embryonic stem cells: comparative behavior of murine, porcine and ovine embryos', *Theriogenology*, **34**(5), 879901.

Pila, J. (2001), 'Methods of medical treatment within Australian and United Kingdom patents law', *University of New South Wales Law Journal*, **24**, 420–61.

Pippen, S. (2006), 'Dollars and lives: finding balance in the patent "gene utility" Doctrine', *Boston Journal of Science and Technology Law*, **12**, 193–226.

Pires de Carvalho, Nuno (2003), *The TRIPS Regime of Patent Rights*, 2nd edn, The Hague: Kluwer Law International.

Pitofsky, R. (2001), 'Antitrust and intellectual property: unresolved issues at the heart of the new economy', Berkeley Center for Law and Technology, http://www.ftc.gov/speeches/pitofsky/ipf301.htm, 1 March.

Pollack, A. (2004), 'Patent on test for cancer is revoked by Europe', *The New York Times*, 19 May.

Pollack, A. (2007), '3 patents on stem cells are revoked in initial review', *The New York Times*, 3 April.

Pollack, M. (2002), 'The multiple unconstitutionality of business method patents: common sense, Congressional consideration, and constitutional history', *Rutgers Computer and Technology Law Journal*, **28**, 61–120.

Power, E. (2001), 'Pharmaceutical companies are "failing to patent new technologies"', Silico Research, London, 2 May.

President's Commission on the Patent System (1967), *To Promote the Progress of Useful Arts*, S. Doc. No. 5, 90th Cong., 1st Sess., 20–21.

Proteome Systems, http://www.proteomesystems.com/.
Public Patent Foundation, http://www.pubpat.org/About.htm.
Public Patent Foundation (2006), 'WARF patents to be re-examined at the request of FTCR and PUBPAT: Patent Office finds "substantial question" regarding the validity of WARF's claims', http://www.pubpat.org/warfstemcellgranted.htm, 3 October.
Public Patent Foundation (2007), 'PTO rejects human stem cell patents at behest of consumer groups: re-examination was initiated by Foundation for Taxpayer and Consumer Rights and Public Patent Foundation', http://pubpat.org/warfstemcellpatentsrejected.htm, 2 April.
Rabin, S. (2006), 'The human use of humanoid beings: chimeras and patent law', *Nature Biotechnology*, **24**, 517–19, 1 May.
Rabinow, P. (2000), 'Learning to fly: a new page in history', *GeneLetter*.
Rabinow, Paul (1996), *Making PCR: A Story of Biotechnology*, Chicago: The University of Chicago Press.
Rabinow, Paul (1998), *French DNA: Trouble In Purgatory*, Chicago: The University of Chicago Press.
Rai, A. (1999), 'Regulating scientific research: intellectual property rights and the norms of science', *Northwestern University Law Review*, **94**, 77–152.
Rai, A. (2001), 'The information revolution reaches pharmaceuticals: balancing innovation incentives, cost, and access in the post-genomics era', *University of Illinois Law Review*, 173–210.
Rai, A. (2002), 'Stem cell research: an NPR special report, a "virtual roundtable" on federal funding', http://www.npr.org/programs/specials/stemcells/viewpoints.rai.html.
Regalado, A. and M. Louis (2001), 'Ethical concerns block patents of useful embryonic advances', *The Wall Street Journal*, 20 August.
Reichman, J.H. (2001), 'A contractually reconstructed research commons for scientific data in a highly protectionist intellectual property environment', *Law & Contemporary Problems*, **66**, 315–440.
Resnik, D. (2000), 'Comments on the United States Patent and Trademark Office utility guidelines', http://www.uspto.gov/web/offices/com/sol/comments/utilguide/dresnick.pdf, 16 March.
Resnik, D. (2003), 'Patents and the research exemption', *Science*, **299**, 821.
Ricketson, Sam and Megan Richardson (2005), *Intellectual Property: Cases and Materials*, 3rd edn, Sydney: LexisNexis Butterworths.
Rifkin, Jeremy (1998), *The Biotech Century: Harnessing the Gene and Remaking the World*, New York: Penguin Putnam Inc.
Rimmer, M. (2002–03), 'Genentech and the stolen gene: patent law and pioneer inventions', *Bio-Science Law Review*, **5**(6), 198–211.

Rimmer, M. (2003), 'The attack of the clones: patent law and stem cell research', *Journal of Law and Medicine*, **10**(4), 488–505.

Rimmer, M. (2003), 'Beyond Blue Gene: intellectual property and bioinformatics', *International Review of Industrial Property and Copyright Law*, **34**(1), 31–49.

Rimmer, M. (2003), 'Blame it on rio: biodiscovery, native title, and traditional knowledge', *The Southern Cross University Law Review*, **7**, 1–49.

Rimmer, M. (2003), 'Myriad Genetics: patent law and genetic testing', *European Intellectual Property Review*, **25**, 20–33.

Rimmer, M. (2004), 'The last taboo: patenting human beings', *Expert Opinion on Therapeutic Patents*, **14**(7), 1061–74.

Rimmer, M. (2004), 'The race to patent the SARS virus: the TRIPS Agreement and access to essential medicines', *Melbourne Journal of International Law*, **5**(2), 335–74.

Rimmer, M. (2005), 'Japonica rice: intellectual property, scientific publishing, and data-sharing', *Prometheus*, **23**(3), 325–47.

Rimmer, M. (2005), 'The freedom to tinker: patent law and experimental use', *Expert Opinion on Therapeutic Patents*, **15**(2), 167–200.

Rimmer, M. (2005), 'The Jean Chrétien Pledge to Africa Act: patent law and humanitarian aid', *Expert Opinion on Therapeutic Patents*, **15**(7), 889–909.

Rimmer, M. (2006), 'Miami heat: patent law, informed consent, and benefit-sharing', *Journal of International Biotechnology Law*, **3**, 177–92.

Rimmer, M. (2006), 'The alchemy of junk: patent law and non-coding DNA', *The University of Ottawa Law and Technology Journal*, **3**(2), 539–99.

Rimmer, M. (ed.) (2006), *Patent Law and Biological Inventions, Law in Context*, **24**(1), 1–163.

Rimmer, M. (2007), 'Chapter 8, remix culture: the creative commons and its discontents', in *Digital Copyright and the Consumer Revolution*, Cheltenham, UK and Northampton, MA, USA: Edward Elgar.

Rimmer, M. (2007), 'The Genographic Project: traditional knowledge and population genetics', *Australian Indigenous Law Review*, **11**(2), 33–55.

Rivers, L. (2002), 'Introduction of *The Genomic Research and Diagnostic Accessibility Act* of 2002 H.R. 3967 and *The Genomic Science and Technology Innovation Act* of 2002 H.R. 3966', Congressional Record, 14 March, E353.

Roberts, Michael Symmons (2004), 'Mapping the genome', *Corpus*, London: Cape Poetry.

Robertson, E. (1987), 'Embryo derived stem cell lines', *Teratocarcinomas and Embryonic Stem Cells; A Practical Approach*, Oxford: 1RL Press, ch. 4:71112.

Robertson, E., M. Kaufman, A. Bradley and M. Evans (1983), 'Isolation, properties and karyotype analysis of pluripotential (EK) cell lines from normal and parthenogenetic embryos', *Teratocarcinoma Stem Cells*, Cold Spring Harbor Laboratory, Cold Spring Harbor, **10**, 647663.

Robertson, S. (2005), 'Re-imagining economic alterity: a feminist critique of the juridical expansion of bioproperty in the Monsanto decision at the Supreme Court of Canada', *University of Ottawa Law and Technology Journal*, **2**, 227–53.

Robinson, William (1890), *Robinson on Patents*, §§133–43.

Rosenwald, M. (2006), 'J. Craig Venter's next little thing: the man who mapped the human genome has a new focus – using microbes to create alternative fuels', *The Washington Post*, 27 February.

Royal Society of the United Kingdom (2003), *Keeping Science Open: the Effects of Intellectual Property Policy on the Conduct of Science*, London: The Royal Society, April; http://www.royalsoc.ac.uk/files/statfiles/document-221.pdf.

Royal Society of the United Kingdom (2005), *Personalised Medicines: Hopes and Realities*, London: Royal Society, http://www.royalsoc.ac.uk/displaypagedoc.asp?id=17049.

Rural Advancement Foundation International (1998), 'Biotech activists oppose terminator technology', http://www.etcgroup.org/en/materials/publications.html?id=418, 13 March.

Rural Advancement Foundation International (1998), 'US patent on new technology will prevent farmers saving seed', http://www.etcgroup.org/en/materials/publications.html?id=420, 11 March.

Rusconi, W. (2005), 'The National Academies Fifth Meeting of the Committee on Intellectual Property Rights in Genomic and Protein-Related Inventions', http://www7.nationalacademies.org/step/Genomics_Committee_Meeting_6_transcript.pdf, 11 February.

Ryan, A. (2000), 'Testimony for hearing on gene patents and other genomic inventions', House of Representatives Subcommittee on Courts and Intellectual Property, http://www.aipla.org/Content/ContentGroups/Legislative_Action/106th_Congress/Testimony4/Statement_at_ Oversight_Hearing_on_Gene_Patents_and_other_Genomic_Inventions,_by_ M_Andrea_Ryan,_July_.htm, 7 July.

Sainsbury, P. (2004), 'Australia–United States Free Trade Agreement and the Australian pharmaceutical benefits scheme', *Yale Journal of Health Policy, Law and Ethics*, **4**(2), 387–99.

Salleh, A. (2000), 'A patent mistake', ABC Science Online, http://www.abc.net.au/science/news/stories/s102681.htm, 23 February.

Sampson, T. (2004), 'Madey, Integra and the wealth of nations', *European Intellectual Property Review*, **26**(1), 1–6.

Sanderson, J. (2006), 'Essential derivation, law, and the limits of science', in M. Rimmer (ed.), *Patent Law and Biological Inventions, Law in Context*, **24**(1), 34–53.
Saviotti, P., M.-A. de Looze, S. Michelland, and D. Catherine (2000), 'The changing marketplace of bioinformatics', *Nature Biotechnology*, **18**(12), 1247–9.
Schmeiser, P. (2003), 'In his own words', Vancouver Central Library, http://www.gmofreemendo.com/press_releases/percyschmeiser_to_mendo.html, 10 December.
Schmeiser, P. (2004), 'Percy Schmeiser claims personal and moral victory in Supreme Court decision', *Grain*, http://www.grain.org/bio-ipr/?id=397, 22 May.
Schneider, K. (1987), 'Agency and Congress face clash over patenting of animals', *The New York Times*, 23 July.
Schneider, K. (1988), 'Life patents: doubts are registering', *The New York Times*, 7 August.
Schubert, S. (2005), 'Europe halts decisions on stem cell patents', *Nature*, **435**, 720–21, 9 June.
Schwartzenberg, R.-G. (2001), 'First large scale analysis of the human genome sequence', Minister of Research, France, http://www.recherche.gouv.fr/english/ministre/discours.htm, 12 February.
Scott, A. (1999), 'The Dutch challenge to the bio-patenting directive', *European Intellectual Property Review*, **21**(4), 212–15.
Scott, Alexander, J. (2003), 'Mouse trap', Canadian Bar Association, October, p. 29.
Scott, R. (2000), 'Testimony for hearing on gene patents and other genomic inventions', House of Representatives Subcommittee on Courts and Intellectual Property, http://judiciary.house.gov/Legacy/scot0713.htm, 7 July.
Securities and Exchange Commission (2003), 'Quarterly report pursuant to Section 13 or 15(*d*) of the Securities Exchange Act of 1934 for the quarterly period ended March 31, 2003 for Applera Corporation', http://media.corporate-ir.net/media_files/NYS/ABI/reports/10q_q3_2003.pdf, March, p. 16.
Sell, Susan (2003), *Private Power, Public Law: The Globalization of Intellectual Property Rights*, Cambridge: Cambridge University Press.
Sevilla, C., C. Julian-Reynier, F. Eisinger, D. Stoppa-Lyonnet, B. Bressac-de Paillerts, H. Sobol and J.-P. Moatti (2003), 'The impact of gene patents on the cost-effective delivery of care: the case of BRCA1 testing', *International Journal of Technology Assessment in Health Care*, **19**(2), 287–300.
Sherman, B. (2002), 'Biological inventions and the problem of passive infringement', *Australian Intellectual Property Journal*, **13**, 146–54.

Sherman, B. and L. Bently (1998), 'The question of patenting life', in S. Maniatis (ed.), *Intellectual Property and Ethics*, London: Sweet & Maxwell, p. 111.

Sherman, Brad and Lionel Bently (1999), *The Making of Modern Intellectual Property Law: The British Experience 1760–1911*, Cambridge: Cambridge University Press.

Sherman, B. and S. Hubicki (2005), 'The killing fields: intellectual property and genetic use restriction technologies', *University of New South Wales Law Journal*, **28**(3), 740–57.

Shi, Q. (2005), 'Patent system meets new sciences: is the law responsive to changing technologies and industries?', *New York University Annual Survey of American Law*, **61**, 316–46.

Shiva, Vandana (2000), *Tomorrow's Biodiversity*, London: Thames and Hudson.

Shreeve, J. (2004), 'Craig Venter's epic voyage to redefine the origin of species', *Wired Magazine*, **12**(8), 107–13, 146–51.

Shreeve, James (2004), *The Genome War: How Craig Venter Tried to Capture the Code of Life and Save the World*, New York: Ballantine Books.

Shulman, S. (2001), 'The morphing patent problem', Owning the Future, *Technology Review*, November.

Shulman, S. (2001), 'Doctors without patents', Owning the Future, *Technology Review*, December.

Shulman, S. (2002), 'Of oncomice and men', Owning The Future, *Technology Review*, September.

Shulman, Seth (1999), *Owning The Future*, New York: Houghton Mifflin.

Schurman, Rachel, Dennis Doyle and Kelso Takahashi (2003), *Engineering Trouble: Biotechnology and its Discontents*, Berkeley and Los Angeles: University of California Press.

Simons, M. (2004), '"Junk DNA" non-coding DNA patents: the inventors' View', *Simons Haplomics*, http://www.simonsjunkdna.com/Full%20Article.htm, 18 May.

Slind For, V. (2001), 'Both sides now: MoFo IP Partners Kate Murashige and Gerald Dodson diverge on the PTO's new guidelines for biotech patents', IP Worldwide, 14 June.

Smith, D. (2000), 'Adventures in Lilliput', *The Sydney Morning Herald*, 11 December.

Smith, D. (2001), 'Patent battle looms over cancer gene', *The Sydney Morning Herald*, 15 March.

Smith Hughes, S. (2001), 'Making dollars out of DNA: the first major project in biotechnology and the commercialization of molecular biology, 1974–1980', *Isis*, **92**, 541–78.

Stanković, B. (2005), 'Patenting the Minotaur', *Richmond Journal of Law and Technology*, **12**(2), 1–39, http://law.richmond.edu/jolt/v12i2/article5.pdf.

Stanley, J., H. Hidu and S. Allen (1984), 'Growth of American oysters increased by polyploidy induced by blocking Meiosis I but not Meiosis II', *Aquaculture*, **37**, 147–55.

Steel, D. and K. Thomson (2001), 'Perspectives: Mark Skolnick, Myriad Genetics', Wasatch Digital iQ, November.

Stolberg, S. (2001), 'Patent laws may determine shape of stem cell research', *The New York Times*, Washington Report, 16 August.

Stop Animal Patents, http://www.stopanimalpatents.org/.

Strauss, J. (2002), 'Genetic inventions and patents: a German empirical study', Organisation for Economic Co-operation and Development Workshop, http://www.oecd.org/dataoecd/36/22/1817995.pdf, 24–5 January.

Sulston, John and Georgina Ferry (2002), *The Common Thread: A Story of Science, Politics, Ethics, and the Human Genome*, London: Bantam Press.

Sung, L. and D. Pelto (1998), 'Bioinformatics may get boost from "State Street": software that can manipulate vast libraries of genetic data may receive patent protection', *The National Law Journal*, **21**(8), 19 October.

Suthers, G. (2004), 'Our genes: humanity's heritage or cash cow?', *Issues*, **67**, 23.

Swanson, R. (1996–97), 'An oral history conducted by Sally Smith Hughes', Bioscience and Biotechnology Archives and Oral Histories, Bancroft Library, http://sunsite.berkeley.edu:2020/dynaweb/teiproj/oh/science/swanson/@Generic__BookView.

Taymor, K., C.T. Scott, and H. Greely (2006), 'The paths around stem cell intellectual property', *Nature Biotechnology*, **24**, 411–13.

ten Kate, Kerry and Sarah Laird (1999), *The Commercial Use of Biodiversity: Access to Genetic Resources and Benefit-Sharing*, Earthscan, London.

Thackray, Arnold (ed.) (1998), *Private Science: Biotechnology and the Rise of the Molecular Sciences*, Philadelphia: University of Pennsylvania Press.

Thoreau, Henry David (1854), *Walden, or Life in the Woods*, New York: Signet Classic.

Thurow, L. (1995), 'Needed: a new system of intellectual property rights', *Harvard Business Review*, September–October 95.

Tokar, Brian (ed.) (2001), *Redesigning Life? The Worldwide Challenge to Genetic Engineering*, London: Zed Books.

Trans National Institutes of Health Mouse Initiatives, http://www.nih.gov/science/models/mouse/.

Transition Therapeutics, http://www.transitiontherapeutics.com/.

Treatment Action Campaign, http://www.tac.org.za/.

Trudinger, M. (2004), 'BIO profile: Mervyn Jacobson, Genetic Technologies', *Australian Biotechnology News*, http://www.biotechnews.com.au/index.php?id=218823635, 1 June.

Trudinger, M. (2004), 'Gene patent system ain't broke, but needs fine tuning: ALRC', *Australian Biotechnology News*, http://www.biotechnews.com.au/index.php?id=1801276438, 6 September.

United Kingdom Patent Office (2003), 'Inventions involving human embryonic stem cells', http://www.patent.gov.uk/patent/notices/practice/stemcells.htm, April.

United Kingdom Stem Cell Bank, http://www.ukstemcellbank.org.uk/.

United States Department of Health and Human Services (2001), 'Memorandum of understanding between the WiCell Research Institute and the Public Health Service', http://www.nih.gov/news/stemcell/WicellMOU.pdf, 5 September.

United States House of Representatives Subcommittee on Courts, the Internet, and Intellectual Property (2007), 'American innovation at risk: "the case for patent reform"', oversight hearing, http://judiciary.house.gov/oversight.aspx?ID=27115, February.

United States Patent and Trademark Office (2004), 'New cross-reference digest for nanotechnology', http://www.uspto.gov/web/patents/biochempharm/crossref.htm.

United States Patent and Trademark Office (1987), 'Notice: animals – patentability', Official Gazette, United States Patent and Trademark Office, **1077**(8), 21 April.

United States Patent and Trademark Office (1998), 'Media advisory: facts on patenting life forms having a relationship to humans', http://www.uspto.gov/web/offices/com/speeches/98-06.htm, 1 April.

United States Patent and Trademark Office (1999), 'Revised interim utility examination guidelines', *Federal Register*, **64**(244), http://www.uspto.gov/web/offices/com/sol/notices/utilexmguide.pdf, 21 December.

United States Patent and Trademark Office (2000), 'Public comments on the United States Patent and Trademark Office revised interim utility examination guidelines', *Federal Register*, **65**, FR 3425, http://www.uspto.gov/web/offices/com/sol/comments/utilguide/index.html, 21 January.

United States Patent and Trademark Office (2001), 'Utility examination guidelines', *Federal Register*, **66**, 1092, http://www.uspto.gov/web/offices/com/sol/notices/utilexmguide.pdf.

United States Patent and Trademark Office (2005), 'Interim guidelines for examination of patent applications for patent subject matter eligibility', Official Gazette of the Patent Office, **1300**, 142, http://www.uspto.gov/web/offices/com/sol/notices/70fr75451.pdf, 22 November.

United States Patent and Trademark Office (2006), 'Regarding interim guidelines for examination of patent applications for patent subject matter eligibility', 'http://www.uspto.gov/web/offices/pac/dapp/opla/comments/ab98/ab98.html, 11 August.

Urban, R. (2007), 'Police seize executives' computers in genetic technologies raid', *The Sydney Morning Herald*, 8 March.

Urquijo, A. (2004), 'The restriction of access to healthcare by patent law: fact or fiction?', *University of New South Wales Law Journal*, **27**(1), 170–97.

Vaidhyanathan, Siva (2006), 'Nanotechnology and the law of patents: a collision course', Geoffrey Hunt and Michael Mehta (eds), *Nanotechnology: Risk, Ethics, Law*, London: Earthscan Publications, http://ssrn.com/abstract=740550.

van Overwalle, G. (2006), 'The implementation of the biotechnology directive in Belgium and its after-effects: the introduction of a new research exemption and a compulsory licence for public health', *International Review of Intellectual Property and Competition Law*, **37**(8), 889–920.

Varmus, H. (2000), 'Testimony for hearing on gene patents and other genomic inventions', House of Representatives Subcommittee on Courts and Intellectual Property, http://judiciary.house.gov/Legacy/varm0713.htm, 13 July.

Venter, J.C. (2000), 'Prepared statement', the Subcommittee on Energy and Environment, United States House of Representatives Committee on Science, http://www.ostp.gov/html/00626_4.html, 6 April.

Venter, J.C. et al. (2001), 'The sequence of the human genome', *Science*, **291**, 16 February, 1301–4.

Verbeure, B., G. Matthijs and G. van Overwalle (2005), 'Analysing DNA patents in relation with diagnostic genetic testing', *European Journal of Human Genetics*, **14**(1), 26–33.

Versitech, see http://www.versitech.hku.hk.

Vise, David (2005), *The Google Story*, New York: Random House.

Vondran, C. and R. Florence (2002), 'Bioinformatics: patenting the bridge between information technology and the life sciences', *The Journal of Law and Technology*, **42**, 93–131.

Wahlberg, D. (2006), 'Was stem-cell advance "obvious"', *Wisconsin State Journal*, 14 October.

Walpole I., H. Dawkins, P. Sinden and P. O'Leary (2003), 'Human gene patents: the possible impacts on genetic services healthcare', *Medical Journal of Australia*, **179**, 203–5.

Walsh, J.P., A. Arora and W.M. Cohen (2003), 'Working through the patent problem', *Science*, **299**, 1021.

Warburg, R., A. Wellman, T. Buck and A. Ligler Schoenhard (2003), 'Patentability and maximum protection of intellectual property in proteomics', *Biotechnology Law Report*, **22**(3), 264–72.
Waring, P. (2003), 'Patenting genetic information', Mutation Detection, International Symposium on Mutations in the Genome, July.
Washington Biohistory (2007), 'The Microsoft effect', http://www.wabio.com/biohistory_microsoft.htm?pp=1.
Wayner, Paul (2000), *Free For All: How Linux and the Free Software Movement Undercut the High-Tech Titans*, New York: Harper Business.
Weck, E. (2005), 'Exclusive licensing of DNA diagnostics: is there a negative effect on quantity and quality of healthcare delivery that compels NIH rulemaking?', *William Mitchell Law Review*, **31**, 1057–91.
Wegner, H. (2005), 'Post-Merck experimental use and the "safe harbor"', *Federal Circuit Bar Journal*, **15**, 1–36.
Wei, T. (2003), 'Patenting genomic technology – 2001 utility examination guidelines: an incomplete remedy in need of prompt reform', *Santa Clara Law Review*, **44**, 307–33.
Weinberg, A. (1961), 'Impact of large-scale science on the United States', *Science*, **134**(3473), 161–4.
Weiss, R. (1988), 'Animal patent report lacks support', *Science News*, 9 April.
West, J. (2001), 'Proteome systems', Harvard Business School, 11 October.
Westerman, H. and R. Urban (2005), 'Genetic wins little fight over DNA work', *The Age*, 16 December.
Westphal, S.P. (2002), 'Your money or your life', *New Scientist*, **175**, 29 at 33.
Wilkie, D. (2004), 'Stealth stipulation shadows stem cell research: a provision in the US Appropriations Bill bans patents on human organisms', *The Scientist*, **18**(4), 42, 1 March.
Williams, J.J. (2005), 'Protecting the frontiers of biotechnology beyond the genome: the limits of patent law in the face of the proteomics revolution', *Vanderbilt Law Review*, **58**, 955–93.
Williams, T. (2006), '*Merck KGAA v. Integra Lifesciences I, Ltd*: Does the Breadth of Safe Harbor Protection toll the Death Knell for Biotech Research Companies?', *Mercer Law Review*, **57**, 917–31.
Williams-Jones, B. (2002), 'History of a gene patent: tracing the development and application of commercial BRCA testing', *Health Law Journal*, **10**, 123–46.
Williamson, J. (1937), 'Scientific property', *Journal of Scientific Instruments*, **14**(3), 73–5, http://ej.iop.org/links/q08/RTvVxpHr8QVNaF2A7hhvVw/siv14i3p73.pdf.
Wilson, Jack (2002), 'Patenting organisms: intellectual property law meets biology', in David Magnus, Arthur Caplan and Glenn McGee (eds), *Who Owns Life?*, Amherst, NY: Prometheus Books, pp. 25–58.

Wisconsin Alumni Research Foundation and Geron Corporation (2002), 'WARF and Geron resolve lawsuit and sign new license agreement', Press Release, 9 January.

Wysocki, B. (2004), 'Cutting edge: a laser case sears universities' right to ignore patents', *The Wall Street Journal*, 11 October, A1.

Ziff, B. (2005), 'Travels with my plant: *Monsanto v Schmeiser* Revisited', *University of Ottawa Law and Technology Journal*, **2**(2), 493–509.

Zovko, N. (2006), 'Nanotechnology and the experimental use defense to patent infringement', *McGeorge Law Review*, **37**, 129–56.

CASE LAW

Aktiebolaget Hässle v Alphapharm Pty (2002), S287/2001, oral proceedings transcripts, http://www.austlii.edu.au/cgi-bin/disp.pl/au/other/hca/transcripts/2001/S287/1.html, 29 May.

Aktiebolaget Hässle v Alphapharm Pty Limited (2002) 212 C.L.R. 411.

The Amalgamated Society of Engineers v The Adelaide Steamship Company Limited and Ors (The Engineers Case) (1920) 28 CLR 129.

Amazon.com Inc. v Barnes and Noble.com, Inc., 239 F.3d 1343 (Fed. Cir. 2001).

American Cyanamid v Berk Pharmaceuticals [1976] RPC 231.

American Fruit Growers v Brogdex Co., 283 U.S. 1 (1931).

Anaesthetic Supplies Pty Ltd v Rescare Ltd (1994) 28 IPR 383.

Animal Legal Defense Fund v Quigg 710 F.Supp. 728, 9 USPQ2d 1816 (N.D.Cal.1989).

Animal Legal Defense Fund v Quigg 932 F.2d 920 C.A.Fed. (Cal.), (1991).

Apotex Inc. v Wellcome Foundation Ltd. [2002] 4 S.C.R. 153, 2002 SCC 77 (CanLII).

Application of Bergy 563 F.2d 1031 Cust. & Pat.App. (1977).

Application of Bergy 596 F.2d 952 Cust. & Pat.App. (1979).

Application of Chakrabarty 571 F.2d 40 Cust. & Pat.App (1978).

Applied Research Systems/Organon (Follicle-Stimulating Hormone) [1996] NJ 463, 28 IIC 558 (1997); (Dutch Supreme Court, affirming a more extensive judgment of The Hague Court of Appeal, 29 IIC 702 (1998)).

Arrhythmia Research Technology, Inc. v Corazonix Corp., 958 F.2d 1053 (Fed. Cir. 1992).

Asgrow Seed Company v Winterboer et al., 513 US 179 (1995).

AT&T Corporation v Excel Communications, Inc., 172 F.3d 1352 (Fed. Cir. 1999), cert. denied, 528 U.S. 946 (1999).

Attorney-General (NSW) v Brewery Employees Union of NSW (the Union Label Case) (1908) 6 CLR 469.

Auchincloss v Agricultural & Veterinary Supplies Ltd [1997] R.P.C. 649; on appeal [1999] R.P.C. 397.
Auckland District Health Board v Genetic Technologies Ltd (2004) (High Court of Auckland).
Australian Tape Manufacturers Association Ltd v Commonwealth (1993) 176 CLR 480.
The Belgian Society of Human Genetics and the Institut Curie v The University of Utah Research Foundation, European Patent Office Opposition Division, Interlocutory Decision in Opposition Proceedings Against EP785216, (29 June 2005).
Bolar Pharmaceutical Co. Inc. v Roche Products Inc., 469 U.S. 856 (1984).
Bonito Boats, Inc. v Thunder Craft Boats, Inc., 489 U.S. 141 (1989).
Brenner v Manson, 383 U.S. 519 (1966).
Bristol-Myers Squibb v FH Faulding (2000) 46 IPR 553.
Brown v Board of Education, 347 U.S. 483 (1954).
Californian Family Bioethics Council v California Institute for Regenerative Medicine (26 February 2007 A114195, Alameda County, Super. Ct. No. HG05 206766), http://www.courtinfo.ca.gov/opinions/documents/A114195.DOC.
Caltech Patent Application, Examining Division, European Patent Office, T522/04–338 (17 October 2003).
Canada: Patent Protection of Pharmaceutical Products: Complaint by the European Communities and their Member States, 17 March 2000, WT/DS114/R.
Cochrane v Deener, 94 U.S. 780 (1877).
Cultivaust Pty Ltd v Grain Pool of WA [2000] FCA 974.
Cultivaust v Grain Pool Pty Ltd [2004] FCA 638.
Cultivaust v Grain Pool Pty Ltd [2005] FCAFC 223.
Darcy v Allen (1602) 11 Co Rep 84 (the Case of Monopolies).
Diamond v Chakrabarty, 447 U.S. 303 (1980).
Diamond v Diehr, 450 U.S. 175 (1981).
Donoghue v Stevenson [1932] All ER Rep 1; [1932] AC 562.
Dow Chemical Co v Ishihara Sangyo Kaisha Ltd., 5 IPR 415.
Eli Lilly & Co v Medtronic, 496 U.S. 661 (1990).
E.R. Squibb & Sons Inc. v Giovannia Aguggini, 12 June 1995, T. Milano.
Ex parte Allen 2 USPQ 2d, p. 1425, 2 (1987).
Ex Parte Raymond H. Boutin, 2006 WL 2822238, *4+ (Bd.Pat.App & Interf. Sep 28, 2006) (NO. APL 2006-1879, APP 10/010,114).
Ex parte Brinkerhoff 24 Dec Comm'n Patent 349 (1883).
Ex parte Fisher 72 U.S.P.Q.2d 1020, 2004 WL 2185929 (United States Patent and Trademark Office Board of Patent Appeals and Interferences), http://patentlaw.typepad.com/patent/Ex_20Parte_20Fisher.pdf.

Ex parte Latimer 1889 (CD 46 OG 1638).
Ex Parte Preeti Lal, Neil Corley et al. 2006 WL 2710996, *3+ (Bd.Pat.App & Interf. Sep 18, 2006) (NO. APL 2006-1035, APP 09/925,140).
Ex Parte Preeti Lal, Jennifer Hillman, et al., 2006 WL 1665364, *3+ (Bd.Pat. App & Interf. Jan 01, 2006) (NO. APL 2005-0102, APP 09/840,787).
Ex Parte Re-examination of US Patent No. 5,843,780 (No. 90/008102, 30 March 2007), http://www.pubpat.org/assets/files/warfstemcell/780 rejected.pdf.
Ex Parte Re-examination of US Patent No. 6,200,806 (No. 90/008319, 30 March 2007), http://www.pubpat.org/assets/files/warfstemcell/ 806rejected.pdf.
Ex Parte Re-examination of US Patent No. 7,029,913 (No. 95/000, 154, 30 March 2007), http://www.pubpat.org/assets/files/warfstemcell/ 913rejected.pdf.
Ex Parte Gary C. Starling, 2006 WL 1665405, *2+ (Bd.Pat.App & Interf. Jan 01, 2006) (NO. APL 2005-2121, APP 09/745,605).
Ex Parte d. Wade Walke, 2006 WL 2711006, *2+ (Bd.Pat.App & Interf. Sep 18, 2006) (NO. APL 2006-2131, APP 10/309,422).
Fertilitescentrum AB and Luminis Pty Ltd [2004] APO 19; (2004) 62 IPR 420.
Folsom v Marsh (1841) 9 Fed. Cas. 342.
Funk Bros. Co. v Kalo Inoculant Co., 333 U.S. 127 (1948).
GE Healthcare Bio-Sciences Corp. v Genetic Technologies, Ltd (2006, United States District Court for the Southern District of New York, 1:06-cv-13172-WHP).
Generics BV v Smith Kline & French Laboratories Ltd (1997) R.P.C. 801, 803 (European Ct. of J. 1997).
Genetic Technologies Limited v Applera Corporation (2003, United States District Court, Northern District of California), No. c-03-1316 PJH, 2003 WL 23796524.
Genetic Technologies Ltd. v Covance Inc. (2003, United States District Court, Northern District of California) 3:2003cv01315.
Genetic Technologies Ltd. v Nuvelo Inc. (2003, United States District Court, Northern District of California).
Gottschalk v Benson, 409 U.S. 63 (1972).
Grain Pool of Western Australia v Commonwealth (6 October 1999) transcripts.
Grain Pool of Western Australia v Commonwealth (2000) 46 IPR 515.
Greenberg v Miami Children's Hospital Research Institute, Inc. 264 F.Supp.2d 1064 (S. D. Fla 2003).
Greenpeace Deutschland e.V. v The University of Edinburgh, Opposition Division, European Patent Office (24 July 2002).

Harvard College v Canada (The Commissioner of Patents) [1998] 3 FC 510 (Fed T.D.).
Harvard College v Canada (The Commissioner of Patents) [2000] 4 FC 528 (Fed C.A.).
Harvard College v Canada (The Commissioner of Patents) [2002] 2 SCR 45.
Harvard/Onco-mouse, 1989 O.J. EPO 451.
Harvard/Onco-mouse, 1990 O.J. EPO 476.
Harvard/Onco-mouse [2003] O.J. EPO 473.
Harvard/Onco-mouse [2004] T 0315/03–3.3.8.
Hoffman v Monsanto (2005) 2005 SKQB 225.
ICI/Pharbia and Medicopharma (Atenolol) [1993] NJ 735 (1993) GRUR Int. 887 (Dutch Supreme Court).
Imazio Nursery Inc. v Dania Greenhouses 69 F 3d 1560 (1995).
In re Adab, No. 05-1013.
In re Allen, 846 F.2d 77 (1988), 1988 WL 23321.
In re Anderson, No. 1012.
In re Bergy, 563 F.2d 1031, 195 USPQ 344 (CCPA 1977).
In re Boukharov, No. 05-1014.
In re Byrum, No. 1011.
In re Fisher, Fed. Cir., No. 04-1465, 5/3/05 oral argument.
In re Fisher 421 F.3d 1365 (C.A.Fed., 2005).
In re Kirk, 54 C.C.P.A. 1119, 376 F.2d 936 (1967).
In re Kovalic, No. 05-1007.
In re Lalgudi, No. 05-1010.
In re Merat, 519 F.2d 1390, 186 USPQ 471 (CCPA 1975).
Inhale Therapeutic Systems v Quadrant Healthcare Plc [2002] RPC 21.
Institut Curie v Myriad Genetics Inc., European Patent Office Opposition Division, Division Revoking the European Patent EP0699754 (3 November 2004).
Institut Curie v The University of Utah Research Foundation, European Patent Office Opposition Division, Interlocutory Decision in Opposition Proceedings Against EP705903 (9 June 2005).
Integra Lifesciences Ltd v Merck KgaA 1999 WL 398180 (1999).
Integra Lifesciences Ltd v Merck KgaA 331 F. 3d 860 (2003).
Integra Lifesciences I, Ltd. v Merck KGaA 421 F.3d 1289 C.A.Fed. (2005).
JEM Ag Supply Inc v Pioneer Hi-Bred International Inc 534 US 124 (2001).
Joos v Commissioner of Patents (1972) 46 ALJR 438.
Kirin-Amgen Inc. v Transkaryotic Therapies Inc. (No. 2) [2002] RPC 3.
Kirin Amgen/Boehringer Mannheim (Erythropoietin), Judgment of February 3, 1994 (docket No. 93/960) (The Hague Court of Appeal). Affirmed on other grounds by the Dutch Supreme Court: [1996] NJ 462.

Klinische Versuche I (Interferon Gamma) [1997] RPC 623.
Klinische Versuche I (Interferon Gamma) 2001 GRUR 43; 1 BvR 1864/95, http://www.bundesverfassungsgericht.de/cgi-bin/link.pl?entscheidungen (in German).
Klinische Versuche II (Erythropoietin) [1998] RPC 423.
KSR International Co. v Teleflex, Inc. 2007 WL 1237837, 82 U.S.P.Q.2d 1385 (2007).
Laboratory Corp. of America Holdings v Genetic Technologies Ltd. (2003, United States District Court, District of New Jersey) 3:2003cv06067.
Laboratory Corp. of America Holdings v Metabolite Laboratories, Inc. Slip Copy United States Supreme Court Official Transcript. Oral Argument, 2006 WL 711253 (U.S.), 74 USLW 3558.
Laboratory Corp. of America Holdings v Metabolite Laboratories, Inc. 126 S.Ct. 2921 (2006).
Le Roy v Tatham, 14 How. 156, 175, 14 L.Ed. 367 (1853).
Lowell v Lewis, Fed. Cas. No. 8568 (C.C. Mass. 1817) (Story, J).
Mackay Radio & Tel. Co. v Radio Corporation of America, 306 U.S. 86 (1939).
Madey v Duke University 266 F. Supp. 2d 420 (2001).
Madey v Duke University 307 F.3d 1351 (2002).
Madey v Duke University WL 2148935 (2004).
McFarling v Monsanto Co. 125 S.Ct. 2956 (Mem) U.S. (2005).
Merck KGaA v Integra Lifesciences I, Ltd. 545 U.S. 193 (2005).
Metabolite Laboratories v Laboratory Corp. of America Holdings Not Reported in F.Supp.2d, 2001 WL 34778749 (D.Colo., 2001).
Metabolite Laboratories, Inc. v Laboratory Corp. of America Holdings 370 F.3d 1354 (C.A.Fed. (Colo.), 2004).
Monsanto Co. v Genetic Technologies Limited (2006, United States District Court, Eastern District of Missouri), No. 3:06-cv-00989-HEA.
Monsanto Co. v Good, 2004 WL 1664013 (2003).
Monsanto Co. v McFarling, 2002 WL 32069634 (2002).
Monsanto Co. v McFarling, 363 F.3d 1336 (2004).
Monsanto Co v Stauffer Chemical Co (NZ) [1984] FSR 559.
Monsanto v Stauffer Chemical Co (UK) [1985] RPC 515.
Monsanto v Stauffer [1987] FSR 57.
Monsanto v Stauffer Japan (1989) 20 IIC 91.
Monsanto Canada Inc. v Schmeiser, 2001 FCT 256 (CanLII), (2001), 202 F.T.R. 78, 12 C.P.R. (4th) 204.
Monsanto Canada Inc. v Schmeiser (2004) SCC 34; 2004 SCC 34.
Morton v New York Eye Infirmary 2 Am. Law Reg. (N.S.) 672 C.C.N.Y. 1862.
Myriad Genetics Inc. v Cancer Research Campaign Technology Inc. European Patent Office, Opposition to EP 0858 467 (11 November 2004).

Myriad Genetics Inc. v Genetic Technologies Limited (2002, United States District Court for the District of Utah, Central Division), 2-02CV-0964 BSJ.
Neilson v Harford, Webster's Patent Cases 295, 371 (1841).
Netherlands v European Parliament (2001) 3 CMLR 49.
Nichols v Universal Pictures Corp., 45 F.2d 119, 122 (C.A.2 1930) (L.Hand, J).
Nintendo v Centronics Systems (1994) 181 CLR 134.
NRDC v The Commissioner of Patents (1959) 102 CLR 252.
O'Reilly v Morse, 56 U.S. (15 How.) 62 (1853).
Oxford Glycosciences v MDS Inc. Case R 397/2000-1 (2000), Office for Harmonization in the Internal Market (Trade Marks and Designs), http://oami.europa.eu/legaldocs/boa/2000/en/R0397_2000-1.pdf.
Pallin v Singer (1995) 36 USPQ (2d) 1050.
Parker v Flook, 437 U.S. 584 (1978).
Peppenhausen v Falke, 19 F. Cas. 1048, 1049 (C.C.S.D.N.Y. 1861).
Pfizer v Commissioner of Patents (NZ) (2004) 60 IPR 624.
Pioneer Hi-Bred International Inc v JEM Ag Supply Inc 1998 WL 1120829.
Pioneer Hi-Bred International Inc v JEM Ag Supply Inc 200 F.3d 1374 C.A.Fed. (Iowa), (2000).
Pioneer Hi-Bred International, Inc. v Ottawa Plant Food, Inc. 283 F.Supp.2d 1018 N.D. Iowa (2003); and 219 F.R.D. 135 N.D. Iowa (2003).
Ranks Hovis McDougall's Application [1976] 46 AOJP 3915.
Re Application for Patent of Abitibi Co. (1982) 62 CPR (2d) 81.
Re Application for Patent of Connaught Laboratories (1982) 82 CPR (2d) 32.
Re Stuart Newman (2005), United States Patent Office decision, http://patentlaw.typepad.com/patent/files/chimera_final_rejection.pdf.
Roche Products, Inc. v Bolar Pharmaceuticals Co., Inc 733 F. 2d 858 (1984).
Rubber-Tip Pencil Co. v Howard, 20 Wall. 498, 507, 22 L.Ed. 410 (1874).
Sample v Monsanto Co. 218 F.R.D. 644 (2003).
Sawin v Guild 21 Fed. Cas. 554 (1813).
Schmeiser v Monsanto Canada Inc., 2002 FCA 448 (CanLII), [2003] 2 F.C. 165, 218 D.L.R. (4th) 31.
Shell Development Co. v Watson, 149 F.Supp. 279, 280 (D.C.1957).
Smith Kline & French Laboratories Ltd v Evans Medical Ltd [1989] F.S.R. 513.
Sozialdemokratische Partei der Schweiz and the Institut Curie v The University of Utah Research Foundation, European Patent Office Opposition Division, Interlocutory Decision in Opposition Proceedings Against EP705902 (19 September 2005).

State Street Bank & Trust Co. v Signature Financial Group, Inc., 149 F.3d 1368 (Fed. Cir. 1998), cert. denied, 525 U.S. 1093 (1999).
Sun World International Inc v Registrar, Plant Breeder's Rights (1995) 33 IPR 106; (1997) 39 IPR 161; (1998) 42 IPR 321 [1998].
Telephone Cases, 126 U.S. 1 (1888).
Tennessee Eastman Co. v Commissioner of Patents [1974] S.C.R. 111.
United States v Dubilier Condenser Corp., 289 U.S. 178, 199 (1933).
Welcome Real Time SA v Catuity Inc [2001] 51 IPR 327.
Wellcome Foundation v Parexel International & Flamel, Tribunal de Grande Instance de Paris, 20 February 2001.
Whittemore v Cutter 29 F. Cas. 1120 (1813).
Wisconsin Alumni Research Foundation v Geron Corporation (2002) Case No. 01-C-0459-C.
Wisconsin Alumni Research Foundation Patent Application in Respect of 'Primate embryonic stem cells', Examining Division, June 2004.
Wisconsin Alumni Research Foundation Patent Application in Respect of 'Primate embryonic stem cells', T 1374/04–3.3.08 Interlocutory decision of the Technical Board of Appeal 3.3.08 of 18 November 2005.
Woo-Suk Hwang [2004] APO 24.

AMICUS CURIAE BRIEFS

Affymetrix Inc. (2004), 'Brief for Amicus Curiae Affymetrix, Inc. in support of appellee *In re Fisher*', 2004 WL 4996615, 14 December.
Affymetrix (2005), 'Brief for Amici Curiae Affymetrix, Inc. and Professor John H. Barton in support of petitioner in *Laboratory Corp. of America Holdings v Metabolite Laboratories, Inc.*', 2005 WL 3597814, 23 December.
American Association for Retired Persons (2005), 'Brief Amicus Curiae of AARP in support of petitioner in *Laboratory Corp. of America Holdings v Metabolite Laboratories, Inc.*', 2005 WL 3597809, 23 December.
American Bar Association (2001), 'Brief for Amicus Curiae American Bar Association in support of respondent in *JEM Ag Supply Inc v Pioneer Hi-Bred International Inc*', 2001 WL 674189, 15 June.
American Clinicial Laboratory Association (2005), 'Brief of the American Clinical Laboratory Association as Amicus Curiae in support of petitioner in *Laboratory Corp. of America Holdings v Metabolite Laboratories, Inc.*', 2005 WL 3543098, 23 December.
American Corn Growers Association and National Farmers Union (2001), 'Brief for Amici Curiae American Corn Growers Association and National Farmers Union in support of the petitioners in *JEM Ag Supply Inc v Pioneer Hi-Bred International Inc*', 2001 WL 490944, 4 May.

American Crop Protection Association (2001), 'Brief of Amicus Curiae American Crop Protection Association in support of affirmance in *JEM Ag Supply Inc v Pioneer Hi-Bred International Inc*', 2001 WL 674199, 15 June.

American Express Company (2005), 'Brief of Amicus Curiae American Express Company in support of neither party in *Laboratory Corp. of America Holdings v Metabolite Laboratories, Inc.*', 2005 WL 3597810, 23 December.

American Heart Association (2005), 'Brief of the American Heart Association as Amicus Curiae in support of petitioner in *Laboratory Corp. of America Holdings v Metabolite Laboratories, Inc.*', 2005 WL 3561169, 23 December.

American Intellectual Property Law Association (2001), 'Brief for Amicus Curiae American Intellectual Property Law Association in support of respondent supporting affirmance in *JEM Ag Supply Inc v Pioneer Hi-Bred International Inc*', 2001 WL 649829, 11 June.

American Medical Association et al. (2005), 'Brief for the American Medical Association, the American College of Medical Genetics, the American College of Obstetricians and Gynecologists, the Association for Molecular Pathology, the Association of American Medical Colleges, and the College of American Pathologists as Amici Curiae in support of petitioner in *Laboratory Corp. of America Holdings v Metabolite Laboratories, Inc.*', 2005 WL 3597812 (Appellate Brief), 23 December.

American Patent Law Association Inc. (1980), 'Appellate brief on behalf of the American Patent Law Association, Inc., as Amicus Curiae in support of the respondents in *Diamond v Chakrabarty*', 1980 WL 339772, 29 January.

American Seed Trade Association (2001), 'Brief of American Seed Trade Association in support of respondent in *JEM Ag Supply Inc v Pioneer Hi-Bred International Inc*', 2001 WL 670055, 13 June.

American Society for Microbiology (1979), 'Appellate brief on behalf of the American Society for Microbiology as Amicus Curiae in support of the respondents in *Diamond v Chakrabarty*', 1979 WL 200007.

Association of American Medical Colleges (2003), 'Brief for the Association of American Medical Colleges et al. petition for a writ of certiorari, *Duke University v Madey*', Supreme Court of the United States, http://www.aamc.org/newsroom/pressrel/patentbrief.pdf.

Bar of the City of New York (2005), 'Amicus Curiae Brief of the Association of the Bar of the City of New York in support of neither party in *Laboratory Corp. of America Holdings v Metabolite Laboratories, Inc.*', 2005 WL 3597808, 23 December.

BIOTECanada (2003), 'Amicus Brief of BIOTECanada in *Monsanto Canada Inc. v Schmeiser*', http://www.biotech.ca/media.php?mid=838, December.
Biotechnology Industry Organization (2001), 'Brief of Amicus Curiae Biotechnology Industry Organization in *JEM Ag Supply Inc v Pioneer Hi-Bred International Inc*', 2001 WL 689273, 15 June.
Boston Patent Law Association (2006), 'Brief of Amicus Curiae Boston Patent Law Association in support of respondents in *Laboratory Corp. of America Holdings v Metabolite Laboratories, Inc.*', 2006 WL 303909, 6 February.
Chakrabarty, A. (1980), 'Brief for the respondent in *Diamond v Chakrabarty*', 1980 WL 339758, 28 January.
Consumer Project on Technology (2005), 'Brief of Amici Curiae Consumer Project on Technology, Electronic Frontier Foundation and Public Knowledge in support of petitioner in *Merck KgaA v Integra Lifesciences I, Ltd*', 2005 WL 435894, 22 February.
Consumer Project on Technology and Public Knowledge et al. (2003), 'Petition for a writ of certiorari by the Consumer Project on Technology and Public Knowledge in *Duke University v Madey*, Supreme Court of the United States', http://jurist.law.pitt.edu/amicus/duke_v_madey_cert_petition_au.pdf, January.
The Council of Canadians (2002), 'Amicus Brief of the Council of Canadians in *Monsanto Canada Inc. v Schmeiser*', http://www.canadians.org/food/documents/COC_Affidavit.pdf, October.
Delta and Pine Land Company (2001), 'Brief of Amicus Curiae Delta and Pine Land Company in support of respondent in *JEM Ag Supply Inc v Pioneer Hi-Bred International Inc*', 2001 WL 689283, 15 June.
Diamond, S. (1980), 'Brief for the petitioner in *Diamond v Chakrabarty*', 1980 WL 339757.
Duke University (2001), 'Appellate Brief in *Madey v Duke University*', 2001 WL 34633573.
Eli Lilly et al. (2004), 'Brief for Amici Curiae Eli Lilly and Company, Association of American Medical Colleges, Baxter Healthcare Corporation, National Academy of Sciences Dow AgroSciences LLC, and American College of Medical Genetics in support of the United States Patent and Trademark Office in support of affirmance *In re Fisher*', 2004 WL 4996616, 14 December.
The ETC Group (2003), 'Amicus Brief of the ETC Group in *Monsanto Canada Inc. v Schmeiser*', http://www.canadians.org/food/documents/ETC_Affidavit.pdf, 24 September.
Federal Circuit Bar Association (2006), 'Amicus Curiae brief of the Federal Circuit Bar Association in support of respondents in *Laboratory Corp.*

of America Holdings v Metabolite Laboratories, Inc.', 2006 WL 303906, 6 February.
Foundation for Taxpayer and Consumer Rights and the Public Patent Foundation (2006), 'Request for re-examination in respect of US Patent No: 5,843,780', http://www.pubpat.org/assets/files/warfstemcell/780Request.pdf.
Foundation for Taxpayer and Consumer Rights and the Public Patent Foundation (2006), 'Request for re-examination in respect of US Patent No: 6,200,806', http://www.pubpat.org/assets/files/warfstemcell/90008139granted.pdf.
Foundation for Taxpayer and Consumer Rights and the Public Patent Foundation (2006), 'Request for re-examination in respect of US Patent No: 7,029,913', http://www.pubpat.org/assets/files/warfstemcell/913Request.pdf.
Franklin Pierce Law Center (2006), 'Brief Amicus Curiae Franklin Pierce Law Center in support of the respondents in *Laboratory Corp. of America Holdings v Metabolite Laboratories, Inc.*', 2006 WL 304571, 6 February.
Genentech Inc. (1980), 'Appellate Brief of Genentech Inc. as Amicus Curiae in support of the respondents in *Diamond v Chakrabarty*', 1980 WL 339766, 28 January.
Genentech Inc. (2004), 'Brief of Genentech Inc. as Amicus Curiae supporting affirmance and supporting the United States Patent and Trademark Office *In re Fisher*', 15 December.
Genentech Inc. and Biogen Idec Inc. (2005), 'Brief of Amici Curiae Genentech, Inc. and Biogen Idec, Inc. in support of petitioner in *Merck KGaA v Integra Lifesciences I, Ltd.*, 2005 WL 435893', 22 February.
Hood, L. (1980), 'Amicus Brief of Dr Leroy E. Hood, Dr Thomas P. Maniatis, Dr David S. Eisenberg, The American Society of Biological Chemists, The Association of American Medical Colleges, The California Institute of Technology, and The American Council on Education as Amicus Curiae in support of the respondents in *Diamond v Chakrabarty*', 1980 WL 339764, 26 January.
Intellectual Property Owners Association (2005), 'Brief of Amicus Curiae Intellectual Property Owners Association in support of neither party in *Laboratory Corp. of America Holdings v Metabolite Laboratories, Inc.*', 2005 WL 3476621, 15 December.
Intellectual Property Professors (2005), 'Brief of Intellectual Property Professors as Amici Curiae in support of neither party in *Merck KgaA v Integra Lifesciences I, Ltd*', 2005 WL 435892, 22 February.
International Center for Technology Assessment (2003), 'Amicus Brief of

the International Center for Technology Assessment in *Monsanto Canada Inc. v Schmeiser*', http://www.canadians.org/food/documents/ICTA_Affidavit.pdf, 26 September.

Laboratory Corp. of America Holdings (2004), 'Petition for a writ of certiorari in *Laboratory Corp. of America Holdings v Metabolite Laboratories, Inc*', 2004 WL 2505526, 3 November.

Loring, J. (2006), 'Declaration on US Patent No: 5,843,780', http://www.pubpat.org/assets/files/warfstemcell/LoringDeclarations.pdf.

Madey (2001), 'Appellant's opening brief in *Madey v Duke University*', 22 October.

Metabolite Laboratories Inc. (2004), 'Brief for respondents in opposition in *Laboratory Corp. of America Holdings v Metabolite Laboratories, Inc.*', 2004 WL 2803464, 3 December.

Metabolite Laboratories Inc. (2006), 'Brief for respondents in *Laboratory Corp. of America Holdings v Metabolite Laboratories, Inc*', 2006 WL 303905, 6 February.

Monsanto Inc. (2001), 'Brief of Amicus Curiae Monsanto Company in support of respondent in *JEM Ag Supply Inc v Pioneer Hi-Bred International Inc*', 2001 WL 674207, 15 June.

Monsanto Inc. (2004), 'Corrected brief for appellants Dane K. Fisher and Raghunath V. Lalgudi *In re Fisher*', 2004 WL 4996614, 27 September, 39–40.

National Farmers Union (2003), 'Amicus Brief of National Farmers Union in *Monsanto Canada Inc. v Schmeiser*', http://www.canadians.org/food/documents/NFU_Affidavit_final.pdf, 26 September.

Patients not Patents, Inc. (2005), 'Brief of Patients not Patents, Inc., as Amicus Curiae in support of petitioner in *Laboratory Corp. of America Holdings v Metabolite Laboratories, Inc.*', 2005 WL 3597811, 23 December.

Peoples Business Commission (1979), 'Appellate Brief of Peoples Business Commission as Amici Curiae in support of the petitioner in *Diamond v Chakrabarty*', 1979 WL 2005, 13 December.

People's Medical Society (2005), 'Brief of Amicus Curiae People's Medical Society in support of petitioner in *Laboratory Corp. of America Holdings v Metabolite Laboratories, Inc.*', 2005 WL 3597702, 22 December.

Perlegen Sciences (2006), 'Brief for Amici Curiae Perlegen Sciences, Inc. and Mohr, Davidow ventures in support of respondents in *Laboratory Corp. of America Holdings v Metabolite Laboratories, Inc.*', 2006 WL 303908, 6 February.

Pharmaceutical Manufacturers Association (1980), 'Appellate Brief of the Pharmaceutical Manufacturers Association as Amicus Curiae in support of the respondents in *Diamond v Chakrabarty*', 1980 WL 339771, 29 January.

Pieczenik, G. (1980), 'Appellate Brief of Dr George Pieczenik as Amicus Curiae in support of the respondents in *Diamond v Chakrabarty*', 1980 WL 339773, 29 January.

Pollack, M. (2001), 'Brief for Amici Malla Pollack and other Law Professors supporting reversal in *JEM Ag Supply Inc v Pioneer Hi-Bred International Inc*', 2001 WL 476088, 2 May.

Public Patent Foundation (2005), 'Brief of the Public Patent Foundation as Amicus Curiae in support of petitioner in *Laboratory Corp. of America Holdings v Metabolite Laboratories, Inc.*', 2005 WL 3597813 (Appellate Brief), 23 December.

The Regents of the University of California (1980), 'Appellate Brief of the Regents of the University of California as Amicus Curiae in support of the respondents in *Diamond v Chakrabarty*', 1980 WL 339770, 28 January.

Research Foundation for Science, Technology and Ecology (2003), 'Amicus Brief of the Research Foundation for Science, Technology and Ecology in *Monsanto Canada Inc. v Schmeiser*', http://www.canadians.org/food/documents/RSTE_Affidavit.pdf, 19 September.

United States Government (2001), 'Brief for the United States Government as Amicus Curiae supporting respondent in *JEM Ag Supply Inc v Pioneer Hi-Bred International Inc*', 2001 WL 689516, 15 June.

United States Patent and Trademark Office (2004), 'Brief and addendum for appellee, Director of the United States Patent and Trademark Office *In re Fisher*', 7 December, 13.

United States Solicitor-General (2003), 'Brief for the United States as Amicus Curiae in petition for a writ of certiorari in *Duke University v Madey*', Supreme Court of the United States, http://www.usdoj.gov/osg/briefs/2002/2pet/6invit/2002-1007.pet.ami.inv.html, May.

United States Solicitor-General (2005), 'Brief for the United States as Amicus Curiae in *Laboratory Corp. of America Holdings v Metabolite Laboratories, Inc.*', 2005 WL 3533248, 23 December.

University of Edinburgh (2003), 'Statement of grounds of appeal under Article 108 EPC', European Patent Office, 8 December.

Wisconsin Alumni Research Foundation et al. (2005), 'Brief of Amici Curiae Wisconsin Alumni Research Foundation, The American Council on Education, Boston University, The Regents of the University of California, Research Corporation Technologies, The Salk Institute for Biological Studies, University of Alberta and University of Oklahoma in support of respondents in *Merck KgaA v Integra Lifesciences I, Ltd*', 2005 WL 682088, 22 March.

LEGISLATION

Administrative Procedure Act 1966 (US).
Australian Constitution 1901.
Bayh–Dole Act 1980 (US).
Consolidated Appropriations Act 2004 (US).
Drug Price Competition and Patent Term Restoration Act 1984 (US) ('The Hatch–Waxman Act').
Gene Technology Act 2000 (Cth).
Genomic Research and Accessibility Act 2007 (US).
Genomic Research and Diagnostic Accessibility Act 2002 (US).
Genomic Science and Technology Innovation Act 2002 (US).
Grain Marketing Act 1975 (WA).
Intellectual Property Laws Amendment Act 2006 (Cth).
Morrill Act 1862 (US).
Patent Act 1793 (US).
Patent Act 1835 (US).
Patent Act 1870 (US).
Patent Act 1874 (US).
Patent Act 1952 (US).
Patent Act 1970 (RSC).
Patent Act 1985 (RSC).
Patent Reform Act 2005 (US).
Patents Act 1953 (NZ).
Patents Act 1977 (UK).
Patents Act 1990 (Cth).
Patents Regulations 1991 (Cth).
Plant Breeders' Rights Act 1990 (RSC).
Plant Breeder's Rights Act 1994 (Cth).
Plant Breeder's Rights Amendment Act 2002 (Cth).
Plant Patent Act 1930 (US).
Plant Variety Protection Act 1970 (US).
Plant Variety Rights Act 1987 (Cth).
Prohibition of Human Cloning Act 2002 (Cth).
Research Involving Embryos Act 2002 (Cth).
Statute of Monopolies 1623 (UK).
Stem Cell Research and Cures Act 2004 (California), California Proposition 71.
Stem Cell Research Enhancement Act 2005 (US).
Trade Practices Act 1974 (Cth).
Transgenic Animal Patent Reform Act 1989 (US).
United States Constitution.

TREATIES

Agreement on Technical Barriers to Trade 1994.
Australia–United States Free Trade Agreement 2004.
The Budapest Treaty 1977.
Community Patent Convention 1989 (EU).
Doha Declaration on Public Health and the TRIPS Agreement 2001.
European Patent Convention 1973 (EU).
European Union Directive on the Community code relating to medicinal products for human use 2004 (EU).
European Union Directive on the Legal Protection of Biotechnological Inventions 1998 (EU).
European Union Regulation on the Compulsory Licensing of Patents Relating to the Manufacture of Pharmaceutical Products for Export to Countries with Public Health Problems 2006 (EU).
Free Trade Area of the Americas.
League of Nations Draft Convention on Scientific Property A 38 1923 XII (League of Nations Documents).
North American Free Trade Agreement 1994.
Paris Convention for the Protection of Industrial Property 1883.
Patent Cooperation Treaty 1970.
Rio Convention on Biological Diversity 1992.
The TRIPS Agreement 1994.
UNESCO Universal Declaration on Bioethics and Human Rights 2005.
United Nations Declaration on the Rights of Indigenous Peoples 2007.
UPOV Convention 1961.
UPOV Convention 1978.
UPOV Convention 1991.
WTO General Council Decision 2003.

PATENTS AND PLANT BREEDERS' RIGHTS APPLICATIONS

Allen, R., S. Stabler and J. Lindenbaum (1986), 'Assay for sulfhydryl amino acids and methods for detecting and distinguishing cobalamin and folic acid deficiency', US Patent No: 4,940,658.

Anderson, D. and D. Stemple (1993), 'Mammalian multipotent neural stem cells', European Patent Application No: EP0658194.

Archer, G., E. Fells, R. Fryer, W. Orange, E. Reeder and L.H. Strenbach (1964), 'Novel 1-and/or 4-substituted alkyl 5-aromatic – 3H – 1,4 –

benzodiazepines and benzodiazepine-2-Ones', US Patent No: 3,299,053.
Bergy, M., J. Coats and V. Malik (1974), 'Process for preparing lincomycin', US Patent Application No: 477,766.
Chakrabarty, A. (1972), 'Microorganisms having multiple compatible degradative energy-generating plasmids and preparation thereof', US Patent No: 4,259,444.
Cohen, S. and H. Boyer (1979), 'Process for producing biologically functional molecular chimeras', US Patent No: 4,237,224.
Courtney, S. (1997), 'Therapeutic compounds with pyrimidine base', US Patent No: 5,945,406.
Department of Primary Industry Tasmania (1989), 'Franklin Barley', *Plant Varieties Journal*, **2**(2), Application No: 1989/018.
Engelgau, S. (2000), 'Adjustable pedal assembly with electronic throttle control', US Patent No: 6,237,565 B1.
Fisher, D. and R. Lalgudi (2001), 'Nucleic acid molecules and other molecules associated with plants', US Patent Application Serial No: 09/619,643.
Furphy, J. (1882), 'Grain stripping machine', Victorian Patent No: 3297.
Futreal, P., R. Wooster, A. Ashworth and M. Stratton (1995), 'Materials and methods relating to the identification and sequencing of the BRCA2 cancer susceptibility gene and uses thereof', European Patent No: EP 0858 467.
Glass, J., H. Smith, C. Hutchison, N. Alperovich and N. Assad-Garcia (2006), 'Minimal bacterial genome', US Patent No: 20070122826.
Hwang, W.S. (2003), 'A method of inter-species nuclear transplantation', Patent Application No: 2003204164.
Leder, P. and T. Stewart (1985), 'Transgenic animals', Canadian Patent No: CA 1341442.
Leder, P. and T. Stewart (1986), 'Method for producing transgenic animals', European Patent No: EP0169672.
Leder, P. and T. Stewart (1988), 'Transgenic non-human mammals', US Patent No: 4,736,866.
MacLean, D. (2000), 'Myself', United Kingdom Patent Application No: GB0000180.0.
Madey, J. and E. Szarmes (1991), 'Free-electron laser oscillator for simultaneous narrow spectral resolution and fast time resolution spectroscopy', US Patent No: 5,130,994.
Madey, J. and G. Westenskow (1984), 'A microwave electron gun', US Patent No: 4,641,103.
Newman, S. (2003), 'Chimeric embryos and animals containing human cells', US Patent Application: 20030079240.

Niebur, W., R. Riley and S. Noble (1996), 'Hybrid corn plant and seed', US Patent No: 5,491,295.

Oliver, M., J. Quisenberry, N. Trolinder and D. Keim (1995), 'Control of plant gene expression', US Patent No: 5,723,765.

Orchard, M. (2003), 'Benzylidene thiazolidinediones and their use as antimycotic agents', US Patent No: 7,105,554.

Orchard, M. and J. Neuss (2003), 'Thiazolidine derivatives and its use as antifungal agent', US Patent No: 6,740,670.

Parekh, R., R. Amess, J. Bruce, S. Prime, A. Platt and R. Stoney (1997), 'Computer-assisted methods and apparatus for identification and characterization of biomolecules in a biological sample', US Patent No: 6,064,754.

Parekh, R., P. Goulding and D. Pfost (1999), 'Assaying and storing labelled analytes', US Patent No: 6,534,321.

Parekh, R. (2000), 'Computer-assisted isolation of proteins', US Patent No: 6,278,794.

Parekh, R., R. Amess, J. Bruce, S. Prime, A. Platt and R. Stoney (2000), 'Methods for computer-assisted isolation of proteins', US Patent No: 6,459,994.

Parekh, R., R. Amess, J. Bruce, S. Prime, A. Platt and R. Stoney (2000), 'Robotic device for removing selected portions of a polyacrylamide gel', US Patent No: 6,480,618.

Pasteur, L. (1873), 'Improvement in the manufacture of beer and yeast', US Patent No: 141,972.

Pennington, M., D. Scopes and M. Orchard (1999), 'Quinolinium- and pyridinium-based fluorescent dye compounds', US Patent No: 6,335,446.

Pierschbacher, M. (1985), 'Vitronectin specific cell receptor derived from mammalian mesenchymal tissue', US Patent No: 4,789,734.

Raguse, B., B. Cornell, V. Braach-Maksvytis and R. Pace (1997), 'Biosensor membranes', US Patent No: 5,798,030.

Robertson, S., M. Wikland and C. Sjoblom (1999), 'Method and medium for in vitro culture of human embryos', Australian Patent Application No: 44916/99, PCT No: WO/1999/067364.

Rosenkranz Jr., C., C. Keefer and M. Sims (1990), 'Bovine embryo medium', US Patent No: 5,096,822.

Ruoslahti, E. and M. Pierschbacher (1985), 'Tetrapeptide', US Patent No: 4,792,525.

Ruoslahti, E. and M. Piershbacher (1985), 'Tetrapeptide', US Patent No: 5,695,997.

Ruoslahti, E., E. Hayman and M. Pierschbacher (1985), 'Use of peptides in control of cell attachment and detachment', US Patent No: 4,879,237.

Ruoslahti, E., E. Hayman and M. Pierschbacher (1987), 'Peptides in cell detachment and aggregation', US Patent No: 4,988,621.
Schatten, G., C. Simerly and C. Navara (2004), 'Methods for correcting mitotic spindle defects associated with somatic cell nuclear transfer in animals', US Patent Application: 20040268422.
Shah, D., S. Rogers, R. Fraley and R. Horsch (1986), 'Glysophate-resistant plants', Canadian Patent No: 1,313,830.
Shattuk-Eidens, D., J. Simard, E. Mitsuru, Y. Nakamura and F. Durocher (1995), 'In vivo mutations in the 17q-linked breast and ovarian cancer susceptibility gene', European Patent No: EP0705902.
Simons, M. (1992), 'Intron sequence analysis method for detection of adjacent and remote locus alleles as haplotypes', US Patent No: 5,612,179.
Simons, M. (1994), 'Genomic mapping method by direct haplotyping using intron sequence analysis', US Patent No: 5,851,762.
Skolnick, M. and D. Goldgar (1995), 'Method for diagnosing a predisposition for breast and ovarian cancer', European Patent No: EP699754.
Skolnick, M. and D. Goldgar (1995), '17q linked breast and ovarian cancer susceptibility gene', European Patent No: EP705902.
Smith, A. and P. Mountford (1996), 'Isolation, selection and propagation of animal transgenic stem cells', European Patent No: 0695351.
Smith, A. and P. Mountford (1997), 'Isolation, selection and propagation of animal transgenic stem cells', Australian Patent No: AU678233.
Tavtigian, S., A. Kamb, J. Simard, F. Couch, J. Rommens and B. Weber (1996), 'Chromosome 13-linked breast cancer susceptibility gene BRCA2', European Patent No: EP 785216.
Thomson, J. (1996), 'Primate embryonic stem cells', European Patent No: EP0770125.
Thomson, J. (1996), 'Primate embryonic stem cells', US Patent No: 5,843,780.
Thomson, J. (1998), 'Primate embryonic stem cells', US Patent No: 6,200,806.
Thomson, J. (2001), 'Primate embryonic stem cells', US Patent No: 7,029,913.
Townsend, R., R. Parekh, S. Prime and N. Wedd (1997), 'Method for de novo peptide sequence determination', US Patent No: 6,582,965.
Townsend, R. and A. Robinson (2001), 'Automated identification of peptides', US Patent No: 6,963,807.
Williams, R., N. Gough and D. Hilton (1989), 'In vitro propagation of embryonic stem cells', US Patent No: 5,166,065.

Index

access to essential medicines vi–x, 8, 11, 16, 21, 207–8, 250, 300
access to genetic resources 7, 11, 21, 33, 58, 299
Adcock, Mike 76, 205, 207
Advisory Council on Intellectual Property 16, 83, 219, 238
Affymetrix 120–21, 149, 284, 302
AgGenomics Pty Ltd 217
AgResearch 230
Agreement on Technical Barriers to Trade 1994 202
agriculture 14, 35, 38, 60–61, 66, 68, 94, 113, 165, 217, 238, 280
Agriculture Victoria Services Pty Ltd 217
Aktiebolaget Hässle v Alphapharm Pty Limited 235–6
Allen, Standish 84
Amanor-Boadu, Vincent 82
Amazon.com Inc. v Barnes and Noble.com, Inc. 283
Ambri 293–4
Ambri Biosensor 293–4
American Anti-Vivisection Society 8, 83, 105
American Association for Retired Persons 123
American Bar Association 60
American Clinical Laboratory Association (ACLA) 122–3
American Council on Education 37
American Crop Protection Association 60
American Cyanamid v Berk Pharmaceuticals 45
American Farm Bureau 89
American Heart Association 122
American Intellectual Property Law Association 60, 145–6
American Medical Association 121–2
American Patent Association 37–8
American Pomological Society 61
American Seed Trade Association 60
American Society for Microbiology 37
American Society for the Prevention of Cruelty to Animals 86
American Society of Biological Chemists 37
AMRAD 261
Anderson, David 271
Andrews, Lori 110, 122, 129, 157–8
Animal Alliance of Canada 91
animal cloning 3, 13, 82
Animal Legal Defense Fund 86–7
Animal Legal Defense Fund v Quigg 84, 86–8
Animals xi, 3, 7–8, 13, 19, 27, 38, 41, 43, 45, 50–51, 67–8, 73, 75, 82–105, 114, 130, 157, 208, 217, 222, 227, 239, 245, 255, 261, 264–9, 299
anti-commons 5–6, 97, 241, 298–300
'anything under the sun made by man' 12, 24, 26, 29, 32, 38–9, 93, 116, 124, 126, 280, 298–9
Aoki, Keith 74
Applera Corporation 15, 217–26, 239
Applera Genome Initiative 219
Application of Bergy 28–30
Applied Biosystems Group 219
Arrhythmia Research Technology, Inc. v Corazonix Corp. 121
Ashcroft, Alan 200
Association of American Medical Colleges 37, 129–30, 148–9, 154, 170
Association of Veterinarians for Animal Rights 86
Associazone Angelasserra per la Ricerca sul Cancro 192
AT&T Corporation v Excel Communications, Inc. 283
Atkinson, Ryan 98

Attorney-General (NSW) v Brewery Employees Union of NSW (the Union Label Case) 53, 56
Auckland District Health Board 16, 217, 219, 226–31
Auckland District Health Board v Genetic Technologies Ltd 16, 217, 219, 226–31
Australia vii, xi–xii, 5, 10–12, 15–17, 51–9, 75, 80, 174, 219, 231–8, 247, 250–57, 273, 280, 287, 290–91, 294–5
Australian Catholic Bishops Conference 249
Australian Government xi-xii, 11–12, 51–3, 56–9, 75, 174, 231–8, 250, 295
Australian Law Reform Commission (ALRC) 5, 11, 16, 219, 231–8, 256–7, 280
Australia legislation
  Australian Constitution 1901 vii, 52–3, 56–7, 59
  Gene Technology Act 2000 (Cth) 80
  Grain Marketing Act 1975 (WA) 55
  Intellectual Property Laws Amendment Act 2006 (Cth) 238
  Patents Act 1990 (Cth) xi, 16, 232–7, 250–55, 273
  Plant Breeder's Rights Act 1994 (Cth) 52
  Plant Breeder's Rights Amendment Act 2002 (Cth)
  Plant Variety Rights Act 1987 (Cth) 52
  Prohibition of Human Cloning Act 2002 (Cth) 254–6
  Research Involving Embryos Act 2002 (Cth) 256
  Trade Practices Act 1974 (Cth) 235
Australia–United States Free Trade Agreement 2004 xi, 11, 233, 237–8, 247
Australian Research Council 295
Australian Securities and Investments Commission 239
Australian Stock Exchange 225

bacteria 3, 12, 25–8, 32, 43, 54, 64, 92, 111–28, 148, 305

Bagley, Margo 103–4
Baltimore, David vi–vii
Banner, Donald 31
Bar of the City of New York 119
Barton, John 288
Becerra, Xavier 156–7
The Belgian Society of Human Genetics 192, 194, 196, 199
The Belgian Society of Human Genetics and The Institut Curie v The University of Utah Research Division 196–9
Belgium 192, 194, 196, 199, 206–7
Bell, Alexander Graham 28, 110
Benefit-sharing ix, 9, 11–12, 21, 236, 247, 300
Bently, Lionel 17, 77, 263
Berman, Helen 298–300
'Big Science' Projects vii, ix, 1, 9, 138–42, 144, 157–8, 218, 224–5, 282, 292
Bill & Melinda Gates Foundation 286
Biogen 10
bioinformatics 3, 16, 56–7, 220, 280–87
bio-mimicry 293, 305
biopiracy 8, 20–21, 287
BIOTECanada 66, 98
biotechnology 1–301
Biotechnology Industry Organization (BIO) 3–4, 60, 146, 248–9
Blair, Tony 140–41
Blake, Sara 67
Blue Gene 285
Blueprint Worldwide 285
Bohrer, Robert 287
'Bolar' Exception 2, 14–15, 165–6, 175–81, 300
Bolar Pharmaceuticals Co., Inc. 173–4
Bolar Pharmaceuticals Co. Inc. v Roche Products Inc. 173–4
Boston Patent Law Association 118–19
Bowbrow, Martin 224
Bowcock, Anne 190
Bowrey, Kathy 17
Boyer, Herb 44
Braach-Maksvytis, Vijoleta 294, 305
Braithwaite, John 8
BRCA1 15, 187–96, 200, 202, 218, 226, 286

BRCA2 15, 187–90, 195–200, 202, 218, 226, 286
Brenner v Manson 142, 149–52
Brin, Sergey 286
Brogan, Jim 154–5
Brown-bag sales 64
Brown v Board of Education vii
Budapest Treaty, The 1977 45–6
Buff, Ernest 283
Burbank, Luther 54
Burk, Dan 4–5, 281
Bush, George W. 258, 295
business methods 45, 113, 117–18, 255, 280, 282–5, 290

California Institute of Technology (Caltech) 16, 37, 250, 271–2
California Institute of Technology Patent Application, Examining Division 271–2
Californian Institute for Regenerative Medicine 260, 262
Cambrosio, Alberto 10, 20
Canada 2, 5, 7, 11–13, 17, 45, 51, 54, 65–74, 82, 84, 90–98, 104, 130, 174, 182, 188, 201, 206, 217, 236, 291, 298
Canada, legislation
    Patent Act 1985 (RSC) 71, 73, 93–4, 96–7
    Plant Breeders' Rights Act 1990 (RSC) 71, 94, 96
Canada-Patent Protection Case 182
Canadian Biotechnology Advisory Committee 5, 11, 69–70, 72, 82, 97–8
Canadian Canola Growers Association 66
Canadian Council of Churches 91
Canadian Government 5, 11, 69–70, 72, 82, 92, 97–8, 174
Canadian Institute for Environmental Law and Policy 91
Canadian Intellectual Property Office 45, 84, 91, 96–8
Canadian Patent Appeal Board 45
Canadian Seed Trade Association 66
cancer 13, 15, 84, 89, 91, 145, 152, 158, 187–202, 217, 218, 225, 228, 230

Cancer Research Campaign Technology 190, 199, 200
Cannell, Michael 86
Caplan, Arthur 19–20, 46
Casale, Carl 73–4
cattle 38, 83, 238, 253–4
Caulfield, Tim 6, 7, 69, 201
Celera Diagnostics 219–20
Celera Genomics 139–41, 219–20, 282, 286
Celltech Group Plc 290
Chaiton, Jonathan 84
Chakrabarty, Ananda 25, 27, 38, 43
Charles River Laboratories 90
chemical inventions 19, 24, 29, 34, 39, 54, 67, 71–2, 82, 85, 92, 120, 142, 144, 150, 165, 191, 280
chimera, human-animal 7, 13, 82, 98–103, 264
Cho, Mildred 18–19, 209
Christie, Andrew 83
clinical use 18, 122, 165, 175, 189, 193, 201, 204–6, 225, 232, 239, 266, 289, 292
Clinton, Bill 140–41
Coast Oyster Company 84
Cockburn, Iain 207, 284
Cohen, Stanley 44
Collins, Francis 138, 140, 144, 190, 218, 224
commercialisation 7, 9, 16, 51, 66, 86–7, 181, 187, 220, 229, 231, 233, 250, 256, 264, 288, 302
Commonwealth Science and Research Organization (CSIRO) 293–4
competition law 2, 9, 57, 61, 77, 87, 155, 206–7, 232, 234, 235, 238, 274, 300
computer science 4, 37, 113, 116, 118, 124, 187, 190, 219, 281–6, 289–90, 296–7
Confirmant Limited 290
constitutional law 29, 42, 51–9, 77, 121, 127, 130, 142, 171, 205, 283
Consumer Project on Technology 8, 170–71, 177
contract law 65, 71–3, 79, 114–15, 167–8, 219, 228
Cook, Trevor 205–6

Cook-Deegan, Robert 6, 20, 158, 201, 281
Cooper Dreyfuss, Rochelle 177–8, 298–300
copyright law 52–3, 56–9, 72, 77, 126–7, 165, 171, 177, 189, 203, 282, 285, 301
Corn Growers Association 60, 62
Cornell, Bruce 294
Cornish, William 17, 204–5, 207
Coulter, John 250–51
Council of Canadians, The 66
Court of Appeals for the Federal Circuit (US) 10, 14, 60, 85, 87, 90, 113–14, 120, 125, 141–2, 148–9, 154–5, 165, 168–70, 174–5, 180–81, 233–4, 283, 296, 299
Covance Inc. 217
Crespi, R. Stephen 6, 265, 271
Crichton, Michael 157
Crick, Francis 24, 156
Critchfield, Greg 188
cross-licensing 218, 235, 300
Crouch, Deborah 100–102
Crown Research Institutes (New Zealand) 218, 229
Curie, Marie 192
curiosity vii, 14, 165, 169, 174, 204, 223, 234, 248
Curtis, George, T. 110
cystic fibrosis 190, 216, 219–21

Dam, Kenneth 113
Daniel, Keith 294
databases 7, 57, 59, 139, 150, 190, 192, 282, 286–7, 289–92, 295, 297, 301
Davidson, James 200
Davies, Kevin 158–9, 209–10
Davis, Paula 154
deBeer, Jeremy 70
DeCODE Genomics 10
Deep Blue 285
Delta and Pine Land Company 60, 72
designs law 2, 11, 52–3, 56, 59
diabetes 217
diagnostic testing 3–4, 10, 15, 19, 104, 117–18, 123, 130, 142, 155–6, 187–9, 191, 193, 196, 198, 199, 201, 206–8, 217–20, 223, 232, 245, 259, 270, 282, 290, 292

Diamond, Sidney 30
Diamond v Chakrabarty vii–x, 4, 7, 12, 24–49, 51, 54, 57, 60, 62, 64, 74, 84–8, 93, 97, 119, 127, 129, 280, 298
Diamond v Diehr 112–14, 118–24, 127
Dickinson, Todd 141, 143
Dinwoodie, Graeme 300
D'Iorio, Helen 98
DNA (deoxyribonucleic acid) 9, 15, 26, 32–3, 35, 37, 44, 93, 98, 120, 141, 144, 145, 158, 189, 192, 197, 216–39, 253–4, 282–3, 288, 291–2
DNA marker 147, 217, 239, 291
DNA sequencing research 14–15, 19, 43–4, 51, 138–47, 152, 154–8, 190–95, 200, 203, 206, 216–39, 282–3, 287–9, 291–2, 294
Dodson, Gerald 146–7
dogs 83
Doha Declaration on Public Health and the TRIPS Agreement 2001 viii, 11, 208
Doll, John 129
Donley, Elizabeth 171, 261
Donoghue v Stevenson vii
Downing, Sandra 84
Drahos, Peter 8, 19–20, 113, 136, 215, 247
drosophila 82
Duffey, William H. 89
Duffy, John 177–8
Duke University 14, 165–73, 181, 199, 292, 296
Duketon Goldfields Limited 216
DuPont 90
Dutfield, Graham 75–6
Dworkin, Peter 226

Earp, David 260
Ebert, Lawrence 255
economics 2, 6–7, 9, 16, 18, 30–31, 33, 35–6, 41, 51, 57, 71, 87, 98, 135, 146, 155–6, 174, 176, 194–5, 207, 219, 226, 251, 261, 293, 296, 299
Edelman, Bernard 50–51
Edison, Thomas 54
Edson, Margaret 187
Edwards, Barry 103
Einstein, Albert 40, 112, 124, 127, 263, 278

Eisenberg, Rebecca 5–6, 10, 25, 31, 43–5, 117, 129, 259, 282, 298
Electronic Frontier Foundation 177
electronics 162, 166, 167, 280, 294, 296
Eli Lilly 10, 149, 154, 178, 191
embryos 84, 99–100, 102, 248–79
energy 25–7, 35, 42, 280, 301
Enerson, Benjamin 263
Engineers case vii
entrepreneurialism 4, 9, 44, 83, 188, 217
environmentalism 4, 8, 32, 41, 50, 65–6, 83–4, 88, 90–91, 93, 97, 104, 152, 172, 227, 238, 293, 301
Epstein, Richard 207
E.R. Squibb & Sons Inc. v Giovannia Aguggini 204–5
ETC Group 8, 66, 91, 287, 298, 301
ethics 1, 5, 7, 12, 16, 26, 36, 38, 58, 68, 92, 97, 99, 121–2, 145, 193, 202, 205, 207, 250–51, 255, 260, 264–5, 268, 270, 273–4, 291, 301
European Breast Cancer consortium 190
European Commission 202–3, 265
European Court of Justice 202–4
European Group on Ethics in Science and New Technologies 16, 250, 264–5, 268, 273
European Patent Office (EPO) 11, 13, 15–16, 76, 84, 90–91, 187–201, 203, 250, 263–73
European Patent Office Opposition Division 90–91, 192–200, 266–8, 272
European Patent Office Technical Board of Appeal 90–91, 268, 271
European Parliament 11, 201–3, 269–70
European Union 10–11, 13, 15–16, 61, 76, 84, 90–91, 130, 166, 174, 182, 187–208, 217, 234, 250, 263–73, 289, 295, 300
European Union, patent law
  Community Patent Convention 1989 (EU) 203–4, 206
  European Patent Convention 1973 (EU) 90, 192–3, 196–8, 200, 266–7, 269–73

European Union Directive on the Community code relating to medicinal products for human use 2004 (EU) 174
European Union Directive on the Legal Protection of Biotechnological Inventions 1998 (EU) 11, 15–16, 90, 190, 201–3, 206, 208, 264, 268, 270, 273
European Union Regulation on the Compulsory Licensing of Patents Relating to the Manufacture of Pharmaceutical Products for Export to Countries with Public Health Problems 2006 (EU) 207
Evangelical Fellowship of Canada 91
Ex Parte Allen 84–7, 90
Ex Parte Latimer 24–5
Ex Parte Re-examination of US Patent No. 5,843,780 259–63
Ex Parte Re-examination of US Patent No. 6,200,806 259–63
Ex Parte Re-examination of US Patent No. 7,029,913 259–63
expressed sequence tags (ESTs) 14, 139–41, 146–55
Ezzell, Carol 288

farmers' rights 8–9, 12, 52, 61–7, 71–3, 75, 82, 92, 94, 97, 103, 300
Farrer, William 54
Faunce, Thomas 136, 215, 247
Federal Circuit Bar Association 118
Federal Court of Australia 11, 59, 130, 255, 283
Federal Court of Canada 65–6, 91
Federal Trade Commission 181
federation cultivar 54
Fertilitescentrum AB and Luminis Pty Ltd 16, 250–54
Fitzgerald, Brian 58
foetal cell recovery 216
Folsom v Marsh 165
food 33, 65, 82, 92, 152, 280, 294
Food and Drug Administration 175, 178–9
Forest Research 230

Foundation for Taxpayer and
   Consumer Rights 16, 250, 260–63,
   273
France 10, 15, 50, 68, 189–92, 195,
   203–4, 206
Franklin Pierce Law Centre 119–20
Free Trade Area of the Americas 12
freedom of speech 57–8, 120
Freeman, Morris 82
frontier technologies 3, 11, 16, 28, 165,
   280–301
Frow, John 30
Fukuyama, Francis 7
fungi 3, 47
Funk Bros. Co. v Kalo Inoculant Co. 38,
   111–12
Furphy, John xi
Furphy, Joseph xi
Futreal, Andrew 191, 200

Gallo, Robert vi
Gates, Bill 286
GE Healthcare Bio-Sciences Corp.
   217, 240
*GE Healthcare Bio-Sciences Corp. v
   Genetic Technologies, Ltd* 217,
   240
GenBank 141, 192
Gene Logic 200
gene patents 5–8, 11, 16, 44, 66–7, 69,
   141–2, 145–6, 151, 155–8, 165,
   201, 207, 219, 227, 231–2, 235–8,
   280, 288
Gene Technology Regulator 71
gene therapy 93, 262
genealogy 187, 190
Genentech 10, 30, 35–6, 44, 49, 149
General Electric Company 25–7, 32,
   34–5
generic pharmaceutical drugs 165,
   173–4, 179, 181, 204
genetic counselling 189, 200
genetic discrimination ix, 196–9, 210
genetic privacy 196–9, 210, 287
Genetic Technologies Limited (GTG)
   15, 191, 216–40
*Genetic Technologies Ltd. v Applera
   Corporation* 15, 217–26, 239
*Genetic Technologies Ltd. v Covance
   Inc.* 217

Genetic Technologies Ltd. v Nuvelo Inc.
   217
genetic testing 18, 155, 187–208,
   217–18, 231, 236, 239
genetic use restriction technologies 72
genetically modified crops (GM crops)
   3, 65–74
GeneType 216
'genius of junk' 216, 218, 224
Genographic Project 21, 285
genomic mapping 1, 15, 120, 138–9,
   147, 149, 190, 206, 216–20, 223–4,
   236, 238–9, 281–2, 285–6
German Constitutional Court 205
German Supreme Court 205
Germany 190, 194, 203, 205, 266
Geron Corporation 16, 250, 257,
   259–60
Gervais, Daniel 17, 20
Gifford, Brooks 113–14, 130
Gilbert, Walter 187
Gitter, Donna 22, 183, 213, 215, 247
Gold, E. Richard 7, 104, 236
Google 285–7
Gottschalk v Benson 112–13
Government of Ontario 66–7, 213
Grain Pool of Western Australia 51–9
*Grain Pool of Western Australia v
   Commonwealth* 12, 51–59, 75
Greely, Henry 273
Greenpeace 8, 91, 194–5, 266
*Greenpeace Deutschland e.V. v The
   University of Edinburgh* 265–9
Gresshoff, Peter 222–3
Griffith, Philip 17
Grubb, Philip 17, 24
Gulbrandsen, Carl 171, 260, 263

Hahn, Robert 280
Haldeman, Donald 89
Hall, Stephen 9–10
haplotype mapping 15, 216–25, 238
Hardin, Garrett 6
Harradine, Brian 250–51, 253
Harvard Business School 287
Harvard College 3, 13, 54, 68–9, 82–98,
   103–4
*Harvard College v Canada (The
   Commissioner of Patents)* 3, 13,
   54, 68–9, 82–98, 103–4

Harvard Medical School 90
Harvard oncomouse 3, 13, 54, 68–9, 82–98, 103–4
Harvard University 3, 13, 54, 68–9, 82–98, 103–4, 164, 287
health-care 5, 8, 11–12, 14, 16, 32, 36, 41, 67, 69, 82, 104, 121–8, 138–9, 141, 145, 152, 155, 171, 177, 188–90, 193–5, 199–202, 205–8, 217, 219, 226–38, 260, 272, 280, 286, 293
Healy, Bernadine 139
Heller, Michael 5–6, 241, 298
Herald, Dave 252–5
Herder, Matthew 272–3
High Court of Australia 11–12, 51–9, 74, 130, 135, 255
'higher' life-forms 13, 19, 34–5, 44–5, 68–9, 82–4, 88, 92–8
history of science 9–10, 20, 24, 26, 28–9, 32–3, 36, 39–40, 44, 46, 48, 50–51, 53, 56–7, 60, 62–4, 74, 77, 110, 114, 131, 136, 138, 141, 161, 165, 167, 170, 174, 187, 189–90, 196, 200, 209, 213, 250, 258, 263, 295
HIV/AIDS vi–x, 208
Hockett, William A. 193
Hoffman v Monsanto 74
Holmes, Jonathon 228–9
Hood, Leroy 30, 222, 286, 289
Hôpitaux de Paris 192, 195
horses 83, 85
HortResearch 230
Hughes, Sally Smith 9, 20, 44
Human Chimera Patent Initiative 7, 82, 98–102
human cloning 3, 16, 19, 93, 99, 251, 253–4, 256
Human Genome Organisation (HUGO) vii–viii
Human Genome Project ix, 1, 138–42, 144, 157, 218, 224, 282, 292
Human Genome Sciences 139, 141
human leukocyte antigen system (HLA complex) 216, 222–5, 239
human proteome project 287
human rights ix, 12, 103, 194–5, 202, 249, 257–8
Humane Farming Association 86
Hwang, Woo-Suk 16, 250–51, 253–6

IBM 285–6, 295
Imazio Nursery Inc. v Dania Greenhouses 53–4
In re Bergy 112
In re Fisher 14, 141–55, 233
In re Kirk 152
incentives 4, 6, 13, 35–7, 41, 54, 83, 102–3, 123, 126–7, 140, 157, 176, 206, 292, 299
Incyte 139, 146
Indigenous Peoples Council on Biocolonialism 8
information 3, 10, 37, 43, 45, 50, 52, 57–8, 60–61, 77, 114, 116, 122–3, 127–8, 130, 135, 138, 140–41, 145, 147–8, 150, 152–3, 155–7, 165, 174, 176, 178–80, 187, 189–90, 195–6, 205, 207, 216, 220, 224–5, 227–9, 232, 235–6, 238, 280–83, 285–7, 289–90, 292, 296
informed consent ix, 9, 11, 12, 21, 236, 247, 249, 255, 300
Ingram, Colin 97–8
innovation 2–3, 5, 6, 35–6, 56–7, 63, 69, 74, 77, 82, 83, 95, 97–8, 103, 118, 122, 130, 142, 155–7, 193, 208, 232–3, 273, 280, 293, 297
Institut Curie 8, 15, 189–91, 199–201, 206
Institut Curie v Myriad Genetics Inc. (EP 699754) 191–4
Institut Curie v The University of Utah Research Foundation (EP705903) 195
Institut Gustave Roussy 192, 194–5
Institute for Molecular Biosciences 238
Institute for Science, Law and Technology 130
Institute of Cancer Research 195
Integra Lifesciences I, Ltd. 14–15, 166, 173–81
Integra Lifesciences I, Ltd. v Merck KGaA 175–7
Intellectual Property Office of New Zealand 227
Intellectual Property Owners Association 118
International Fund for Animal Welfare 91
Internet 140, 297

inventors 1, 25, 36, 46, 54–5, 84–5, 94, 98, 111, 126, 143–5, 166, 192, 200, 223, 229, 261, 294, 301
investment viii, 3–4, 16, 35, 44, 74, 82, 95, 171, 188, 201–2, 232–3, 239, 261, 272, 295
IP Australia 11, 45, 235, 251–3, 256–7, 263, 275
Irons, Edward 36
Italy 192, 204, 266

Jacobson, Mervyn 216–40
Jaffe, Adam 299
James Hunter Hospital 190
Janis, Mark 65, 75, 183
Japan 188, 295
Jarmul, David 172–3
Jefferson, Thomas 39, 96
JEM Ag Supply Inc. (Farm Advantage) 12, 51, 54, 59–65, 75, 96
JEM Ag Supply Inc v Pioneer Hi-Bred International Inc 12, 51, 54, 59–65, 75, 96
Jepson, Craig 119–20
Joly, Yann 154
Jones, Keith 170
Judge, Elizabeth 17

Kass, Leon 34
Kastenmeier, Robert 82, 84, 88–9
Kelley, James 154
Kesan, Jay 65
Kevles, Daniel 10, 36
Keyes, Jospeh 170
Kieff, F. Scott 6, 10, 201
Kiley, Thomas 35
King, Annette 226
King, Mary-Claire 190, 199, 286
Kinsman, John 86
Klinische Versuche I (Interferon Gamma) 205
Klinische Versuche II (Erythropoietin) 205
Kloppenburg, Jack 60–61, 78
Koppikar, Vivek 295
Korn, David 148–9

La Trobe University 217
Laboratory Corporation of America (LabCorp) 13, 110–31, 217
Laboratory Corp. of America Holdings v Genetic Technologies Ltd. 217
Laboratory Corp. of America Holdings v Metabolite Laboratories, Inc 13, 110–31, 217
Lambert, Janet 98
land grab 288
Large Scale Biology Corporation 291
League of Nations 131
League of Nations Draft Convention on Scientific Property 1923 131
Leder, Philip 90
Leiss, William 92–3
Lemley, Mark 4–5, 17, 281, 296–7
Lerner, Josh 299
Lesavich, Stephen 283
Lessig, Lawrence 57
Lewontin, Richard 138
licences 6, 9, 15, 18, 52, 59, 70, 89, 121, 127, 140, 169–71, 173, 177, 190–91, 199, 206–8, 217–18, 225–32, 234–5, 237–9, 249, 259, 261, 273, 297, 300, 302
Little, Peter 224–5
Livestock Improvement Corporation 230
living organisms 25, 28–34, 37, 41, 43–5, 62, 65, 67, 74, 85–7, 226
Llewelyn, David 17
Llewelyn, Margaret 50, 75–6, 203, 205, 207
Loring, Jeanne 262–3
'lower' life-forms 19, 34–5, 44–5, 92–3, 96
Luddites 35, 98
Ludlam, Charles 146
Lyman, Howard 88–9
Lynn, Rebecca 181

Mackay Radio & Tel. Co. v Radio Corp. of America 111
MacLean, Donna 1
Madey, John 14, 165–73, 181, 292, 296
Madey v Duke University 14, 165–73, 181, 292, 296
Magnus, David 19–20, 46
Maki, David 84–5
Manhattan Project vii
Mann, Jonathan vi
Marconi 290

Marin Humane Society 86
Martin, Larry 82
Massachusetts General Hospital 271
mathematical algorithms and formula 45, 111–13, 126–7, 129, 281–3, 286
Matthijs, Gert 199
Mattick, John 224, 238
McGee, Glenn 19–20, 46, 260–61
McKeough, Jill 17, 59
MDS Proteomics 290–91
mechanical inventions x–xi, 9, 19, 24, 32, 35, 39, 54, 66–8, 83, 93, 95, 111–12, 116, 124, 139, 156, 164–5, 280, 282, 285, 293, 294, 296
Médecins Sans Frontières 8
medicine 3–4, 8, 11, 16, 21, 35, 118–19, 122–3, 125, 138, 156, 170, 188, 207–8, 218, 250, 260, 262, 281, 286, 292, 294, 300
Meldrum, Peter 188
Mennell, Peter 17
Merck KGaA 14–15, 166, 173–81
Merck KGaA v Integra Lifesciences I, Ltd. 14–15, 166, 173–81
Merges, Robert 17, 131, 285
Merton, Robert 10
Merz, Jon 18–19, 209
Metabolite Laboratories Inc. 13, 110–31, 217
methods of human treatment xi, 3, 13, 19, 110, 116, 121–3, 128–31, 232, 236, 246, 255
mice 3, 13, 54, 68–9, 82–98, 103–4, 220, 261–2, 266–7
microarrays 120, 147, 282, 284, 302
micro-organisms xi, 3, 11–12, 19, 24–46, 54, 64, 85, 92, 105, 245, 299
microscopes 14, 148–53, 161, 295, 298
Microsoft 285–6
Millman, Robert 139
Misrock, Leslie 288–91
Modzelewski, Mark 295
Monsanto 12, 14, 51, 60, 65–75, 89, 96, 147–55, 204, 217, 233, 239
Monsanto Co. v Genetic Technologies Limited 217, 239
Monsanto Canada Inc. v Schmeiser 8, 12, 51, 65–75, 96–7
Montagnier, Luc vi

morality vi–x, 8, 13, 26, 43, 73, 83, 99–100, 102–4, 156, 194–5, 198, 227, 252, 257, 263–4, 266–8, 270–71
Morrow, David 97–8
Morse, Samuel 100
Morton v New York Eye Infirmary 111
Mountford, Peter 265–6
mouse genome 220
Mueller, Janice 17, 169–70
Mullally, Veronica 293
Myriad Genetics Inc. 15, 67, 187–208, 217–18, 228, 231
Myriad Genetics Inc. v Cancer Research Campaign Technology Inc. (EP 0858 467) 199–200
Myriad Genetics Inc. v Genetic Technologies Limited 217–18

Nader, Ralph 8, 170
Naimark, Arnold 69
NanoBusiness Alliance 295
nanotechnology xi, 3, 16, 19, 165, 280–81, 293–8
National Academy of Sciences 5, 188
National Farmers Union 60, 62, 66, 88–9
National Institutes of Health (NIH) 129, 139, 144–5, 190–91, 260
National Research Council 5
Nature 8, 24–5, 27–9, 34, 37–8, 40, 43, 51, 67, 85–6, 98, 101, 110–13, 119–30, 144, 156–7, 168, 258, 293, 298–9, 301
Nature 141, 196, 287–8
Nature Biotechnology 65, 102, 283
Netherlands 190, 194–5, 202–4, 266
Netherlands v European Parliament 202–3
Network of Concerned Farmers 8
Neville, Warwick 249
New York Medical College 98
New Zealand 5, 10–11, 15–16, 18, 21, 131, 188, 217–19, 224, 226–31
New Zealand Government 11, 226–31
New Zealand High Court 219, 229–31
New Zealand, legislation
  Patents Act 1953 (NZ) 227
New Zealand Royal Commission on Genetic Modification 5, 226
Newman, Stuart 13, 84, 98–103, 264

Nicol, Dianne 18, 154, 162, 209, 215, 241, 276
Nielsen, Jane 18, 215, 241
non-coding dna ('junk dna') 15, 216–40
Northey, Bruce 229–31
Nottenburg, Carol 223
NRDC v The Commissioner of Patents 54, 135
nuclear science and technology 42, 56, 235, 298
Nuffield Council on Bioethics 5, 145, 205, 207, 291
Nuvelo Inc. 217

OncorMed 199–200
O'Reilly v Morse 110
Organisation for Economic Co-operation and Development (OECD) 6, 18
ownership 70, 72, 144, 194, 196, 200, 216, 249, 284, 302
Oxford Gene Technology 302
Oxford Glycosciences 288–91
Oxford Glycosciences v MDS Inc. 290
Oxford University 288

Page, Larry 286
Pallin v Singer 121, 130, 155–6
Parekh, Raj 289
Paris Convention for the Protection of Industrial Property 1883 2, 11
Parker Elmer Corporation 139
Parker v Flook 112–13
Pasteur, Louis 24, 40
patent abolitionists 7, 281
patent bargain 2, 6, 282
patent clearinghouses 234, 249
Patent Cooperation Treaty 1970 11
patent law 1–301
  claims 2, 13–14, 24–5, 27–30, 32, 34, 40–41, 51, 53, 55, 63, 66–9, 84–6, 91, 99–102, 110–30, 135, 139, 145, 147–54, 162, 190–91, 193, 195–200, 206, 220–25, 227, 229–33, 238–9, 248, 251–6, 258–9, 261–3, 266, 268–73, 282, 284, 288

composition of matter 29, 32, 38–40, 51, 54, 63, 85–6, 91, 93, 95–6, 112, 144, 187, 191, 290
compulsory licensing 9, 18, 190, 206–8, 232, 234, 237–8, 249, 273, 297, 300
contrary to law 1, 254–5
crown use/state use 9, 207, 232, 234, 237, 238, 249, 297, 300
exclusive rights 2–3, 5–12, 15, 31, 38, 42, 60–62, 66, 69, 72, 75, 95, 97, 110, 126, 131, 165, 170, 177, 180, 182, 191, 193, 196, 198, 200, 205, 207–8, 218, 220–21, 229–30, 235, 245, 248–9, 259–61, 270, 272, 282, 288, 291
experimental use 2, 5, 9, 16, 61–4, 164–82, 190, 203, 206, 208, 219, 233–8, 249, 297, 300
generally inconvenient proviso 1, 255
information function 176, 282–3
infringement 5, 9, 15, 59, 65, 67, 69–70, 72, 94, 104, 114–16, 118, 121–2, 130, 145, 155, 164–5, 167–70, 172, 174–6, 178–80, 204, 208, 218–22, 226, 228, 230, 236, 239, 300, 302
innocent bystanders 9, 12, 52, 66–7, 69–71, 82, 94, 97, 104, 300
invention xi, 1, 2, 10, 12, 14–15, 25–8, 31–2, 38, 41, 44, 53–5, 65, 67–8, 73, 88, 90–91, 93–6, 98–102, 104, 110–12, 115–16, 120, 124, 126–7, 129, 135, 140, 142–4, 147–51, 156–7, 162, 166, 174, 176, 181–2, 191, 193–5, 198, 200, 204–5, 216, 223, 225, 227, 230, 232–5, 251–5, 262, 265–6, 271–2, 294, 301
inventive step (non-obviousness) 1–2, 4–5, 9, 16, 50, 55, 61, 65, 73, 82, 85–6, 99–101, 115, 127, 139–41, 154, 162, 192–5, 200, 219, 221–4, 226–7, 230, 232, 235, 261–2, 265, 272, 284, 299
limited liability for medical practitioners 121, 128, 130, 155–6, 208, 236, 300
manner of manufacture x, 1, 7, 24, 26, 28–9, 34, 36–40, 45, 51, 54,

63, 85–6, 91, 93, 95, 96, 112, 116, 124, 130, 135, 237
novelty (anticipation) 5, 9, 11, 16, 50, 55, 82, 99, 115, 139–41, 154, 192–5, 200, 218, 219, 221, 222, 224, 232, 235, 255, 265, 272, 284, 299
person skilled in the art 5, 14, 101, 154, 197, 236, 262, 299–300
prior art 11, 55, 100–101, 148, 154, 162, 192–4, 221, 223, 238–9, 261–2, 284–5
remedies 5, 67, 72–3, 75, 87, 115, 130, 175, 219, 221
safe harbour 2, 14–15, 165–6, 175–81, 300
specifications 2, 13, 34, 101, 164, 219, 221, 253, 288, 292
utility 5, 9, 14, 26, 28, 40, 70, 82, 99–100, 100–102, 113, 139–55, 193, 224, 227, 232–3, 297, 299
written description 4, 5, 16, 45, 115, 143, 145–6, 148, 219, 221
patent loyalists 3, 280
patent pools 235, 249
patent trolls 299
paternity testing 217
Patients not Patents 123
Pattinson, Barry 230–31
Peace, Nicholas 83
People for the Ethical Treatment of Animals 86–7
People's Medical Society 122
Peoples Business Commission 8, 30, 33–5, 41
Peppenhausan v Falke 165
Periodic Table 298
Perlegen 123
personalised medicine 3, 16, 280–81, 288, 291–3
Peter MacCallum Cancer Institute 225
pharmaceutical drugs xi, 2–3, 12, 14–15, 19, 44–5, 89, 102, 116, 118, 123, 146, 157, 165–6, 173–82, 187, 204–5, 207, 208, 250, 273, 280, 291–4
Pharmaceutical Manufacturers Association 36
pharmacogenomics 3, 16, 280–81, 288, 291–3

Phillipson, Martin 98
Phillips, Valeria 99
philosophy 1, 7, 14, 26, 33–6, 38–9, 50, 102, 142, 149, 164–5, 167–9, 172, 174, 234
physics 110, 166, 235, 293
Piccinini, Patricia 248
Pieczenik, George 30, 37
Piedrahita, Jorge 261–2
pigs 38, 82, 100, 261
Pioneer Hi-Bred International Inc. 12, 51, 54, 59–65, 75, 96
plant breeders' rights 3, 7, 11–12, 25, 31–33, 40, 42, 50–76, 79, 83, 94, 96, 165
   breeders' exemption 62, 64–5, 79, 165
   distinctiveness 55, 63
   essential derivation 76
   farm-saved exemption 62, 64–5, 79
   newness 55, 63
   stability 55, 63
   uniformity 55, 63
plants xi, 3, 7, 8, 11–12, 14, 19, 25, 27–9, 31–3, 38, 40, 42, 45, 50–76, 82–4, 88, 90, 92, 94, 96–7, 104–5, 112–13, 138, 147, 149, 165, 217, 222, 227, 239, 245, 266, 299
Pollack, Malla 60, 283
polymerase chain reaction (PCR) 10, 147, 197, 222
polyploid oysters 3, 13, 82, 84–6, 103
Pompidou, Alain 265
population genetics 10, 187, 190, 197–8
preclinical research 15, 175, 178–9
primates 82–3, 101, 259–62, 269
products of nature 8, 24–5, 27, 37–8, 40, 51, 98, 110, 113, 119–21, 127–30, 144, 156, 298–9
Protana Inc. 291
Protein Atlas of the Human Genome 287–91
proteins 3, 16, 22, 44, 98, 146–7, 154, 188, 204, 220, 224, 281, 280–82, 285, 287–91, 299
Proteome Systems 291
proteomics 3, 16, 220, 280–81, 287–91, 299
public domain 2, 9–10, 57, 92, 110, 119, 140–41, 192, 289, 299–300

Public Knowledge 170–71, 177
Public Patent Foundation 16, 113, 120, 250, 260–63, 273, 299

Quigg, Donald 89

Rabin, Sander 102–3
Rabinow, Paul 10, 20
Rai, Arti 6, 177–8, 249, 292–3
Ranks Hovis McDougall's Application 45
Ravicher, Dan 263
Re Application for Patent of Abitibi Co. 45
Re Application for Patent of Connaught Laboratories 45
Re Stuart Newman 13, 84, 98–103, 264
Reagan, Ronald 166
reagents 85, 217, 219, 226
recombinant DNA technology 9, 26, 32, 33, 35, 37, 44, 191
religious concerns 28, 91, 138
Research Foundation for Science, Technology and Ecology 66
research tools xi, 3, 14, 19, 95, 141, 148–53, 164–82, 205, 259–60, 273
Resnik, David 155
Richardson, Megan 17
Ricketson, Sam 17
Rifkin, Jeremy 7–8, 13, 30, 33, 84, 88, 98, 100, 102–3, 264
Risconi, William 188
Rivers, Lynn 155–6
Rio Convention on Biological Diversity 1992 11, 202
Roberts, Michael Symmons 138
Robertson, Elizabeth 261–2
Robertson, Sean 92
Robinson, William 110
Roche Products Inc. 173–4
Roche Products, Inc. v Bolar Pharmaceuticals Co., Inc. 173–4
Royal Society of the United Kingdom 205, 292
Rubinstein, Pablo 222–3
Rutgers University 37
Ryan, Andrea 145–6

Sage, David 228
Sambrook, Joe 225

Sampson, Tim 182
Sarnoff, Joshua 171, 177
Saviotti, Paolo 284
Sawin v Guild 164
Schatten, Gerald 256
Schwartzenberg, Roger-Gérard 206
Science 6, 141, 255, 263
scientific discoveries 7, 8, 13, 35, 39, 110–14, 117–19, 126, 131, 135, 143, 233, 237, 263, 298, 299
scientific knowledge 2, 98, 110–12, 119, 126, 129, 142, 148, 152, 155, 167–8, 171, 176, 208, 218, 258, 286, 294, 299–300
scientific publishing 6, 20, 120, 122–3, 141, 165, 176, 192, 196–7, 222, 224, 261, 283, 290, 301
scientific progress 3, 12, 29, 42–3, 46, 50, 67, 118, 120–21, 123, 127, 130, 144, 149, 152–3, 170–71, 176–7, 248, 258–9, 286, 296
scientific research 1–3, 5–12, 14–16, 19, 26, 32–3, 35–7, 40–44, 51, 54, 61–3, 65, 67, 75, 79, 82, 84, 88–92, 95, 98–9, 102, 114, 120–25, 127, 129, 131, 141–3, 146, 148–58, 164–82, 188–208, 216–40, 248–74, 280, 283–8, 290–95, 298–301
Scott, Christopher Thomas 273
Scott, Randall 146, 289
Scott, Rodney 190
Sears, Mary 36
Seoul National University 253, 255–6
Sequenom Inc. 239
sheep 2, 38, 83, 261–2
Sherman, Brad 17, 67, 263–4
Shi, Qin 291–2
Shiva, Vandana 66
Shreeve, James 139
Shulman, Seth 121, 208, 258
sickle-cell anaemia 216, 224, 243
Sierra Club of Canada, The 66
Silico Research 283
Simons, Malcolm 15, 216–25, 229
Simpson, John 263
single nucleotide polymorphisms (SNPs) 219
Skolnick, Mark 187–91, 195–6
slavery 257
Smith, Austin 265

Smith, Hamilton 301
Social Democratic Party of Switzerland 194–5
Social Democratic Party of Switzerland and the Institut Curie v Myriad Genetics Inc. (EP 705902 and EP 705903) 194–5
sociology of science 1–3, 5–12, 14–16, 19, 26, 32–3, 35–7, 40–44, 51, 54, 61–3, 65, 67, 75, 79, 82, 84, 88–92, 95, 98–9, 102, 114, 120–25, 127, 129, 131, 141–3, 146, 148–58, 164–82, 188–208, 216–40, 248–74, 280, 283–8, 290–95, 298–301
Solicitor-General of the United States 31, 38, 60, 117–18, 127, 171–2
South Australian Clinical Genetics Service 232–3
South Korea 253–6
Stanford University 44, 166
Stankovic, Bratislav 103
State Street Bank & Trust Co. v Signature Financial Group, Inc. 114, 283
Statute of Monopolies 1623 (UK) 1, 77
Stem cell banks 249
Stem Cell Sciences 265–6
stem cells xi 2, 8, 11, 16, 19, 82, 93, 99, 104, 165, 171, 248–79
  modified 259, 265, 273
  multipotent 257
  pluripotent 256–7, 259, 262, 269, 271, 273
  totipotent 256–7, 271, 273
  unmodified 259, 265, 273
Stemple, Derek 271
Stewart, Andrew 39
Stewart, Timothy 90
Stoppa-Lyonnet, Dominique 193, 199
'storehouse of knowledge' 110–12, 126, 299
Strandburg, Katherine 177–8
Stratton, Michael 195–6, 200
Strauss, Joseph 18
sui generis forms of intellectual property 7, 11, 31, 52, 58, 61, 71, 75–6, 83, 94, 131, 165, 281, 291, 297
Sulston, John 8, 140, 218, 224

Supreme Court of Canada 2, 11–13, 51, 54, 65–74, 84, 91–8, 104, 130
Supreme Court of the United States 10, 12–14, 25–6, 30, 25–43, 45, 51, 54, 57, 59–65, 74, 86–8, 93, 96, 110–29, 142, 149, 151, 154, 162, 164, 166, 170–81, 299–300
surgical methods 3, 104, 111, 113, 121, 128, 130, 155, 208, 245
Suthers, Graeme 232–3, 237
Swanson, Robert 35
Swiss-type claims 130, 137
Switzerland 190, 194, 216, 263, 278
synthetic biology 301
Synthetic Genetics 301

Tait, Brian 222–3
Taymor, Kenneth 273
technology neutrality 4–5, 281
Technology user agreements 65, 71–3, 79
telegraph 110–11
Telephone Cases 110–11
terminator technologies 72
thalassemia 217, 222
Then, Christopher 266
therapeutic cloning 3, 16, 248, 250, 253, 255
therapeutic methods 3, 16, 44, 89, 104, 130, 140, 158, 175, 187–8, 191, 206, 208, 218, 220, 245, 248, 250, 253, 255, 259, 270, 284, 288, 291
Thomson, James 259–63
Thoreau, Henry David 164
Thurow, Lester 9
Tizard, Judith 226
trademark law 2, 11, 52–3, 59, 61, 290
transgenic animals 13, 51, 68, 82, 84, 88–91, 96, 103–4, 265–6, 269
Treatment Action Campaign 8
TRIPS Agreement 1994 viii, xi, 2, 11, 45, 104, 130, 182, 202, 208, 232, 236, 245–6, 300

UNAIDS viii
UNESCO Universal Declaration on Bioethics and Human Rights 2005 ix–x, 12

United Kingdom 1, 12, 16–17, 35, 45, 53, 119, 127, 201, 204–7, 249, 256–7, 261, 271, 273, 288, 290, 292, 302
United Kingdom Department of Health 205, 207
United Kingdom National Health Service 200
United Kingdom, patent legislation
  Patents Act 1977 (UK) 204, 256–7
United Kingdom Patent Office 1, 16, 256–7, 271, 273
United Kingdom Stem Cell Bank 249
United Nations vii–x, 12, 21, 58, 67, 182, 297
United Nations Declaration on the Rights of Indigenous Peoples 2007 21
United Nations Economic, Scientific, and Cultural Organization (UNESCO) vii–x, 12, 58
  International Bioethics Committee vii–ix
United Nations Permanent Forum On Indigenous Issues 12
United States Congress 10, 14, 25, 33, 37–8, 43, 61, 65, 82, 84, 88, 89, 103, 121, 142, 155–6, 166, 174, 258, 299
United States District Courts 14, 60, 87, 114–15, 167–9, 173, 175, 179, 219, 225, 239, 277, 302
United States Patent and Trademark Office (USPTO) 7, 10, 12, 14–16, 24–6, 29–32, 35, 40, 43, 84, 86–7, 89–90, 99–100, 102–4, 114, 118, 121, 125, 129–30, 139–41, 143–6, 148–9, 151, 153–5, 216, 222, 224, 233, 250, 257–8, 261–3, 273, 283–4, 288, 292, 295
  Board of Patent Appeals and Interferences 14, 25, 27, 29, 32, 85–6, 142, 148–9, 154–5
  Eligible patentable subject matter guidelines 129–30
  Utility guidelines 102, 141–8, 151, 155
United States, legislation
  Administrative Procedure Act 1966 (US) 87

Bayh–Dole Act 1980 (US) viii, 173
Consolidated Appropriations Act 2004 (US) 101, 109, 258
Drug Price Competition and Patent Term Restoration Act 1984 (US) ('The Hatch–Waxman Act') 174–81
Genomic Research and Accessibility Act 2007 (US) 11, 142, 156–8
Genomic Research and Diagnostic Accessibility Act 2002 (US) 10, 142, 155–6
Genomic Science and Technology Innovation Act 2002 (US) 10–11, 142, 156
Morrill Act 1862 (US) 61
Patent Act 1793 (US) 39, 96
Patent Act 1835 (US) 39
Patent Act 1870 (US) 39
Patent Act 1874 (US) 39
Patent Act 1952 (US) 24, 29, 39, 54, 62, 128, 156, 174–5
Plant Patent Act 1930 (US) 3, 25, 31–3, 40, 42, 51, 54, 60–64, 74
Plant Variety Protection Act 1970 (US) 25, 31–2, 40, 42, 51, 54, 60–64, 74, 79
Stem Cell Research and Cures Act 2004 (California), California Proposition *71*, 260
Stem Cell Research Enhancement Act 2005 (US) 258
Transgenic Animal Patent Reform Act 1989 (US) 82, 88–9
United States Constitution 29, 42, 57, 106, 121, 127, 130, 136, 142, 171, 283
United States v Dubilier Condenser Corporation 33
Universities 9–10, 14, 30, 36–7, 44, 84, 90, 146, 164–73, 181, 184, 194–7, 199, 217, 249, 253, 258, 259, 264–9, 286, 288, 292, 296
University of California, The 9–10, 30, 36–7, 44
University of Edinburgh, The 264–9
University of Hawaii 165, 173
University Patents Incorporated 114
University of Pittsburgh 256

University of Utah Research Foundation 15, 194–7
University of Washington 84, 286
University of Wisconsin 258–9
UPOV Convention 1961 11, 50
UPOV Convention 1978 11
UPOV Convention 1991 11, 52, 58
Utah Cancer Registry 190

Vaidhyanathan, Siva 297–8
van Overwalle, Geertrui 18, 206–7
Varmus, Harold 145, 191
Venter, Deon 236
Venter, J. Craig 139–41, 286–7, 301
Vereniging van Stichtingen Klinische Genetica Leiden 195

Wald, George 34
Walden 164
Walsh, John 6, 18, 201
Warburg, Richard 287–8
Watson, James 24, 139, 156
Waxman, Seth 149
Welcome Real Time SA v Catuity Inc 283
Weldon, Dave 156–7, 258
Wellcome Foundation v Parexel International & Flamel 204
White House Office of Science and Technology Policy 156
Whittemore v Cutter 164, 167
WiCell Research Institute 260
Wilkins, Marc 287
Williams, Daryl 231
Williams, J. Jason 291
Williams, R. Sanders 172

Williams, Robert Lindsay 261–2
Wilson, Jack 26
Winn, David 293
Wisconsin Alumni Research Foundation (WARF) 16, 171, 250, 259–64, 269, 273
Wisconsin Alumni Research Foundation v Geron Corporation 259–60
Wisconsin Alumni Research Foundation Patent Application in Respect of 'Primate embryonic stem cells', Examining Division, 269–70
Wisconsin Alumni Research Foundation Patent Application in Respect of 'Primate embryonic stem cells', Technical Board of Appeal 270–71
Wisconsin Family Farm Defense Fund 86
Wise, Steven M. 88
Wiseman, Roger 191
Woo-Suk Hwang 16, 250–51, 253–6
Wooster, Richard 200
World Intellectual Property Organization (WIPO) 12, 182
World Trade Organization (WTO) 2, 45, 130, 182, 245, 300
WTO General Council Decision 2003 11, 208

yeast 3, 24, 40, 92

Zareski, Terry 74
zebra fish 82
Ziff, Bruce 65
Zoocheck Canada 91

IV/158